Advances in Computed Tomography for Geomaterials

Advances in Computed Tomography for Geomaterials

GeoX 2010

Edited by
Khalid A. Alshibli
Allen H. Reed

Associate Editors
Les Butler, Joanne Fredrich,
Jeffrey Nunn, Karsten Thompson
and Clinton Willson

Library of Congress Cataloging-in-Publication Data

GeoX 2010 (2010 : New Orleans, La.)
 Advances in computed tomography for geomaterials / GeoX 2010 ; edited by Khalid A. Alshibli, Allen H. Reed.
    p. cm.
 Papers presented March 1-3, 2010 in New Orleans, La. sponsored by Louisiana Sate University and the Naval Research Laboratory, Stennis Space Center, Mississippi.
 Includes bibliographical references and index.
 ISBN 978-1-84821-179-7
 1. Soil mechanics--Research--Congresses. 2. Rock mechanics--Research--Congresses. 3. Tomography--Congresses. 4. Three-dimensional imaging in geology--Congresses. 5. Materials--Testing--Congresses. 6. Concrete--Analysis--Congresses. 7. Radiography--Industrial--Congresses. I. Alshibli, Khalid. II. Reed, Allen H. III. Louisiana State University (Baton Rouge, La.) IV. Naval Research Laboratory (John C. Stennis Space Center) V. Title.
 TA710.A1G475 2010
 625.1'22--dc22

                                                                                        2009048641

British Library Cataloguing-in-Publication Data
A CIP record for this book is available from the British Library
ISBN 978-1-84821-179-7

Printed and bound in Great Britain by CPI Antony Rowe, Chippenham and Eastbourne.

Organizing Committee

Prof. Khalid A. Alshibli, Louisiana State University, Co-Chair
Dr. Allen Reed, Naval Research Laboratory, Co-Chair
Prof. Clinton Willson, Louisiana State University
Prof. Karsten Thompson, Louisiana State University
Dr. Joanne Fredrich, BP America, Inc
Prof. Les Butler, Louisiana State University
Prof. Jeffrey A. Nunn, Louisiana State University

International Advisory Committee

Table of Contents

Foreword

Geomaterials are often the fundamental building blocks of infrastructure. They are the soil, sediment and rock upon which manufactured geomaterials, such as asphalt, composites and concrete are laid or poured. Geomaterials are also a fundamental foundation of modern society, providing energy through coal, gas, oil, etc. Working with these materials provides interesting, complex and difficult challenges, such as modification, construction, maintenance and repair of the building blocks as along with extraction of energy and sequestration of carbon dioxide. In this book, numerous techniques are presented to address issues that stem from the use and evaluation of geomaterials with computed tomography (CT) imagery.

CT imagery provides a basis by which many complex structures/feature within geomaterials can be visualized and evaluated. CT sections the scanned material into small parts and then reconstructs these parts into three-dimensional images. This process has seen widespread used in medical fields and has grown increasingly common in diagnosing ailments in humans. At the same time, CT has been applied to geomaterials, which are being studied for industrial and research purposes.

In this book, advances in CT are presented that are built upon petroleum research conducted in the late 1980s and was first addressed by a collective international group of researchers at GeoX2003 workshop (Japan) and then again addressed by a international effort at GeoX2006 (Aussois, France). GeoX2010 follows in the tradition of this great research by applying the latest tools and techniques to computed tomography in studies of geomaterials.

This book is a compilation of 49 papers presented at GeoX2010 in New Orleans, Louisiana, USA, March 1-3, 2010. These papers address geomaterials from many perspectives by: 1) using advanced software and numerical methods to address complex geometries efficiently and more completely; 2) applying novel imaging techniques, such as neutron and nanometer scale tomography as well as traditional x-ray computed tomography; 3) addressing issues related to energy exploration and climate change; 4) flow through porous media and 5) coupling computed tomography with geotechnical testing methods to address deformations and progress of failure in sand, rock, asphalt and concrete.

Overall, this compilation is a broad-based address of CT applications to geomaterials that has been made possible by the efforts of faculty members from Louisiana State University and the Naval Research Laboratory, Stennis Space Center, Mississippi and due to the innovation and sustained research efforts by the authors, their support and their staff.

Khalid A. ALSHIBLI
Allen H. REED
All the chairs and reviewers that helped out with these papers

Sand Deformation at the Grain Scale Quantified Through X-ray Imaging

G. Viggiani — P. Bésuelle — S. A. Hall — J. Desrues

Laboratoire 3S-R
University of Grenoble – CNRS
38041 Grenoble
France
cino.viggiani@grenoble-inp.fr
pierre.besuelle@grenoble-inp.fr
stephen.hall@grenoble-inp.fr
jacques.desrues@grenoble-inp.fr

ABSTRACT. *This paper presents a study of localized deformation processes in sand with grain-scale resolution. Our approach combines state-of-the-art x-ray micro tomography imaging with 3D Volumetric Digital Image Correlation (3D V-DIC) techniques. While x-ray imaging and DIC have in the past been applied individually to study sand deformation, the combination of these two methods to study the kinematics of shear band formation at the scale of the grains is the first novel aspect of this work. Moreover, we have developed an original grain-scale V-DIC method that enables the characterization of the full kinematics (i.e. 3D displacements and rotations) of all the individual sand grains in a specimen. We present results obtained with both "continuum" and "discrete" DIC on Hostun sand, and a few preliminary results (continuum DIC only) recently obtained on ooid materials, which are characterized by spheroidal, layered grains.*

KEYWORDS: *strain localization, granular media, in-situ x-ray tomography, 3D volumetric digital image correlation*

1. Introduction

Shear banding, the localization of deformation into thin zones of intense shearing, is a phenomenon commonly observed in sand and other granular materials. It has quite a practical relevance from an engineering standpoint, and has been thoroughly investigated in the laboratory for decades. However, it should be kept in mind that in the presence of localized deformations, the meaning of stress and strain variables derived from boundary measurements of loads and displacements is only nominal. Therefore, the most valuable experimental contributions to the understanding of shear banding are those measuring, in one way or another, the *full field* of deformation in the specimen – which is the only means by which test results can be appropriately interpreted (Viggiani and Hall 2008). Full-field analysis of strain localization in sand possibly started in the late 1960s in Cambridge (e.g. Roscoe *et al.* 1963), and was continued over the last decades in the work of a number of groups, including Grenoble; see Desrues and Viggiani (2004) for a review. Most of these works were performed using specifically designed plane strain devices, and used a range of full-field methods, the more advanced of which allowed observation of the specimen throughout loading by optical methods, thereby permitting measurement of the evolving strain field. In the 1960s, x-ray radiography was first used to measure 2D strain fields in sand (e.g. Roscoe 1970). From the early 1980s, x-ray tomography was used by Desrues and coworkers (see Desrues 2004 for a review) and later by Alshibli *et al.* (2000). These studies provided valuable 3D information on localization patterning in sand, and also demonstrated the potential of x-ray tomography as a quantitative tool, e.g. for measuring the evolution of void ratio inside a shear band and its relation to critical state (Desrues *et al.* 1996).

The recent advent of x-ray *micro* tomography, originally with synchrotron sources and now with laboratory scanners, has provided much finer spatial resolution, which opens up new possibilities for understanding the mechanics of granular media (in 3D) at the scale of the grain. For example, Oda *et al.* (2004) presented micro tomography images of sand grains inside a shear band, showing organized structures that would not have been seen in standard x-ray tomography images (because of insufficient resolution) and that had only previously been observed in 2D thin sections (Oda and Kazama 1998).

It should be noted that the images by Oda *et al.* (2004) were obtained *post-mortem*, i.e. after testing. However, a full understanding of the mechanisms of (localized) deformation can only be achieved if the entire deformation process is followed throughout a test while the specimen deforms. This is possible by using *in-situ* x-ray tomography (*in-situ* meaning x-ray scanning at the same time as loading). A number of such *in-situ* studies for triaxial tests on sand were performed using medical or industrial tomography systems (e.g. Desrues *et al.* 1996, Alshibli *et al.* 2000, Otani *et al.* 2002). More recently, Matsushima *et al.* (2006, 2007) have used synchrotron x-ray *in-situ* micro tomography, which allowed them to identify individual sand grains and track their displacements throughout a triaxial test – note that this tracking was carried out only in 2D for a section through the specimen.

The aim of the study presented in this paper was both to observe the material evolution under loading with grain-scale resolution and to image the deformation processes. In recent work, presented at the previous GeoX workshop (Bésuelle *et al.* 2006), we applied 3D Volumetric Digital Image Correlation (V-DIC) to a sequence of x-ray tomography images taken during a triaxial test on a clay-rock specimen (see also Lenoir *et al.* 2007). In the present paper, we show results of a similar DIC-based analysis of deformation for sand specimens under triaxial compression. Two different granular materials were tested: *Hostun sand*, a fine-grained, angular siliceous sand with a mean grain size (D_{50}) of about 300 µm, and *Caicos ooid*, a material characterized by spheroidal grains with D_{50} of about 420 µm. In addition, we have developed a grain-scale V-DIC method that permits the characterization of the full kinematics (i.e. 3D displacements and rotations) of all the individual grains in a specimen. So far, such a method has been applied only to Hostun sand.

The structure of the paper is as follows. First, we describe the experimental setup for triaxial testing with concurrent x-ray micro tomography. We then describe the main features of the two V-DIC methodologies (continuum and discrete) used in this study. Results obtained with both methods are presented and discussed for a triaxial compression test on Hostun sand. For the tests on Caicos ooid, the analysis is still ongoing, and the evolution of full-field incremental kinematics has been obtained only from the continuum V-DIC. For both materials, distinct features of localized deformation are identified and their spatial and temporal development is characterized.

2. Experimental setup, testing program and materials tested

The experimental results presented in this work come from two testing programs. The former, on Hostun sand, was carried out on beamline ID15A at the European Synchrotron Radiation Facility (ESRF) in Grenoble; results of such program have already been presented elsewhere (Hall *et al.* 2009, 2010a). The latter, on Caicos ooid, was carried out using the multi-scale x-ray CT scanner recently acquired at Laboratoire 3S-R, which was designed and manufactured by RXSolutions at Annecy, France (see Figure 1). In this laboratory scanner, a large cabin allows for the flexible working space that is needed to perform *in-situ* scanning. Spatial resolution can be adjusted by changing the spot size and by moving the rotation stage, thus changing the distance between the x-ray source and the object to be scanned (the distance between source and detector remaining the same). Both at the ESRF and at 3S-R, x-ray micro tomography allowed for high spatial resolution (in the order of a few microns), which is crucial for understanding mechanics down to the grain scale. It should be noted that, for a given spatial resolution, using a synchrotron source provides very fast scanning thanks to the high photon flux (12 minutes for a scan of the entire sample, as opposed to about 3 hours for our laboratory x-ray scanner).

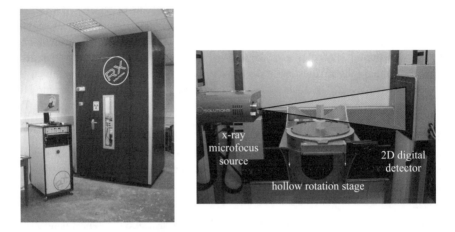

Figure 1. *X-ray CT scanner at Laboratoire 3S-R. Overall view of the large cabin (internal dimensions: height 290 cm, width 175 cm, depth 135 cm) and remote computer control (left photograph), and main components of the system (right photograph)*

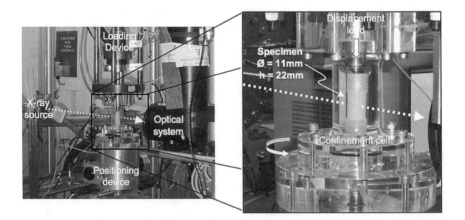

Figure 2. *Tomography set-up for triaxial testing at the beamline ID15A at ESRF. Complete set-up on the beamline (left photograph) and zoom on the specimen inside the triaxial cell (right photograph)*

The tests were conducted using a specifically built *in-situ* setup that could be placed in the x-ray beam allowing the specimens to be scanned under load, see Figure 2. The triaxial apparatus, made from Plexiglas (very transparent to x-rays), is practically the same as a conventional system, except the much smaller size and the shape of the confining cell. As opposed to conventional triaxial systems, the tensile reaction force is carried by the cell walls and not by tie bars (which avoids having any obstacles to the x-rays). The axial load and hence the stress deviator are applied using a motor-driven screw actuator, which also does not interfere with the tomographic x-ray scans. See Lenoir (2006) and Viggiani *et al.* (2004) for full details. Note that essentially the same setup was used in the experiments at the ESRF and at 3S-R, except for the fact that the loading system is above the cell in the former case, and below in the latter (i.e. the axial piston moves upward during deviatoric loading).

Several triaxial compression tests were performed on dry specimens of Hostun sand and Caicos ooid. Results from just two tests are discussed herein, both performed under a confining pressure of 100 kPa. Both materials were tested starting from an initially dense packing, obtained by dry pluviation. Deviatoric loading was strain controlled, with a screw driven piston moving at 60 μm/min, which corresponds to quite a low strain rate (0.27%/min for a 22 mm high specimen). As in conventional triaxial testing, the specimen slenderness was equal to 2, i.e. the diameter of the specimen was 11 mm. It should be noted that despite the small sample size (in comparison to standard triaxial tests on sands), the specimen can be considered large enough to be mechanically pertinent (i.e., its response can be considered representative of that of a larger mass of the material); in fact the sample comprises roughly 50,000 grains for Hostun sand and 20,000 for Caicos ooid. These reduced dimensions were imposed by the x-ray imager width at the ESRF, which was just 14 mm (the sample needed to be smaller than this to not risk passing out of the field of view, although this did occur by the end of the test on Hostun sand; see later). The spatial resolution (i.e. the voxel size) was set to 14x14x14 μm^3 at the ESRF, while it was slightly larger for the experiment on Caicos ooid performed in the scanner at 3S-R (where the side of the voxel was 14.7 μm). Such a resolution was enough to clearly identify the individual grains of both materials tested. As an example, Figure 3 shows two tomographic slices obtained from scanning a specimen of Hostun sand at the ESRF and at 3S-R (recall that the mean grain size of Hostun sand is around 300 μm or about 20 voxels; therefore each grain contains about 5500 voxels in the tomography images). Interestingly, the quality of the two slices looks very much comparable at first glance. However, important differences may exist between them (for example in terms of the signal-to-noise ratio, which is lower in the ESRF images because unfortunately scanning parameters were not optimized), which are all but "minor details" when it comes to quantitative image processing, e.g. digital image correlation.

As for the materials tested, Hostun sand can be considered relatively well-known, as it has been the reference sand for our laboratory as well as many other research groups in Europe. It is a fine-grained, angular siliceous sand coming from the Hostun quarry (Drôme, France). A uniform gradation of the natural sand (so-called

Hostun RF, or S28 Hostun) was used in the testing program, with 100% of the material passing a 0.63 mm sieve and retained on a 0.16 mm sieve. The mean particle diameter is 0.30 mm and the coefficient of uniformity ($C_u = D_{60}/D_{10}$) is 1.70.

The ooid material has been provided by the Resource Sciences Laboratory of ExxonMobil Research and Engineering Co., Anandale (USA). The material was collected from the Caicos platform, which is the southernmost platform of the Bahamian archipelago that has significant emergent islands. More precisely, the material tested comes from a region there described as Ambergris Shoal, which is a marine shoal depositional environment. Ooids are rounded, "coated" (layered) sedimentary grains that are essentially pure CaCO3. X-ray diffraction indicates that they are >96% aragonite, the remainder being calcite and high magnesium calcite. The grains are composed internally of concentric spherical layers that build out from a single core. The gradation used in this testing program is relatively uniform, with 100% of the grains passing a 0.60 mm sieve and retained on a 0.18 mm sieve. The mean particle diameter is 0.42 mm and the coefficient of uniformity is 1.90.

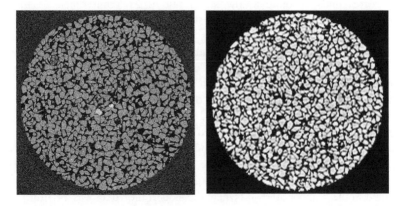

Figure 3. *Horizontal slices extracted from 3D x-ray micro tomography images of a specimen of Hostun sand obtained at the beamline ID15A at ESRF (left) and at 3S-R (right)*

3. Continuum and discrete volumetric digital image correlation

Digital Image Correlation (DIC) is a method that is being used increasingly in experimental solid mechanics to measure displacement and deformation fields over surfaces (e.g., Pan *et al.* 2009) or through a volume (e.g., Bay 2008). While the method was pioneered in the 1980s, its extension to measure displacement and strain fields within solid objects, e.g., using 3D images acquired by x-ray tomography, is more recent (e.g., Bay *et al.* 1999, Bornert *et al.* 2004, Lenoir *et al.* 2007; see Bay 2008 for a review).

Whatever dimension (2D or 3D), DIC is a mathematical tool to define the best mapping of an image into another. More precisely, the aim is to determine the transformation Φ that relates reference and deformed configurations of an evolving system. The method is based on the fundamental assumption that at any point x the grey levels in the first image, f(x), are convected into the grey levels of the second image, g(x), by the transformation Φ, that is, g(Φ(x)) = f(x). In practice this relation is never fully satisfied, because of systematic and random noise. For the case of images acquired by x-ray micro tomography, random noise can be high and systematic reconstruction artifacts are often present.

Implementations of DIC usually involve local evaluations of the transformation Φ over cubic (for the volume case) subsets that are regularly distributed over the reference image. The evaluation requires solving an optimization problem for each subset, in which essentially some measure of the similarity of f(x) and g(Φ(x)) in the considered subset is maximized over a parametric set of transformations. As a digital image is a discrete representation of grey levels, any integral over subsets is in fact discretized into a sum over voxels. Some interpolation is therefore necessary to evaluate the transformation with subvoxel accuracy; see Lenoir *et al.* (1997) for further details.

It should be noted that standard implementations of the approach described above assume a continuous displacement field, at least within each subset. Locally, the transformation is assumed to be a rigid translation, or a low order (usually linear or quadratic) polynomial expansion of the actual transformation. When deriving strain from the displacements of separate subsets, continuity between subsets is assumed. For this reason, we refer to this DIC analysis as "continuum DIC". Such a procedure can be applied to study the deformation of a granular material such as sand as long as the spatial scale of the investigation remains large with respect to the grain size. It may also be used at somewhat smaller scales (a few grains within the correlation subsets) under the condition that only small deformation increments are considered. However, a different DIC approach is possible, which recognizes the granular character both of the images and the mechanical response, and has therefore the specific objective of investigating the kinematics of individual sand grains. In this work, a "discrete DIC" procedure has been developed with the specific aspect that the regularly shaped and spaced subsets are replaced by subsets centered on each individual grain, with a shape following the actual shape of the grain. In practice, the subsets include a grain plus a small surrounding layer a few voxels thick. The reason for this layer is that, possibly because of the relatively high noise level in the x-ray images and almost uniform x-ray absorption of the sand grains, the grey level variation within a grain was not enough for DIC. Adding a layer provided the extra information of grain shape, which is characteristic of each individual grain. If the grains are assumed to be rigid, then the transformation of each subset is a rigid motion, i.e. it involves a three component translation vector plus a rotation. The latter is represented by a rotation axis and a positive angle of rotation about this axis (the axis is parameterized by two polar angles, a longitude with respect to the specimen axis, and a latitude in the cross-sectional plane).

The practical implementation of this Discrete DIC comprises four consecutive steps. First, the image of the undeformed specimen is segmented in order to identify and label individual grains. Then, a mask is defined for each grain, covering the grain plus a three-voxel wide layer around the grain. Standard DIC procedures of CMV-3D are then applied to determine a first evaluation of the translation of each grain, making use of sufficiently large cubic subsets centered on the grains. Finally, starting from these initial estimates, the translation and rotation of each grain are determined using the discrete DIC algorithm. See Hall *et al.* (2010a) for further details.

In the following, we present results obtained using both continuum and discrete V-DIC. The former uses the code TOMOWARP, which is based on the work of Hall (2006) (see also Hall *et al.* 2010b for a 2D application to a granular material). Discrete V-DIC has been integrated into CMV-3D, a code developed by Bornert *et al.* (2004) (see also Lenoir *et al.* 2007 for further details and an application to geomechanics).

4. Selected results

4.1. *Hostun sand*

X-ray tomography scans were carried out at key moments throughout the test, which are marked by (small) relaxations in the loading curve in Figure 4. The sample stress-strain response shows a roughly linear initial trend followed by a curvature to the peak stress at around 11% nominal axial strain, after which the stress drops, to what is probably the beginning of a plateau, after which the test was stopped and the sample unloaded.

The image on the right in Figure 4 shows in 3D the grain detail which is possible to obtain for Hostun sand through x-ray tomography. For the sake of clarity, in the following we will show only 2D slices through this volumetric data and the subsequent V-DIC results. The top row of Figure 5 shows a series of vertical slices through the x-ray tomography images at different stages in the test (see stress-strain curve in Figure 4). These slices, which are roughly perpendicular to the planar band of localized strain that developed during the test, show that the specimen gradually leans to one side, with a rotation of the upper platen in the latter part of the test. However, there is no clear evidence of localized deformation in these images. Porosity maps shown in the bottom row of Figure 5 were obtained from the grey-scale images based on overlapping cubic windows of side 61 voxels (854 μm) throughout the sample volume. From these porosity fields an evolving inclined zone of localized dilation can be seen.

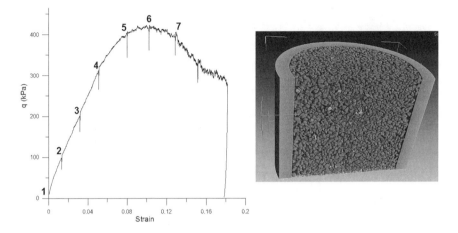

Figure 4. *Stress deviator versus axial strain curve for the deviatoric loading part of the triaxial compression test on Hostun sand (left) and 3D rendering of the sand specimen showing the grain detail (right)*

Continuum V-DIC has been carried out on consecutive pairs of 3D images to provide the incremental displacement and strain fields (the results are thus averages over the given time interval). The key DIC parameters are the distance between the calculation nodes (which also represents the reference length for subsequent strain calculation) and the correlation window size; in this analysis these were, respectively, 20 voxels (or 280 µm) and a cube with sides of 21 voxels (or 294 µm) reduced to 11 voxels (or 154 µm) for the sub-voxel derivation. Results from this analysis indicate that, despite the granular nature of the material, smooth and relatively continuous displacement fields are measured. Figure 6 shows vertical slices through the 3D field of maximum shear strain $(\varepsilon_1 - \varepsilon_3)/2$ (where ε_1 and ε_3 are the major and minor principal strains) for increments 3-4, 4-5, 5-6 and 6-7. These strain images clearly show the evolution of a localized band that traverses the sample diagonally from top left to bottom right. It is worth noting that this is an incremental analysis, therefore indicates the deformation active in each strain increment. This is different from what can be seen with accumulated porosity changes shown in Figure 5. As such it is seen from these incremental maps that the localization possibly initiated in increment 4-5, and was clearly developed in 5-6, i.e. before the peak load. Note that localization is visible in these maps before it becomes clear in the porosity images (bottom row of Figure 5). The general picture is of a localization of shear strain and dilatancy which starts as a broad zone and then progressively thins with loading. In increment 6-7, this zone has a width of about 5 mm (i.e. about 17 D_{50}). It is also clear that the localized zone is not uniform, showing a degree of structure.

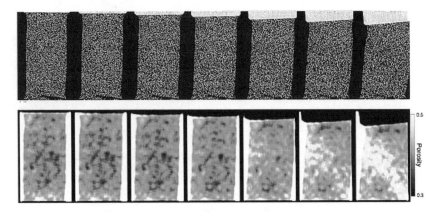

Figure 5. *Vertical slices extracted from the seven 3D x-ray micro tomography images of the sand specimen acquired throughout the triaxial compression (top row) and equivalent slices through the 3D volumes of calculated porosity (bottom row). For scale, remember that the initial sample diameter was 11 mm.*

Discrete V-DIC has been applied to provide incremental analysis of grain kinematics. Following the procedure detailed earlier, for each sand grain in the specimen a set of six scalar quantities (three displacements and three rotations) describing the kinematics of the grain has been determined. From these results, displacement components at any position within a grain can be deduced. Despite these results having been derived from a discrete analysis, they indicate a relatively continuous field of displacements, even in the presence of strain localization, which explains why continuum V-DIC performs well; see Hall *et al.* (2010a).

However, *locally* (i.e. at the scale of the grain contact) the field can be discontinuous. It should kept in mind that individual grains are identified through image segmentation, and it is well possible that grains that are in contact will not appear so as they have been artificially separated. Therefore, we cannot differentiate from such segmented images grains that are in contact from those that are not. A more detailed study of grain contact evolution in space and time would require defining contacts based on the original, non-segmented images (see the paper by Hall *et al.* in this volume, which gives a snapshot of work in progress in our group).

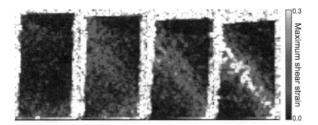

Figure 6. *Continuum V-DIC derived incremental maximum shear strains (as defined in text) for increments 3-4, 4-5, 5-6 and 6-7 (previous increments showed much the same picture as 3-4). The images show vertical slices through the shear strain volume near the middle of the specimen at an equivalent position to Figure 5.*

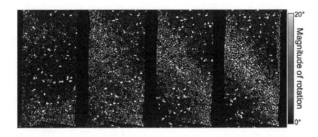

Figure 7. *Discrete V-DIC derived incremental grain rotations for increments 3-4, 4-5, 5-6 and 6-7. The magnitude of grain rotations is plotted for vertical slices through the middle of the specimen at an equivalent position to Figure 6; note that the grains are plotted in the configuration at the start of the test for all increments. Grains colored grey are those for which the image correlation was not successful and those colored white are those with a rotation above a threshold value of 20°.*

Figure 7 shows, for increments 3-4, 4-5, 5-6 and 6-7, the magnitude of rotation for each grain about its rotation axis (recall this is specific for a grain) in a vertical slice corresponding to the middle of the specimen, as in Figure 6. Note that the grains in Figure 7 are represented in the configuration that existed at the beginning of the test, and not in their displaced positions. These images indicate that grain rotations become progressively more intense into a zone that roughly corresponds to where shear strain localizes (see Figure 6).

4.2. *Caicos ooid*

As for the test on Hostun sand, x-ray tomography scans were carried out at key moments throughout the test – marked by relaxations in the loading curve in Figure 8. The sample stress-strain response shows a well-defined peak stress at around 6%

nominal axial strain, after which the stress drops to a plateau; the test was stopped at about 17% axial strain and the sample unloaded.

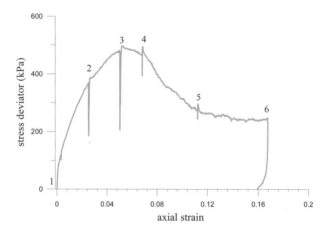

Figure 8. *Stress-strain response from the triaxial compression test on Caicos ooid*

The top row of Figure 9 shows a series of vertical slices through the x-ray tomography images at different stages in the test (see stress-strain curve in Figure 7). These slices are roughly perpendicular to the planar band of localized strain that developed during the test. Differently from what was observed for the test on Hostun sand (see Figure 5), the shear band is clearly visible on these images – starting from post-peak slice 5, and possibly already in slice 4, i.e. immediately after the stress deviator peak. However, the corresponding porosity maps (shown in the bottom row of Figure 9) clearly indicate an evolving inclined zone of localized dilation in the specimen.

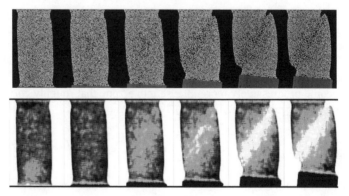

Figure 9. *Vertical slices extracted from the six 3D x-ray micro tomography images of the specimen acquired throughout the triaxial compression (top row) and equivalent slices through the 3D volumes of calculated porosity (bottom row)*

Continuum V-DIC was carried out for this test, the values of the key DIC parameters (distance between the calculation nodes and correlation window size) being the same as for the analysis on Hostun sand. Figure 10 shows vertical slices through the 3D field of maximum shear strain for increments 1-2, 2-3, 3-4 and 4-5. The bottom row results have been smoothed, which yields lower values of the maximum shear strain but also better shows some details of the inner structure of the shear band prior to and around the peak. As it was the case for the test on Hostun sand, these strain images clearly show the evolution of a localized band that traverses the sample diagonally. Again, localization initiated in increment 2-3, i.e. before the peak load; the general picture is of a localization of shear strain and dilatancy which starts as a broad zone and then progressively thins with loading. It is also clear that the localized zone in this test on Caicos ooid shows a lower degree of structure as compared to that observed in the test on Hostun sand.

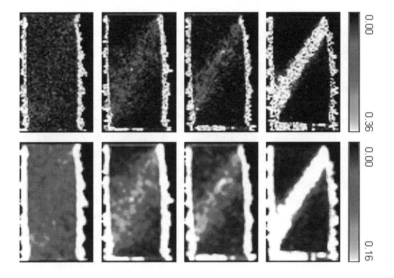

Figure 10. *Continuum V-DIC derived incremental maximum shear strains for increments 1-2, 2-3, 3-4 and 4-5. The images show vertical slices through the shear strain volume near the middle of the specimen at an equivalent position to Figure 9. Results shown on bottom row have been smoothed*

5. Conclusions

The objective of this work was to observe and quantify the onset and evolution of localized deformation processes in granular materials with grain-scale resolution. The key element of our approach is combining state-of-the-art x-ray micro tomography imaging with 3D Volumetric Digital Image Correlation techniques.

This allows not only to view the grain scale details of a deforming sand specimen, but also and more importantly to assess the evolving 3D displacement and strain fields throughout loading.

The application of continuum V-DIC has allowed the development of a localized shear band to be characterized throughout a test. Incremental analysis of consecutive steps reveals that strain localization begins before the peak stress, and indicates a diffuse, wide band progressively thinning after peak. It also appears that the shear band contains a narrower internal core of much higher strain, and that within the band there are aligned zones of either reduced or elevated strains at angles "conjugate" to the main band direction. The degree of such structure seems related to the nature/shape of the grains.

The results obtained so far using the discrete V-DIC confirm the importance of grain rotations associated with strain localization. A clear correspondence can be established between the zones of the specimen experiencing localization of (continuum) shear strain and the zones where grain rotations are more intense. A deeper analysis of continuities/discontinuities at grain contacts and their evolution is now possible, and will be investigated in future work.

In constitutive modeling, one must understand the physics governing material behavior – from the micro scale to the continuum scale. This is particularly true when modeling emergent fine-scale mechanisms whose characteristic length scales are only a few particles wide, e.g. shear bands. With the tools we have presented in this paper, we now have the capability to capture experimentally and at a pertinent level of resolution the details of grain-scale processes, including those that underlie the localization phenomena of interest here. However, the potential of the approach has still to be fully exploited. Future directions of this research include more detailed analysis of: effects of the grains shape (e.g. angularity and aspect ratio of the grains); kinematics across grain contacts and its evolution with strain localization at the macro scale; emergence of grain-scale structures inside a shear band (e.g. the "columns" of aligned grains observed by Oda *et al.* (2004) and also advocated by Rechenmacher (2006) based on continuum 2D DIC); organized kinematics, in particular grain rotation, at the onset of shear banding and through its evolution.

This work was carried out within the framework of the project MicroModEx funded by the French research agency, ANR (contract number: ANR-05-BLAN-0192). We acknowledge M. Di Michiel from the ESRF for his invaluable contribution to the x-ray CT experimental program, and M. Bornert and Y. Pannier at LMS (Paris) who carried out the discrete DIC analysis. C. Rousseau and N. Lenoir (formerly at Laboratoire 3S-R) are also acknowledged for their contributions to the experimental program. Finally, H.E. E. King from ExxonMobil Research and Engineering Co. is acknowledged for providing the ooid material and financial support of the experiments carried out at 3S-R.

6. References

Alshibli, K.A., Sture, S., Costes, N.C., Franck, M.L., Lankton, M.R., Batiste, S.N., Swanson, R.A., "Assessment of localized deformation in sand using X-ray computed tomography", *Geotechnical Testing J.*, vol. 23, 2000, p. 274-299.

Bay, B.K., "Methods and applications of digital volume correlation", *J. Strain Analysis*, vol. 43, 2008, p. 745-760.

Bay, B.K., Smith, T.S., Fyhrie, D.P., Saad, M., "Digital volume correlation: three-dimensional strain mapping using X-ray tomography", *Experimental Mechanics*, vol. 39, 1999, p. 217-226.

Bésuelle, P., Viggiani, G., Lenoir, N., Desrues, J., Bornert, M., "X-ray Micro CT for Studying Strain Localization in Clay Rocks under Triaxial Compression", *Advances in X-Ray Tomography for Geomaterials*, 2006, ISTE, London, p. 35-52.

Bornert, M., Chaix, J.M., Doumalin, P., Dupré, J.C., Fournel, T., Jeulin, D., Maire, E., Moreaud, M., Moulinec, H., "Mesure tridimensionnelle de champs cinématiques par imagerie volumique pour l'analyse des matériaux et des structures", *Instrum. Mes. Metrol.*, vol. 4, 2004, p. 43-88.

Desrues, J., "Tracking strain localization in geomaterials using computerized tomography", *Proc. 1st Int. Workshop on X-ray Tomography for Geomaterials*, 2004, Balkema, p. 15-41.

Desrues, J., Chambon, R., Mokni, M., Mazerolle, F., "Void ratio evolution inside shear bands in triaxial sand specimens studied by computed tomography", *Géotechnique*, vol. 46, 1996, p. 527-546.

Desrues, J., Viggiani, G., "Strain localization in sand: an overview of the experimental results obtained in Grenoble using stereophotogrammetry", *Intern. J. for Num. and Anal. Methods in Geomechanics*, vol. 28, 2004, p. 279-321.

Hall, S.A., "A methodology for 7D warping and deformation monitoring using time-lapse seismic data", *Geophysics*, vol. 71, 2006, p. 21-31.

Hall, S.A., Bornert M., Desrues J., Pannier Y., Lenoir N., Viggiani, G., Bésuelle, P., "Discrete and Continuum analysis of localised deformation in sand using X-ray micro CT and Volumetric Digital Image Correlation", *Géotechnique*, in print, 2010a.

Hall, S.A., Lenoir N., Viggiani G., Bésuelle P., Desrues J., "Characterisation of the evolving grain-scale structure in a sand deforming under triaxial compression", *this volume*, 2010.

Hall, S.A., Lenoir N., Viggiani G., Desrues J., Bésuelle P., "Strain localisation in sand under triaxial loading: characterisation by x-ray micro tomography and 3D digital image correlation", *1st Int. Symp. on Computational Geomechanics*, 2009, IC^{2}E, p. 239-247.

Hall, S.A., Muir Wood, D., Ibraim, E., Viggiani, G., "Image analysis and internal patterning in tests on an analogue granular material", *Granular Matter*, accepted, 2010b.

Lenoir, N., Comportement mécanique et rupture dans les roches argileuses étudiés par micro tomographie à rayons X, PhD thesis, Joseph Fourier University, Grenoble, 2006.

Lenoir N., Bornert M., Desrues J., Bésuelle P., Viggiani G., "Volumetric digital image correlation applied to X-ray micro tomography images from triaxial compression tests on argillaceous rocks", *Strain*, vol. 43, 2007, p. 193-205.

Matsushima, T., Uesugi, K., Nakano, T., Tsuchiyama, A., "Visualization of grain motion inside a triaxial specimen by micro X-ray CT at SPring-8", *Advances in X-Ray Tomography for Geomaterials*, 2006, ISTE, London, p. 35-52.

Matsushima, T., Katagiri, J., Uesugi, K., Nakano, T., Tsuchiyama, A., "Micro X-ray CT at SPring-8 for granular mechanics", *Soil Stress-Strain Behavior: Measurement, Modeling and Analysis*, 2007, Springer, Netherlands, p. 225-234.

Oda, M., Kazama, H., "Microstructure of shear band and its relation to the mechanism of dilatancy and failure of granular soils". *Géotechnique*, vol. 48, 1998, p. 465-481.

Oda M., Takemura T., Takahashi, M., "Microstructure in shear band observed by microfocus X-ray computed tomography", *Géotechnique*, vol. 54, 2004, p.539-542.

Otani, J., Mukunoki, T., Obara, Y., "Characterization of failure in sand under triaxial compression using an industrial X-ray scanner", *Int. J. of Physical Modelling in Geotechnics*, vol. 1, 2002, p. 15-22.

Pan, B., Qian, K., Xie, H., Asundi, A., "Two-dimensional digital image correlation for in-plane displacement and strain measurements: a review", *Meas. Sci.Technol.*, vol. 20, 2009, doi:10.1088/0957-0233/20/6/062001.

Rechenmacher, A.L. "Grain-scale processes governing shear band initiation and evolution in sands", *J. of the Mech. and Phys. of Solids*, vol. 54, 2006, p. 22-45.

Roscoe, K.H., "The influence of strains in soil mechanics", *Géotechnique*, vol. 20, 1970, p. 129-170.

Roscoe, K.H., Arthur, J.R.F., James, R.G., "The determination of strains in soils by an x-ray method", *Civ. Eng. Public Works Rev.*, vol. 58, 1963, p. 873-876 and 1009-1012.

Viggiani, G., Lenoir, N., Bésuelle, P., Di Michiel, M., Marello, S., Desrues, J., Kretzschmer, M., "X-ray micro tomography for studying localized deformation in fine-grained geomaterials under triaxial compression", *Comptes rendus mécanique*, vol. 332, 2004, p. 819-826.

Viggiani G., Hall S. A., "Full-field measurements, a new tool for laboratory experimental Geomechanics", *Deformation Characteristics of Geomaterials*, IOS Press, Atlanta, USA, 2008, p. 3-26.

Quantitative Description of Grain Contacts in a Locked Sand

J. Fonseca* — C. O'Sullivan** — M. R. Coop***

Imperial College
Geotechnics Section
London SW7 2AZ
UK
*joana.fonseca04@ic.ac.uk
** cath.osullivan@ic.ac.uk
*** m.coop@ic.ac.uk

ABSTRACT. *Quantifying the fabric of intact soil is of great importance in both geomechanics and geology. A unique and interesting example of fabric can be found in "locked sands". These geologically old sands are characterized by significant grain interlocking and a low cement content. They can be sampled with minimal fabric disturbance. This study analyzes images acquired by x-ray microtomography of resin impregnated samples of a natural sand, Reigate Silver Sand part of the Folkestone Bed formation from southeast England. 2D and 3D image analyses were carried out to identify the grain-grain contacts and quantify individual contact areas. In contrast to earlier studies that have focused on the coordination number, this work demonstrates that for non-punctual contacts a measure of fabric that considers the contact area may be more appropriate.*

KEYWORDS: *granular material, fabric, contacts, image analysis, locked sands*

1. Introduction

This paper considers fabric characterization of a type of natural sand formation referred to as a "locked sand". The term "locked sand" was introduced by Dusseault and Morgestern (1979) to describe sands that, due to particular deposition conditions, have very low porosity values and a fabric comprising of inter-locked grains. In addition to being a very interesting material to study, the locked fabric of these sands allows undisturbed samples to be obtained. Previous studies on the fabric of natural soils have been limited due to the difficulties in sampling and examining undisturbed samples. A block sampling method was used in this project to obtain intact sand samples. Samples were carefully trimmed in the laboratory and impregnated with a resin prior to extracting sub-volumes for μCT scanning and microscopy. A methodology that makes use of image analysis together with x-ray micro tomography is presented for a three dimensional (3D) characterization of the grain to grain contacts. Combining 2D and 3D characterization will give insight into the applicability of existing two dimensional (2D) to real 3D materials.

2. Locked sand

Locked sands constitute an extreme example of fabric, where grains with irregular and complex morphologies are interlocked with other grains as can be seen in Figure 1(a). As a consequence of these arrangements, in their natural state these sands exhibit relative densities greater than 100%, i.e. the *in situ* porosities are less than can be obtained following the standard ASTM method to obtain the maximum material density. Regarding the type of contacts, there is a predominance of straight contacts and concavo-convex contacts giving high grain-grain contact areas, as shown in Figure 1(b) and (c) respectively. Tangential or point contacts are less frequent and floating particles (i.e. particles with no contacts) are very rare. The fact that this type of fabric cannot be reproduced in laboratory makes it particularly attractive to micro-scale characterization and above all requires intact samples to be used in investigations of the mechanical response. While the original strata to which the term "locked sands" are found in North America (Dusseault and Morgestern, 1979), there are similarities between these formations and the Lower Greensand Formation located in South-Eastern England.

To understand the way the components of fabric influence the yielding and stiffness of these materials, the behavior of intact and reconstituted samples (i.e. samples produced in laboratory following destructuration of the material), of Greensand were compared by Cuccovillo & Coop (1997) and Cresswell & Powrie (2004). The intact sand samples (with undisturbed fabric) were found to have higher shear stiffnesses, higher rates of dilation and higher peak strengths than equivalent reconstituted samples.

Figure 1. *Thin sections of Reigate Silver Sand under cross polarized light (a) illustrating the interlocking fabric (b) concavo-convex contact (c) straight contact*

3. Material and methods

3.1. *Reigate Silver Sand*

Reigate Silver Sand is part of the Folkestone Beds – Lower Greensand Formation, a marine shallow-water deposit from Cretaceous Age. It is a quartzitic sand, classified as being fine to medium graded with mean grain diameter of about 270 μm. It is virtually free of cementation with the exception of some localized deposition iron oxide evidenced by staining.

3.2. *Field sampling*

The intact samples of Reigate Silver sand were sampled at Park Pit Quarry Buckland, Reigate (SE England). The block sampling methodology used consisted of obtaining undisturbed blocks of the material for transportation to the laboratory. The ability of this sand to remain intact in the absence of a confining pressure made it possible to shape blocks of undisturbed soil to a slightly smaller size than the pre-fabricated wooden boxes (35 x 35 x 25 cm). Each block was then wrapped in cling-film and waxed to create an effective barrier against moist loss. Once the wooden box was placed around the block, expansive foam was injected between the soil and the box walls to provide a confinement, minimizing the disturbance to the sample during transportation. The foam was left to set, the base of the block was cut with a wire saw and the block was gently turned over prior to being closed.

3.3. *X-ray micro CT imaging*

The degree of resolution required to analyze the grains in detail necessitated the use of small CT samples only a few millimeters diameter. As a consequence of the friable character of this sand, obtaining this size of sample without significant fabric disruption required the sand to be impregnated with resin.

Cylindrical samples, 38mm in diameter and 76mm high were impregnated in a modified triaxial cell. During impregnation, a constant rate of resin flow was maintained while keeping the flow rate low enough to minimize the disturbance of the soil fabric and to prevent air bubbles from being trapped in the soil pores.

Epo-Tek 301, a very low viscosity resin was used. The CT samples were then drilled from these impregnated triaxial samples, to sizes of about 4 mm to 6 mm, mounted on a rotating steel sample holder and imaged using the commercial microCT systems. A 3D grayscale volume was reconstructed from the series of 2D projections. The final volume was a binary 16 bit dataset. Different magnifications were used resulting in resolutions ranging from 0.8 μm to 7.0 μm. It was found that for this material and for the purpose of the analysis, resolutions of around 3μm give good results. The wide range of particle sizes and the complex morphology of the grains were found to be the primary factors controlling the resolution for this analysis. In fact for resolutions of 7μm details are lost and the grains become blurred. This happens because properties of different materials are merged within a single voxel, i.e. the grain-void transitions are defined by voxels with intermediate gray values. Therefore grain edges are not clearly defined and this inhibits contact detection.

4. Image analysis

Image analysis involves the extraction of measurements from an image. Prior to analysis, μCT images are converted from grayscale to images with meaningful structures. Image processing algorithms were used in order to obtain segmented images where the pixels belonging to each grain are assigned grey level that clearly differentiates them from both the void space and the other grains. The segmentation technique used is based on the 3D watershed algorithm and further details can be found in Fonseca *et al.* (2009).

Figure 2 shows the result of morphological segmentation on a 3D image developed from an original dataset with 2.5 μm resolution. These images are then converted into MATLAB matrices, with 2D arrays representing 2D images and 3D arrays representing 3D images. Each pixel in the image corresponds to a single element in the matrix and the value of this element is the pixel intensity or grey level. Figure 3 is a magnified sub-volume of Figure 2, displaying the individual pixels forming the grain surfaces.

Figure 2. *Representative image processing output, 3D*

Figure 3. *Image analysis input, 3D*

A MATLAB code based on the image processing toolbox (Mathworks, 2004) was developed to identify the contacts between two grains and measure the contact areas as well as the particle surface areas (in both cases these are lengths in 2D). In the algorithm used the elements in the matrices representing the images are assigned a value *0* if they correspond to the background (or pore space) pixels. The pixels representing the solid particle volume are given integer values ranging from 1 to N_p, where N_p represents the number of particles. The contacts between particles are then identified in an automated process of finding for each pixel the neighboring pixels meet the criteria of having an intensity value that is both different from the primary pixel and is not *0*. The contact pixels are labeled with a distinct grey value (that is greater than N_p) and the contact related information is stored using the cell array data structure available within MATLAB.

4.1. *Grain contact description*

Prior grain contact characterization of locked sands has been mainly 2D and qualitative using descriptions like tangential, long or short straight and long or short concavo-convex, (e.g. Palmer and Barton, 1987). Some indices have been proposed to quantitatively characterize the degree of interlocking on a locked sand including the Tangential Index (TI) that gives the proportion of tangential (or point) contacts amongst all grain-grain contacts (Barton, 1993). In general, the most basic particle scale metric to characterize the packing granular media is the coordination number (CN). The average coordination number is defined as the average number of contacts per particle in the system, i.e. twice the number of contacts divided by the number of particles. The coordination number for an individual particle is simply the number of contacts in the system that involves that particle. An additional particle scale measure of contact intensity, the average Contact Index (CI) is introduced here. The CI for an individual particle is the ratio between the contact surface area (S_C) and the particle surface areas (S_P). The average CI for the system is given by equation [1]. In 2D S_C is calculated as a contact length while S_P is the particle perimeter.

$$CI_{ave} = 2\sum_{Nc} S_C \Big/ \sum_{Np} S_p \qquad\qquad [1]$$

5. Results and discussion

The results of the grain contact investigation for two images, one 2D and the second 3D, are presented here. Figure 4 shows the 2D segmented image with the particles displayed in different colors, while Figure 5 shows the contacts between particles. The focus of the research to date has been an assessment of the sensitivity of the microCT data to the imaging approach adopted, as well as the development of the segmentation algorithms (Fonseca *et al.*, 2009). The analysis of the system is computationally expensive and requires significant computer memory. The data presented here are therefore the preliminary results following development of the necessary image processing and analysis algorithms and only a limited volume is considered. Despite the small dataset considered useful insight into the material fabric is attained.

   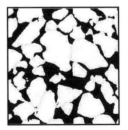

Figure 4. *Image after segmentation, 2D*   **Figure 5.** *Contacts identification, 2D*

The particle coordination numbers are presented in Figure 6(a) and (b) for the two and three dimensional analyses respectively. The range of coordination number values observed in the 3D case (1-10) greatly exceeds the 2D range (1-4). The average CN values were 2.16 in 2D and 3.67 in 3D. The range of coordination numbers will always be greater in 3D in comparison with the 2D case; uniform 2D disks can attain a maximum coordination number of 6, while uniform 3D spheres can attain a value of 12. In this case the 2D image considers only contacts intercepted by an (arbitrary) planar surface. In contrast, the 3D data takes the entire particle volume into account and all the contacts are counted, not just those intercepted by the section plane. While there are no particles with a CN value of 0 (in these images), in both analyses there are noticeable number of particles with

CN=1. For the 3D data, particles with a CN of 1 cannot participate in stress transmission through the material. Particles with a high CN values are less frequent than particles with low CN values. This is a consequence of the range of particle sizes, there are fewer larger particles and the CN values tend to increase with increasing particle size.

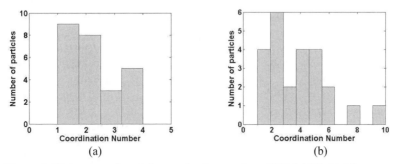

(a)   (b)

Figure 6. *Histograms of particle coordination numbers (CN) (a) 2D (b) 3D*

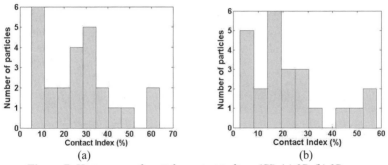

(a)   (b)

Figure 7. *Histograms of particle contact indices (CI) (a) 2D (b) 3D*

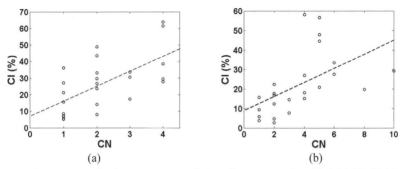

(a)   (b)

Figure 8. *Contact index (CI) variation with Coordination Number (CN) (a) 2D (b) 3D*

The contact index data is presented in Figure 7(a) (2D) and Figure 7(a) (3D). The data is rather erratic and it is difficult to identify a dominant or representative CI value, it is however clear from both the 2D and 3D analysis that most of the particles have CI values in that are less than 30%. However, in 3D the CI value can exceed 50%. The average CI values were 27% in 2D and 22% in 3D. One would expect a linear relationship between the CI and CN values, i.e. the proportion of the particle surface participating in a contact should increase as the number of contacts increases. The CI values for each particle are plotted against the corresponding CN values in Figure 8. There is indeed a general trend for the CI values to increase as the CN numbers increase; however there is a very large amount of scatter. Upon fitting a straight line to the data using a least squares approach, the correlation coefficient for the 2D data was only 0.39 and the 3D data showed an even poorer fit, with a correlation coefficient of only 0.27. It is particularly interesting to consider the 3D data. From this data it is seen that particles with the highest CN values have relatively low CI values, this is a consequence of the presence of tangential contacts increasing the CN to particularly high values. There is also a particle size effect. A large grain may be surrounded by smaller grains and in this case, the CN value is high because of the number contacts, but the CI can be low as the contact area in each case will be small relative to the size of the large grain. On the other hand, a small grain can potentially participate in a relatively large conforming contact. Consequently, in 2D CI values of 40% are observed in cases where the CN is 1. It seems that such large CI values are not attainable for the low CN values in 3D, however it is notable that the highest CI values observed in 3D correspond with CN values that were just slightly higher than the average CN value of 3.67.

6. Conclusions

This paper has presented a methodology to characterize the fabric of natural locked sands based on the qualitative description of particle contacts. Using μCT images of intact soil, an algorithm was developed to identify and measure the contacts. Rather than classifying the contact intensity simply by using the coordination number (whose use is very common in discrete element analyses), the contact index parameter that considers the contact area was introduced. While the contact index tended to increase with coordination number, the two parameters are poorly correlated. The energy required to cause slip along the contacts (and hence deformation of the material) will be related to the contact area, rather than simply the coordination number. Use of the contact index parameter is particularly important for materials with non-punctual contacts. Furthermore the 3D analysis has also revealed information that cannot be obtained by looking at 2D sections.

7. References

Barton, M.E., "Cohesive sands: The natural transition from sands to sandstones", *Geotechnical Engineering of hard Soils-Soft Rocks*, Balkema, Rotterdam, 1993.

Cresswell, A., Powrie, W., "Triaxial tests on an unbonded locked sand", vol. 54, no. 2, pp.107-115, *Géotechnique*, 2004.

Cuccovillo, T., Coop, M.R., "Yielding and pre-failure deformation of structured sands", vol. 47, no. 3, pp.491-508, *Géotechnique,* 1997.

Dusseault, M.B., Morgensten, N.R., "Locked sands", vol. 12, pp. 117-131, *Q. J. Engng. Geol.* 1979.

Fonseca, J., O'Sullivan, C., Coop, M.R., "Image Segmentation Techniques for Granular Materials" *Powders and Grains*, Nakagawa and Luding (eds.), Melville, New York, 2009.

Mathworks, Image Processing Toolbox 7.0 (2004).

Palmer, S.M., Barton, M.E., "Porosity reduction, microfabric and lithification in UK uncemented sands", no. 36, pp. 29-40, *Geological Society Special Publication*, 1987.

3D Characterization of Particle Interaction Using Synchrotron Microtomography

K. A. Alshibli* — A. Hasan**

* Dept. of Civil & Env. Engineering
Louisiana State Univ.-Southern Univ
Baton Rouge, LA 70803
USA
Alshibli@lsu.edu

** Centre National de la Recherche Scientifique
Domain Universitaire
38041 Grenoble Cedex 9
France
Ahasan2@tigers.lsu.edu

ABSTRACT. *Granular particles experience sliding and rolling as they are sheared. X-ray Synchrotron Microtomographic (SMT) was used to acquire 3D scans of a triaxial specimen of sand at eight axial strain levels. The specimen measures 9.5 mm in diameter and 20 mm in height. Several particles within and outside the shear band were identified and tracked as shearing progressed. The analysis reveals that sliding of particles within the shear band is much more significant than particles outside the shear band. Particles within the shear band continue to rotate throughout the experiment while particles outside the shear band exhibit insignificant rotation.*

KEYWORDS: *sand, triaxial, computed tomography*

1. Introduction

The mechanical behavior of granular materials is highly dependent on the arrangement of particles, particle groups and associated pore space. These geometric properties comprise the so-called *structure* or *fabric* of a material. The literature lacks three-dimensional (3D) experimental measurements of fabric changes of sheared granular materials at the particle level. As a result, many researchers used the Discrete Element Method (DEM) to quantify particles sliding and rotation (e.g. Oda *et al.* 1997). Experimental measurements of particles interaction provide key answers to calibrate models for better understanding on the behavior of such materials. In the literature, internal structure analysis techniques are mainly classified as destructive (e.g. specimen stabilization and thin-sectioning) and nondestructive techniques (e.g. magnetic resonance imaging (MRI), ultrasonic testing, x-ray radiography, and computed tomography (CT)). Recently, x-ray CT was utilized to conduct qualitative analysis of sheared soils such as shear band visualization (e.g., Alshibli *et al.*, 2000), however still only few studies extended the analysis to include quantitative measurements of physical properties and tracking the movements of particles. This might be due to the lack of image resolution to visualize particles for some x-ray CT systems. Furthermore, such quantitative analysis requires challenging particle identification techniques and computational power. For example, Chang *et al.* (2003) developed algorithms to track the movement of glass beads at different loading stages from 3D CT images. They embedded a steel ball in the mix of glass beads, which served as a reference point to track relative locations of the other beads. Similarly, Alshibli and Alramahi (2006) tracked the interaction of spherical plastic beads in a cylindrical triaxial specimen at different axial strain increments. The particles have holes which helped to track their relative movement, rotation, and interaction with each other. The data were then used to calculate particles rotation and local strains at the particle-to-particle microscopic level during shear.

Synchrotron x-ray microtomography (SMT) has emerged as the most powerful non-destructive scanning technique that can produce a high-resolution 3D of scanned objects (e.g. Matsushima *et al.*, 2006; Hasan *et al.*, 2008). This paper presents a quantification of 3D particles sliding and rotation within a sheared sand specimen. The analysis focuses on comparing the behavior of particles within and outside the shear band during shear.

2. Materials and methods

A miniature triaxial apparatus was especially fabricated to conduct the *in situ* axisymmetric triaxial experiment. It should be light weight and small in size to facilitate mounting it on the stage of the SMT scanner and must rotate freely to acquire the SMT scans. The small size also helps in increasing the resolution of the SMT scans. The triaxial cell was mounted on the SMT scanner stage using a clamping chuck via a pin attached to the base of the cell. The miniature triaxial apparatus system has similar capabilities as the conventional triaxial cell except that the specimen is confined using vacuum. It consists of a stepper motor to drive the loading ram at a constant displacement rate, a load cell located inside the test cell, a cylindrical acrylic chamber, two endplates, top and base plates, a latex membrane, and a tubing line to connect the specimen to the vacuum. The motor was powered by 12 volts electricity and automatically controlled by a computer. A data acquisition card acquired the signal from load cell with an interface to a computer.

The specimen is cylindrical in shape and measures 9.5 mm in diameter and 20 mm in height. Dry F-75 Ottawa sand was used to prepare the specimen. It is natural uniform silica sand that was mined from Ottawa, Illinois, USA and marketed by the American Silica Company. It has a mean particle size (d_{50}) of 0.16 mm, specific gravity of solid particles of 2.65, minimum and maximum void ratio values of 0.486 and 0.805, respectively. The specimen has a relative density of 90% (dry bulk density of 1.75 g/cm^3). Vacuum was used to apply a confining pressure (σ_3) of 58 kPa. The specimen was compressed axially (σ_1) at a displacement rate of 0.1 mm/min. Compression was paused at 0%, 2.5%, 5%, 7.5%, 10%, 12.5%, 15%, and 17.5% nominal axial strain (ε_1) to acquire the SMT scans. It took approximately 2 hours to acquire each scan.

The CT scans were acquired using Sector 13-BMD synchrotron microtomography beamline at the Advanced Photon Source (APS), of the Argonne National Laboratory (ANL), Illinois, USA. An x-ray beam with energy of 33 keV was used to scan the specimen which produced a good contrast between sand particles and the void space. The voxel size of images was 10.26 µm at which individual sand particles were clearly seen by a naked eye. Algorithms were developed using the Interactive Data Language (IDL) software to reconstruct the image projections, to build 3D renderings, and to further process the scan images (Thompson *et al.* 2006).

3. Particle sliding and rotation during shear

A group of particles within and outside shear band regions were visually identified, tracked, and analyzed. They were randomly selected at the initial stage (before compression/shearing). Then, the same particles were tracked in the subsequent shearing stages. The displacement fields (Δx, Δy, and Δz), and horizontal and vertical rotation angles ($\Delta \kappa$ and $\Delta \lambda$) were calculated as follows:

$$\Delta x = x_n - x_o \qquad\qquad\qquad [1]$$

$$\Delta y = y_n - y_o$$
$$\Delta z = z_n - z_o$$
$$\Delta \kappa = \kappa_n - \kappa_o$$
$$\Delta \lambda = \lambda_n - \lambda_o$$

where subscript n denotes the coordinate at the nth strain stage and subscript o denotes the coordinate at the initial stage (i.e. when $\varepsilon_1 = 0\%$). The rotation of each particle was calculated as the change in the orientation of the particle major axis. Angles λ and κ define the orientation with respect to the z-axis (axial direction) and x-y plane (axisymmetric direction), respectively. Two rotation angles can fully describe the rotation of particles since the x-y plane represents the axisymmetric plane in triaxial testing.

Figure 1 shows the visualization of the evolution of the tracked particles displacements and rotations within and outside shear band. They are displayed in y-z plane (axial plane) with each figure frame serves as a reference for particles location. Note that there are some particles that might exist between these particles, but they were not displayed in the figure. There are two primary mechanisms of particle interaction visualized here: interparticle sliding and rolling. Particles within the shear band exhibit a significant sliding as the test continued due to the intensive shearing within the shear band where particles slide together globally as well as slide relative to each other. Globally, all particles within the shear band move downward to the right parallel to the shear band inclination angle. To some extent, some particles lose their interlocking which permits them to slide/shear (either face-to-face or face-to-edge) with their neighboring particles. On the other hand, particles outside the shear band exhibit insignificant sliding or rolling. Some particles within the shear band show a considerable rotation while sliding. They exhibit rotation in a continuous or an oscillatory mode either in the horizontal (axisymmetric) or vertical (axial) directions.

Figure 2 displays the displacement in the x, y and z directions of the tracked particles within and outside shear band at each stage. The curves show that particles within the shear band continue to slide throughout the test. Little change in curvature is shown for the x- and y-displacements at 5% axial strain. No change in curvature is shown at 5% axial strain for the z-displacement curve. The x-, y- and z-displacements increase as axial strain increases and the rate decreases after 12.5% axial strain. At $\varepsilon_1 = 17.5\%$, some particles slide as high as 1.1 mm (11.6% of specimen initial diameter (D_o) or $6.9d_{50}$) for the x-displacement, 0.35 mm (3.7% of D_o or $2.2d_{50}$) for the y-displacement, and 1.1 mm (5.5% of specimen initial height (H_o) or $6.9d_{50}$) for the z-displacement. On the other hand, particles outside shear band show much less sliding when compared to those within the shear band. At $\varepsilon_1 = 17.5\%$, particles slide only 0.07 mm (0.07% of D_o or $0.4d_{50}$) for the x-displacement, 0.05 mm (0.05% of D_o or $0.3d_{50}$) for the y-displacement, and 0.1 mm (0.05% of H_o or $0.9d_{50}$) for the z-displacement.

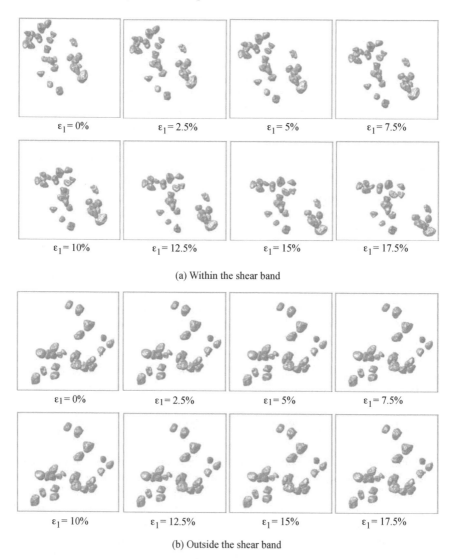

(a) Within the shear band

(b) Outside the shear band

Figure 1. *Illustrative example of particles evolution during shear*

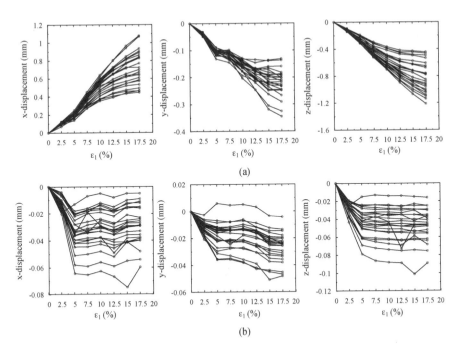

Figure 2. *Displacements of tracked particles*
(a) within the shear band; (b) outside the shear band

However, the trends show that these particles experience a small amount of sliding until 5% axial strain. Then, they become almost flat and stabilize after 5% axial strain, which implies particles no longer slide. Note that the 5% axial strain is the axial strain close to the peak stress state. The curves demonstrate that before the peak stress state, particles in both locations experience sliding due to the axial loading. After the peak stress state, deformations localize into the shear band when particles outside the shear band experience little change as shearing continue along the shear band.

4. Conclusions

The SMT was successfully utilized to visualize the evolution of particles sliding and rolling and to calculate their orientation distributions in the axisymmetric and axial directions. Particles within and outside the shear band region slide during shearing until peak stress state is reached. Then, only particle within the shear band continue to slide during the post peak stress regime. Sliding of particles within the shear band is much more significant than particles outside the shear band. Particles within the shear band continue to rotate throughout experiment while particles outside the shear band exhibit insignificant rotation.

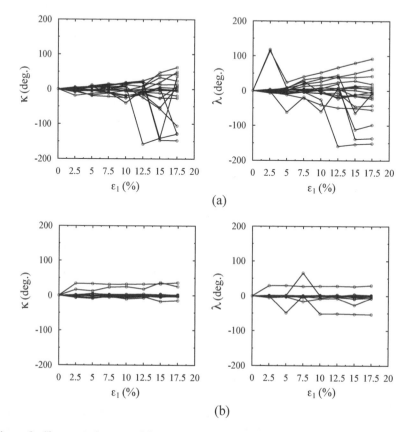

(a)

(b)

Figure 3. *Changes in horizontal (axisymmetric) rotations (Δκ) and vertical (axial) rotations (Δλ) of tracked particles (a) within the shear band; (b) outside the shear band*

5. Acknowledgments

The authors gratefully acknowledge the financial support of the National Science Foundation grant number CMMI- 0653957. The data presented in this paper were collected using the x-ray Operations and Research Beamline Station 13-BMD at Argonne Photon Source (APS), Argonne National Laboratory. We thank Mark Rivers of (APS) for help in performing the scans. Use of the Advanced Photon Source was supported by the U. S. Department of Energy, Office of Science, Office of Basic Energy Sciences under Contract No. DE-AC02-06CH11357.

6. References

Alshibli, K. A., and Alramahi, B. A., "Microscopic Evaluation of Strain Distribution In Granular Materials during Shear", *ASCE J. Geotech. Geoenviron. Eng.*, vol. 132, no. 1, pp. 483-494, 2006.

Alshibli, K. A., Sture, S., Costes, N. C., Frank, M., Lankton, M., Batiste, S., and Swanson, R., "Assessment of Localized Deformations in Sand Using x-ray Computed Tomography", *ASTM Geotech. Testing J.*, vol. 23, no. 3, pp. 274-299, 2000.

Chang, C.S., Matsushima, T., and Lee, X., "Heterogeneous strain and bonded granular structure change in triaxial specimen studied by computer tomography", ASCE *J. Eng. Mech.*, vol. 129, no. 11, pp. 1295-1307, 2003.

Hasan, A., Alshibli, K., Heinrich, J., Rivers, M., and Eng, P., "Visualization of Shear Band in Sand Using Synchrotron Micro-Tomography", *Proc. GeoCongress 2008, Characterization, Monitoring, and Modeling of GeoSystems (GSP 179)*, ASCE, New Orleans, pp. 1028-1035, 2008.

Matsushima, T., Uesugi, K., Nakano, T., and Tsuchiyama, A., "Visualization of grain motion inside a triaxial specimen by micro X-ray CT at SPring-8", *Advances in X-ray Tomography for Geomaterials*, J. Desrues, G. Viggiani, and P. Besuelle (eds.), ISTE Ltd., London, pp. 255-261, 2006.

Oda, M., Iwashita, K., and Kakiuchi, T., "Importance of particle rotation in the mechanics of granular materials", *Powder & Grains 97*, Balkema, Rotterdam, 1997.

Thompson, K.E., Willson, C.S., and Zhang, W., "Quantitative computer reconstruction of particulate materials from microtomography images", *Powder Technology*, vol. 163, pp. 169-182, 2006.

Characterization of the Evolving Grain-Scale Structure in a Sand Deforming under Triaxial Compression

S. A. Hall[*] — N. Lenoir[**] — G. Viggiani[*] — P. Bésuelle[*] — J. Desrues[*]

*Laboratoire 3S-R,
CNRS – Grenoble Universities
Domaine Universitaire
BP 53
38041 Grenoble
France
stephen.hall@hmg.inpg.fr
cino.viggiani@hmg.inpg.fr
pierre.besuelle@hmg.inpg.fr
jacques.desrues@hmg.inpg.fr

**Previously at Laboratoire 3S-R,
Now at Institut Navier - LMSGC
2, allée Kepler
77420 Champs-sur-Marne

ABSTRACT. This paper is a snapshot of work in progress in which the aim is to analyze possible evolving grain-scale structures associated with localized deformation in a sand specimen undergoing triaxial compression. 3D digital image analysis of in-situ-acquired x-ray tomograms provides a characterization of porosity, contact density and grain coordination number distributions. These characteristics are compared with strain fields derived by 3D-volumetric digital image correlation (DIC), which reveal the development of a localized shear band. Structures conjugate to the main shear band have been observed clearly in incremental DIC-derived shear strain images and also in porosity images. New results on grain contact and grain coordination number distributions suggest that lower values of both exist in the localized strain zone, but bridges of higher coordination-number grains appear to traverse this zone.

KEYWORDS: localization, granular media, in-situ x-ray tomography, digital image analysis, digital image correlation

1. Introduction

To understand the mechanisms of deformation and failure of materials, and in particular strain localization phenomena, requires an entire deformation process to be followed in 3D, at the appropriate scale, while the test specimen deforms under load. This can be achieved using *in-situ* x-ray tomography (i.e. x-ray scanning at the same time as loading). In recent work (Hall *et al.*, 2010; see also Viggiani *et al.* "Sand deformation at the grain scale quantified through x-ray imaging", in this volume) in-situ x-ray micro-tomography with grain-scale resolution of triaxial compression tests on sand has been presented. In this previous work the development of strain localization was studied, based on the tomography images taken at key stages throughout the loading, using two types of 3D-volumetric Digital Image Correlation (DIC). In the first instance "continuum" DIC was used to derive incremental 3D displacement and strain fields for the deforming specimen by essentially ignoring the granular aspect of the specimen and images and treating the medium/image as piece-wise continuous. Despite the neglect of the grains, this analysis provided interesting insight into the development of a localized shear band, with some indications of the grain-scale mechanics.

A second 3D-DIC analysis was carried out that allowed the tracking of the individual displacements and rotations of all the grains in the specimen. This refined grain-scale analysis demonstrated the significant role of grain rotations in the localization process. A key observation from both types of DIC analyses, but perhaps more evident in the continuum results (see later), is the emergence of grain-scale structures (zones of lower or elevated shear strains conjugate to the main band direction) inside the shear band. These structures resemble the "columns" of aligned grains observed in sand by Oda *et al.* (2004) in post-mortem (i.e. after test) x-ray micro-tomography images and also advocated by Rechenmacher (2006), based on continuum 2D-DIC. Furthermore, Tordesillas and Muthuswamy (2009), amongst others, suggest that the underlying mechanism for shear banding in granular materials is by the buckling of such columns of grains. These aspects need to be further investigated and this work aims to analyze the grain-scale structures that might exist and their evolution by 3D digital image analysis of the x-ray tomograms.

In the following, the experimental data are briefly described before presenting analyses of the structure and structural evolution of the sample based on previous results (porosity analysis and 3D-volumetric DIC) and new analyses to assess the grain and grain-contact organization. This paper represents just a snapshot of work in progress with preliminary results and analyses. For example, only two load steps have currently been analyzed in terms of the grains and grain-contacts – the first, before deviatoric loading, and the seventh, well after the peak in stress.

2. Experimental set-up and material

2.1. *Material studied and mechanical response*

The experimental results analyzed in this work come from an *in-situ* triaxial compression test performed on a dry specimen of S28 Hostun sand, under a confining pressure of 100 kPa carried out at the European Synchrotron Radiation Facility (ESRF) in Grenoble on beamline ID15A. The specific loading system used is described in more detail in Hall *et al.* (2010) (see also Lenoir *et al.*, 2006).

The S28 Hostun sand is a fine-grained, angular siliceous sand with a mean grain size (D_{50}) of about 300 μm. For this test, the sample was 11 mm in diameter and 22 mm high with an initially dense packing (initial void ratio ~0.6). Figure 1(a) shows the sample stress-strain response, in which a roughly linear initial trend is followed by a curvature to the peak stress at around 11% axial strain. After the peak load, the stress drops and appears to begin to plateau before the sample unloaded. Relaxations can be seen at the moments at which x-ray tomography scans were carried out.

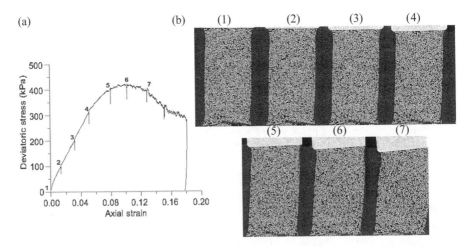

Figure 1. *(a) Sample stress-strain response: deviatoric stress (q = the maximum minus the minimum principal stress) versus axial strain. (b) Vertical slices through the x-ray micro-tomography image volumes at the 7 levels of strain indicated in (a). The slices are cut through the middle of the specimen roughly perpendicular to the eventual band of localized strain. For scale: the sample was originally 11 mm diameter and 22 mm high.*

2.2. *In-situ x-ray micro-tomography*

X-ray tomography scans were carried out at key moments throughout the test as indicated in Figure 1(a). Figure 1(b) shows a series of vertical slices through each of these x-ray tomography image volumes. The vertical slices shown are cut through the middle of the volumes roughly perpendicular to the "plane" of localization that developed during the test. The voxel size of these images is $14 \times 14 \times 14 \ \mu m^3$, which means the individual grains can be clearly identified (the mean grain diameter is around 300 μm or 21 voxels so there are several thousand voxels per grain).

These *in-situ* images show that in the latter part of the test the sample starts to lean to the right and the upper platen rotates, which is probably the result of a non-perfect sample. From these images, there is no indication of localized deformation is immediately obvious and enhanced analysis is required to visualize and quantify the localization structure as discussed in the following.

3. Data analysis: characterization of structural evolution

3.1. *Porosity analysis*

The porosity fields presented in Figure 2 were calculated from the x-ray tomography image volumes described in section 2.2. This calculation was based on a binarization of the images into "grain" and "pore-space" voxels (i.e. a threshold gray-scale value has been defined such that voxels with a gray-scale value above this threshold are considered to be within a grain and those below the threshold, pore space). Porosities are given for overlapping cubic windows of side 141 voxels (1974 μm, about 7 x D_{50}) at every 10th voxel throughout the sample volume. The window size used is a compromise between a good representative elementary volume size and the need for spatial resolution. Note also that the derived values represent the accumulated porosity from the start of the test.

From the porosity fields, an evolving inclined zone of localized porosity increase (dilation) can be seen. Also, there are indications of structure in the band in the form of high and low porosity striping conjugate to the main band orientation.

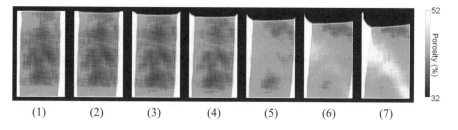

Figure 2. *Vertical slices through the porosity volumes for the 7 load steps indicated in Figure 1(a) for the same positions as in Figure 1(b)*

3.2. *3D-volumetric DIC*

Continuum 3D-volumetric DIC was carried out on consecutive pairs of the 3D x-ray tomography image volumes to provide the 3D displacement and strain fields for each increment. The details of this procedure are presented in Hall *et al.* (2009) and Hall *et al.* (2010). Figure 3 shows vertical slices through DIC-derived maximum shear strain volumes; as in the previous images, these vertical slices are cut roughly perpendicular to the "plane" of localization that developed during the test.

The strain images in Figure 3 clearly show the evolution of a localized band that traverses the sample diagonally from top-left to bottom-right. The localization appears to initiate in the increment 4-5, i.e. well before the peak load. Furthermore it can be seen that the band starts as a broad zone and converges towards a narrow band of localization in step 6-7 (the zone is around 5 mm, about 17 D_{50}, at this stage) with a narrower, high strain core. It is also clear that the localized zone is not uniform and shows some structure, including aligned zones of either reduced or elevated strains at angles "conjugate" to the main band direction.

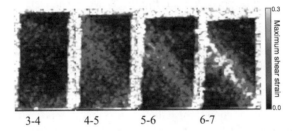

Figure 3. *Vertical slices through the incremental maximum-shear-strain volumes for load increments indicated (see Figure 1a); slice positions as previous figures*

3.3. *Grain contact analysis*

The DIC results show that the localized band is not uniform and, while it is well defined (and quite planar), there are aligned zones of reduced and elevated strains at an angle "conjugate" to the main band. The orientations of these zones are similar to those of "columns" of aligned grains identified by Oda *et al.* (2004) in a shear band in sand. Therefore the shear-strain structures might indicate the presence of columnar structures in the grain assemblage. In order to investigate if indeed these structures are related to some grain-scale organization, e.g. columns, and if such structures evolve with loading, in this section procedures for analyzing grain and grain contact distributions are investigated. At present, just the first and last x-ray tomography images have been analyzed. The first step in this analysis is the identification and separation of the individual grains, which is followed by the identification of the contacts between neighboring grains; these steps are described in the first subsection below. Once the images have been segmented and the contacts identified, it is then possible to characterize the distribution of the contacts, which is discussed in the subsequent section.

3.3.1. *Image segmentation and contact recognition*

The x-ray tomography volume images of the specimen are segmented in order to identify and label individual grains. This is performed using an image binarization followed by the application of a watershed algorithm using the image-processing package VISILOG (©Noesis, http://www.noesisvision.com/). This yields an image volume in which each grain is assigned a unique label (number) such that they can subsequently be identified and distinguished.

Once the grains are separated and labeled, contacts between grains might be identified. Here, this is achieved by defining a contact by black voxels (i.e. those that are not within a grain) that have direct-neighbor voxels (i.e. those in the surrounding 26 voxels) that belong to two different grains, identified by their unique labels. Applying this approach over the whole volume thus provides a binary or labeled volume of contacts. In fact these are the potential contacts and likely an over-estimate of the number of contacts for two reasons. First, contacts can only be resolved within the resolution of the images (sub-voxel size grain separations will appear as contacts). Second, segmented images are used, in which contacts have been cut to separate the grains, so differentiating cut contacts from grains originally separated by a single voxel is not possible. An alternative approach is to subtract the segmented image from the unsegmented one - the non-zero voxels that remain will be contact voxels, but currently the segmented images are being used.

3.3.2. *Contact distribution – contact density and grain coordination number*

Given the volumes of contacts it is now possible to quantify their distribution. Currently this has been carried out in terms of the contact density distribution (number of contact pixels per unit volume, a "continuum" field measure like porosity) and the grain coordination number (number of contacts per grain, a "discrete" grain-based measure).

In a similar way to the porosity fields, the calculation of contact density is based on the ratio of black and white voxels (in a binary volume where contacts are the white voxels) in overlapping cubic windows of side 61 voxels (854 μm or about $3 \times D_{50}$) throughout the sample volume. This procedure yields a 3D map of contact densities throughout the sample in terms of the number of voxels in contacts (i.e. the contact area) per unit volume; an alternative measure would be simply the number of contacts. Figures 4 and 5 show results of the contact density distribution and coordination number calculations for tomography images 1 and 7, respectively.

(a)

(b)

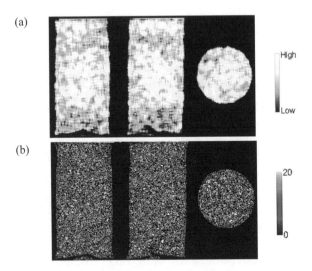

Figure 4. *Slices through the (a) contact density and (b) coordination number volumes at step 1. The three slices in each case are cut through the middle of the volume along the three image axes – the middle slice is the same slice position as in previous figures*

(a)

(b)

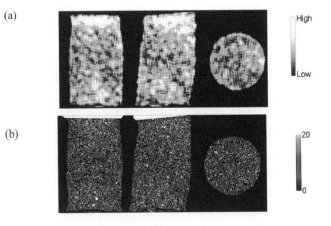

Figure 5. *As Figure 4, but for step 7*

From the images in Figures 4-6 it can be seen that the zone of localized shear strain (see Figure 3) is associated with a lower density of contacts and is populated with lower coordination number grains, which are both consistent with the observed increase in porosity in this zone. It can also be seen that, in general, the coordination numbers were higher and more homogenously distributed at the start of the test, although the contact density maps do indicate some degree of heterogeneity in the contact distribution even before the deviatoric loading. Thus is appears that the localization is associated with a loss of contacts, but so far no clear structures are observed, although, from Figure 6, it can be seen that the zone of generally lower contacts in the localized strain band is traversed by higher coordination number grains. Further quantitative analysis is needed to assess the possible structures in these images.

4. Conclusions

This paper has given a snapshot of work in progress in which the aim is to analyze possible evolving grain-scale structures associated with localized deformation using 3D digital image analysis of *in-situ*-acquired x-ray tomograms. So far structures conjugate to the main shear band have been observed clearly in incremental DIC-derived shear strain images and also in porosity images. New results on grain contact and grain coordination number distributions have been presented and initial analyses suggest that lower contact densities and lower coordination-number grains characterize the localized strain zone. Additionally, bridges of higher coordination number grains appear to traverse this zone. So far, only two load steps have been analyzed, one before deviatoric loading and the other after the shear band is well developed (and so potentially after structures have been broken). Analysis of the other load steps and of other structural features will yield more insight. It should be noted that much of what is being looked for is related to force chains, which cannot be seen in tomography images. However, for a force chain to exist requires contacts (although the presence of a contact does not indicate force communication nor does the size of contact indicate its importance in terms of force transfer). Therefore, the identification of lines of contacts and changes in contacts with time might indicate some details of the underlying force communication evolution; the continuation of this work should shed new light on such aspects.

Figure 6. *3D rendered view (image J) through the coordination number volumes (left – step 1, right – step7) with thresholding to remove grains with low coordination numbers*

5. References

Hall S.A., Lenoir N., Viggiani, G., Desrues J., Bésuelle, P., "Strain localisation in sand under triaxial loading: characterisation by x-ray micro tomography and 3D digital image correlation", *Proceedings of COMGeo09*, IC^2E, 2009, p 239-247.

Hall S.A., Bornert M., Desrues J., Pannier Y., Lenoir N., Viggiani, G., Bésuelle, P., "Discrete and Continuum analysis of localised deformation in sand using X-ray micro CT and Volumetric Digital Image Correlation", *Géotechnique*, in print, 2010.

Lenoir N., Bornert M., Desrues J., Bésuelle P., Viggiani G., "Volumetric digital image correlation applied to X-ray micro tomography images from triaxial compression tests on argillaceous rocks", *Strain*, vol. 43, 2007, p. 193-205.

Oda M., Takemura T., Takahashi, M., "Microstructure in shear band observed by microfocus X-ray computed tomography", *Géotechnique*, vol. 54, 2004, p.539-542.

Rechenmacher A.L. "Grain-scale processes governing shear band initiation and evolution in sands", *J. of the Mech. and Phys. of Solids*, vol. 54, 2006, p.22-45.

Tordesillas A., Muthuswamy M., "On the modeling of confined buckling of force chains", *J. of the Mech. and Phys. of Solids*, vol. 57, 2009, p. 706-727.

Visualization of Strain Localization and Microstructures in Soils during Deformation Using Microfocus X-ray CT

Y. Higo* — F. Oka* — S. Kimoto* — T. Sanagawa** — M. Sawada*** — T. Sato* — Y. Matsushima*

* *Department of Civil and Earth Resources Engineering, Kyoto University*
C1 Bd., Kyotodaigaku-katsura 4, Nishikyo-ku, Kyoto, 615-8540, Japan
higo@mbox.kudpc.kyoto-u.ac.jp
foka@mbox.kudpc.kyoto-u.ac.jp
kimoto@mbox.kudpc.kyoto-u.ac.jp
tomoya@earth.mbox.media.kyoto-u.ac.jp
yoshiki@handballer.mbox.media.kyoto-u.ac.jp

** *Department of Civil and Earth Resources Engineering, Kyoto University*
(Presently, Railway Technical Research Institute)
Kokubunji-shi, Tokyo, 185-8540, Japan
sanagawa@rtri.or.jp

*** *Department of Civil and Earth Resources Engineering, Kyoto University*
(Presently, Taisei Corporation, Technology Center)
Totsuka-ku, Yokohama, 245-0051, Japan
swdmi-00@pub.taisei.co.jp

ABSTRACT. It is well known that strain localization is an important issue for the onset of failure problems. In order to clarify the mechanism of failure, it is necessary to visualize the strain localization and microstructure changes in detail. The aim of this paper is to observe strain localization behavior and microstructures of soils during deformation process using a microfocus X-ray CT. Strain localization of unsaturated Toyoura sand specimen and Maruyamagawa clay specimen during compression tests have been observed and discussed. In addition, microstructures of sands have been visualized by partial CT scan. In the partial CT images, we can clearly visualize each soil particle distinguished from the others.

KEYWORDS: strain localization, microstructure, microfocus X-ray CT, unsaturated sand, clay

1. Introduction

It is well known that the microstructure such as arrangement of soil particles and void distribution is a key to understand the strain localization and failure of geomaterials. The strain localization of geomaterials has been studied by many researchers since the strain localization phenomenon is an important issue for onset of failure problems. For partially saturated soil, the pore water exists as meniscus with suction force which behaves as capillary force between the soil particles. The water meniscus strengthens the soil as a capillary force while the collapse of the water meniscus causes drastic loss of strength by shearing or infiltration of water.

Strain localization of geomaterials has been studied by many researchers. In particular, investigations using x-ray CT have achieved a lot of outcomes (e.g. Desrues *et al.* 1996; Alshibli *et al.* 2000; Otani *et al.* 2000; Kodaka *et al.* 2006), and, recently, microstructure changes due to shearing have been studied by micro computed tomography (e.g. Oda *et al.* 2004; Kikuchi 2006; Matsushima *et al.* 2006). To the author's knowledge, however, the published researches are mainly for dry sand although most of the natural geomaterials are fully saturated or unsaturated porous media.

In the present study, strain localization and microstructures in unsaturated sand and clay were investigated using a microfocus x-ray CT. The microfocus x-ray CT system used in this study has very small focus size and has high voltage and current; hence, it is possible to visualize the microstructures in rather larger specimens with high spatial resolution. In addition, since the triaxial cell can be mounted on the rotation table, x-ray CT scan during triaxial compression tests can be performed. In the present paper, using the microfocus x-ray CT, visualization of the strain localization and microstructure during compression tests for unsaturated Toyoura sand and Maruyamagawa clay specimens were conducted. Moreover, we have performed partial CT scan with very high magnification for sand specimens, in order to visualize each sand particle distinguished from others, i.e. the microstructures of sand specimens.

2. Microfocus x-ray CT

In this study, we used a microfocus x-ray CT system, TOSCANER-32250µHDK assembled by TOSHIBA IT & Control Systems Corporation, which was newly introduced in Dept. Civil & Earth Resources Engineering of Kyoto University. The specification of the system is listed in Table 1. The focus size of the microfocus x-ray tube is very small, 4 µm, which provides very high spatial resolution of 5 µm. The system also has a high penetrating ability with the maximum voltage of 225 kV and the maximum current of 1 mA. The voltage and the current can be controlled independently and the maximum electric power consumption is 200 W.

Figure 1 depicts a schematic figure of the microfocus x-ray CT system. The microfocus x-ray tube generates a cone shape of white x-ray beam. A two dimensional x-ray image intensifier (I.I.) records the x-ray attenuation at different angles equally spaced 360° by rotating the object on a computer-controlled rotation table (rotate-only method). The x-ray I.I. converts an X-ray photon to visible light, and then the light is transformed to the digital data by a CCD camera. Using the data, the computed tomography technique provides a spatial distribution of CT value with a gray scale, i.e. CT image. The total number of the colors is 256.

X-ray source	Maximum voltage	225 kV
	Maximum current	1 mA
	Maximum electric power consumption	200 W
	Min. focus size	4 μm
	Maximum projection views	4,800 (0.075°)
	Maximum accumulation per 1 view	50
Work table	Maximum size of specimen	φ700 mm, h700 mm
	Maximum scanning area	φ200 mm
	Maximum weight of specimen	441 N (45 kgf)
Image Intensifier	Size of detector	9/6/4.5 inch
	Image matrix	$512^2/1024^2/2048^2$
CT image	Spatial resolution performance	5 μm
	Scanning method	single scan (2D CT image)/ cone-beam scan (3D CT image)

Table 1. *Specifications of the microfocus X-ray CT*

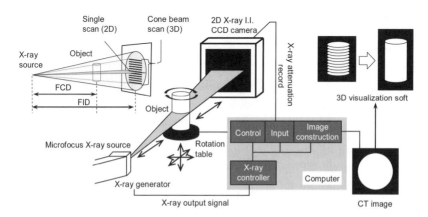

Figure 1. *Schematic diagram of the microfocus X-ray CT*

The cone beam technique gives several horizontal CT images in the vertical direction with one scanning. Finally, we can obtain a complete three-dimensional CT image constructed by a visualization software (VGStudio MAX2.0, Volume Graphics GmbH). The distance between the x-ray source and the rotation table (FCD) and that between the x-ray source and the x-ray I.I. (FID) can be changed manually so that the magnification can be determined arbitrarily. The magnification also depends on the x-ray image intensifier with the user-selectable size of 4.5, 6 and 9 inches in diameter. In addition, since the triaxial cell can be mounted on the rotation table, the specimens during triaxial tests can be scanned.

3. Visualization of strain localization in soil specimens

3.1. *Unsaturated sand during triaxial compression tests*

The test sample used in this study was Toyoura sand which is classified as semi-angular sand. The average diameter D_{50} was 0.185mm, the uniformity coefficient was 1.6, the maximum void ratio was 0.975, the minimum void ratio was 0.614, and the fine content was 1%.

The specimen was prepared by the moist-tamping method using a rammer and a mould. Prior to performing the compaction, the dry Toyoura sand and the water were mixed; the initial water content was 8.14%. Then, the mixed wet Toyoura sand was compacted in several layers in a cylindrical mould with 35 mm in diameter and 70 mm in height. The initial void ratio is 0.66 and the relative density is 87%. The triaxial cell used in the present study was made by lucid acryl so that x-ray attenuation by the triaxial cell is as small as possible. Minus 20kPa of air pressure was applied inside the specimen as a confining pressure during the compression test, while the cell pressure was equal to atmospheric pressure. Hence, this test was carried out under drained conditions for water and air. After setting the specimen, the triaxial cell was placed on the rotation table. Axial pressure was loaded by displacement control system of DC servomotor. The applied axial strain rate was 0.5%/min.

The specimen was scanned at four different steps: initial state, before and after the peak deviator stress, and after visible formation of shear bands. Cone-beam CT method has been used for scanning, in which the voltage is 150 kV, the current was 200 µA, the size of the x-ray I.I. was 6 inch, the number of projection views was 1200, the accumulation per one view was 10, the matrices of image was 1024×1024, and the voxel size was 41 µm × 41 µm × 300 µm.

The stress-strain relation obtained by the test is demonstrated in Figure 2. The specimen exhibits strain-hardening behavior until an axial strain of 5.5%, and then it shows strain-softening. At an axial strain of 15%, shear bands developed clearly enough to be visible by naked eyes. We have scanned at four stages: at the initial state, before the peak stress, after the peak stress, and after the visible formation of shear bands. At each stage the axial loading was stopped just before the scanning, and then the specimen was scanned, namely, the axial strain was held constant during the scanning. Every scanning in this study spent about two hours; hence, stress relaxation can be seen in the stress-strain relation. After the scanning, axial loading was restarted with the same axial strain rate.

Figure 2. *Stress-strain relation and the scanning steps (Toyoura sand)*

Figure 3. *Cross-sections of CT image*

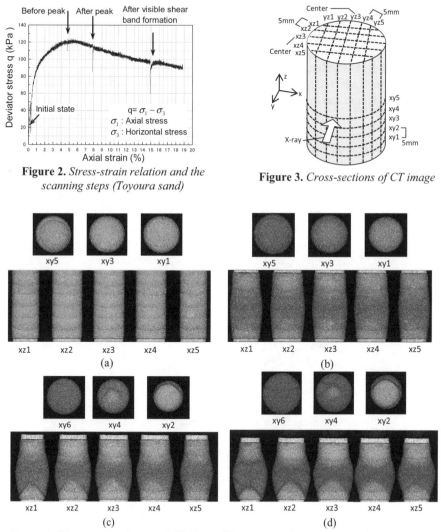

Figure 4. *Horizontal and vertical CT slices of Toyoura sand at four stages during triaxial compression test: (a) initial state, (b) before peak (axial strain of 5%), (c) after peak (axial strain of 8%), and (d) after visible shear bands formation (axial strain of 15%)*

Figure 4 shows reconstruction of CT images at four different stages. In the CT images, darker regions denote lower densities and lighter regions denote higher densities. The location of the sections is illustrated in Figure 3.

At the initial state several curved lines at almost equal distance can be seen in the vertical sections, and circular line appears in the image of xy5 section. These lines correspond to the boundaries between compaction layers. The lower density region at the center of the specimen can be seen before the peak stress, at an axial strain of 5%. This is because the volume expansion due to dilatancy occurs in the specimen.

After the peak stress, shear bands can be seen in the lower part of the specimen in the CT image at an axial strain of 8% although the shear bands could not be seen with naked eyes. The localization pattern involves two types of shear band, which is similar to that observed in air-dried sand by Desrues *et al.* (1996) (one of them is an axial conical band and the other is a radial-planar band).

3.2. Water-saturated clay during unconfined compression test

The clay used in this study was Maruyamagawa clay sampled at Kyoto Pref., which was almost fully saturated and very soft clay: the wet density was 1.545 g/cm^3, natural water content was 64.99%, the degree of saturation was 99.4%, and the initial void ratio was 1.6. The clay sample was trimmed to be a cylindrical specimen with 35 mm in diameter and 70 mm in height. Then the specimen was put into the triaxial cell same as above, but no confining pressure was applied, namely, this test was conducted under the unconfined compression conditions. The axial strain rate was 0.5%/min.

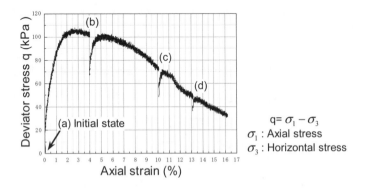

Figure 5. *Stress-strain relation and the scanning steps (Maruyamagawa clay)*

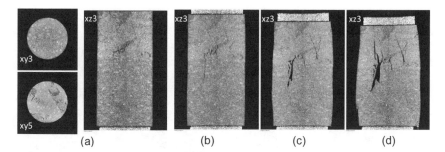

Figure 6. *Horizontal and vertical CT images of Maruyamagawa clay at four steps during unconfined compression test: (a) Initial state, (b) Axial strain of 4%, (c) Axial strain of 10%, and (d) Axial strain of 13%*

The stress-strain relation is shown in Figure 5. X-ray scanning at four steps was carried out as shown in Figure 5. The CT image demonstrates that the density of the lower part of the specimen was smaller than that of the upper part at the initial state; however, initial cracks can be seen in the upper part while the lower part was rather homogenous; shown in Figure 6. It can be seen in Figures 6(b), 6(c), and 6(d) that the initial cracks trigger the strain localization and the shear band develops in the downward direction.

4. Microstructures in sand specimens

In order to visualize microstructures in sand specimen, we performed a partial CT scan by which the volume of interest of the specimen was partially scanned non-destructively. The partial CT scans for Toyoura sand with D_{50} of 0.185 mm and Silica sand with D_{50} of 0.372 mm were performed. The voltage was 150 kV, the current was 200 μA, the size of the detector was 4.5 inch, the projection views are 2400, the accumulation was 20, the matrices of image was 2048 × 2048, and the voxel size was 1.7 μm × 1.7 μm × 52 μm. The sand sample was dropped into a glass bottle with 32 mm in diameter as shown in Figure 7(a), then placed on the rotation table and was scanned by an x-ray fan-beam.

Figures 7(b) and 7(c) show images of microstructures constructed by microfocus x-ray CT for Toyoura sand and Silica sand, respectively. As shown in the figures, we can clearly distinguish each particle from others. It was confirmed that the spatial resolution of this system was very high enough to visualize non-destructively the microstructures in sand specimens with rather larger specimen and that the microstructural changes in sand specimen during triaxial tests could be studied in the future.

| (a) Test sample | (b) Toyoura sand | (c) Silica sand |

Figure 7. *Microstructures of sand samples visualized by the microfocus X-ray CT*

5. Conclusions

The strain localization behavior in unsaturated Toyoura sand during and after the triaxial compression and Maruyamagawa clay during unconfined compression test was visualized using a microfocus x-ray CT with high spatial resolution. The development of shear bands in the unsaturated sand specimens was clearly observed and the strain localization mode was similar to that of published results for air-dried sands. The microstructures of sands were successfully observed by the partial CT scan, namely, soil particles and pore air can be seen independently. It was confirmed that the spatial resolution and the x-ray energy of the microfocus x-ray CT used in this study is high enough to visualize the strain localization behaviors and microstructures in soil specimens with the usual sizes of triaxial specimens. We will study the microstructure changes in the shear bands of soil specimens through comparison with air-dried or fully saturated soils and qualitative evaluation of local void ratio or local water content by the CT images, in the future.

6. References

Alshibli K.A., Sture S., Costes N.C., Frank M.L., Lankton M.R., Batiste S.N. and Swanson R.A., "Assessment of localized deformations in sand using X-ray computed tomography", *Geotechnical Testing Journal*, Vol.23, No.3, 2000, p. 274-299.

Desrues J., Chambon R., Mokni M. and Mazerolle F., "Void ratio evolution inside shear bands in triaxial sand specimens studied by computed tomography", *Géotechnique*, Vol.46, No.3, 1996, p. 539-546.

Kikuchi Y., "Investigation of engineering properties of man-made composite geo-materials with micro-focus X-ray CT", *Advances in X-ray Tomography for Geomaterials, Proc. of the Second International Workshop on X-ray CT for Geomaterials*, GeoX 2006, Aussois, France, 2006, p. 255-261.

Kodaka T., Oka F., Otani J., Kitahara H. and Ohta H., "Experimental study of compaction bands in diatomaceous mudstone", *Advances in X-ray Tomography for Geomaterials, Proc. of the Second International Workshop on X-ray CT for Geomaterials*, GeoX 2006, Aussois, France, 2006, p. 255-261.

Matsushima T., Uesugi K., Nakano T. and Tsuchiyama A., "Visualization of grain motion inside a triaxial specimen by micro X-ray CT at SPring-8", *Advances in X-ray tomography for geomaterials, Proc. of the Second International Workshop on X-ray CT for Geomaterials*, GeoX 2006, Aussois, France, 2006, p. 255-261.

Oda M., Takemura T. and Takahashi M., "Microstructure in shear band observed by microfocus X-ray computed tomography", *Géotechnique*, Vol.54, No.8, 2004, p. 539-542.

Otani J., Mukunoki T. and Obara Y., "Characterization of failure in sand under triaxial compression using an industrial X-ray CT scanner", *Solis and Foundations*, Vol.40, No.2, 2000, p. 111-118.

Determination of 3D Displacement Fields between X-ray Computed Tomography Images Using 3D Cross-Correlation

Problems and Solutions

M. Razavi* — B. Muhunthan**

*Mineral Engineering Department
New Mexico Institute of Mining and Technology
801 Leroy Place
Socorro, NM 87801
USA
mehrdad@nmt.edu

**Civil Engineering Department
Washington State University
Pullman, WA 99164
USA
muhuntha@wsu.edu

ABSTRACT. *The difficulties associated with applying the method of cross correlation to determine the 3D displacement fields between x-ray computed tomography (x-ray CT) images are discussed in this study. The high-resolution 3D x-ray CT images of Silica sand specimens were obtained using the x-ray CT facility at Washington State University. A computer code, M-DST, was developed determine the 3D displacement fields between the x-ray CT images of the soil specimen before and after moving the specimen. Displacement fields obtained by changing the x-ray CT scan parameters as well as different computer programming approaches in M-DST code show that the effect of x-ray source fluctuation is reduced by increasing the number of images to generate radiographs. In addition, computing time is reduced significantly by defining a limited search region around the template and using running sum to calculate the 3D cross correlation.*

KEYWORDS: *x-ray computed tomography, cross-correlation, displacement field, M-DST*

1. Introduction

Detection of the internal displacement fields for a specimen under loading in real time is important to characterize material behavior. Deformations may be observed and photographed directly for simple cases such as plane strain problems. Use of easily detectable material as markers, such as colored sand or tiny metallic spheres is commonly used to find internal displacements. Colored sand layers are usually used behind a transparent sheet and the movement of the colored particles is traced from photographs that are taken continuously as the specimen is loaded. On the other hand, instead of markers, a rectangular grid, made of horizontal and vertical lines, is plotted on the transparent sheet and the displacements of the grains are obtained by tracing the relative movement of the particles respect to the grid lines (White and Bolton 2004). Very small metallic balls or wires have also been placed inside the specimen at certain locations to quantify the internal displacement fields using radiography or X-ray CT (Nemat-Nasser and Okada 2001; Wood 2002; Alshibli and Alramahi 2006). Disturbance and changes in material properties by this method limit the calculation of 3D displacements to only small number of points.

Several different methods using image processing and computer vision techniques have been developed to quantify the 2D displacement fields. The cross-correlation method remains the most popular among them. In signal processing, cross-correlation number is a measure of the similarity between two signals. It can also be used to compare the similarity between two digital images before and after application of displacements.

Sadek *et al.* (2003), Liu and Iskander (2004), and several other researchers have applied the cross-correlation technique to find the 2D displacement fields on the exterior boundaries of soil specimens. This method has not been applied for 3D x-ray CT images due to some difficulties such as fluctuations of x-ray intensity with time, large sizes of 3D images, and extremely long processing time.

In this study the cross-correlation technique is extended to three dimensions and practical methods are developed to obtain 3D displacement fields from successive 3D x-ray CT images. An interactive computer program (M-DST) is developed to determine the 3D displacement fields using these techniques.

2. Cross-correlation technique for pure displacements

A small box-shaped subvolume is taken from the current image $F(X_i)$, which is represented by $f(X_i)$. It is assumed that after applying a displacement d_i (with three components u_i, v_i, w_i in x, y, and z directions, respectively) to $f(X_i)$ there is another box shaped subvolume with the same size of the template in the current image like target $g(X_i)$, which is the same as $f(X_i)$. Normalized cross-correlation (NCC) is a measure of the similarity between f and g defined by (Lewis 1995):

$$NCC(d_i) = \frac{\sum_{i=1}^{m}\left[g(X_i)-\bar{g}\right]\left[f(X_i-d_i)-\bar{f}\right]}{\left(\sum_{i=1}^{m}\left[g(X_i)-\bar{g}\right]^2 \sum_{i=1}^{m}\left[f(X_i-d_i)-\bar{f}\right]^2\right)^{0.5}}$$ [1]

where:

$NCC(d_i)$ = normalized cross-correlation as a function of displacement d_i

$\bar{f}$ = average of template f

$\bar{g}$ = average of target g

Correlation coefficients range from −1 to +1, in which +1 shows 100% similarity between template and target or both are the same; 0 means no similarity and −1 shows 100% similarity in the reverse direction. In any case, the maximum absolute value resulting from equation [1] is always considered to compare two images.

2.1. Determination of 3D displacement fields

The reference image is divided into small box-shaped subvolumes to find the displacement fields using cross-correlation techniques. The size of the subvolumes depends on many different factors such as image size, available computer memory, and accuracy. For each template f in the reference image F a target g in the current image G is searched so that the normalized cross-correlation (equation [1]) is a maximum. The displacement d_i is determined based on the maximum similarity between two blocks. In case of significant rotations different orientations of the target must be examined to include the effect of affine deformation (displacement and rotation). This is done by using optimization techniques or robust statistical methods (Clocksin et al. 2002). However, for small rotations, rotation of the target blocks may be neglected to reduce processing time.

2.2. NCC issues and solutions

There are several issues with NCC (equation [1]) that need to be resolved before programming the method:

1. X-ray intensity fluctuates, whereas the main assumption in NCC is to use the same x-ray intensity for both reference and current images.

2. NCC is not invariant and for the repeating patterns and affine deformation the results may fail or incur large errors.

3. As shown in the last example, for every block in the template all the voxels in the target must be examined. Thus, NCC technique is extremely time consuming. For a $M \times N \times P$ voxel reference image and a $Q \times R \times S$ voxel current image, NCC is calculated $M \times N \times P \times Q \times R \times S$ times. For instance for a $300 \times 300 \times 500$ voxel reference image and the same current image size, NCC is calculated 2.025×10^{15} times to find the displacement at every voxel. Besides, several times this number is needed for the total suite of operations, which requires a tremendous amount of CPU time.

The following solutions are provided to overcome the above issues:

1. Use of the average of several different frames (x-ray snap shots) for each single digital radiograph and increasing the number of averaged frames smoothes the x-ray intensity fluctuation so that it can be assumed to be of constant intensity with a good accuracy. To reduce the effect of x-ray beam fluctuations, it is recommended to use at least an average of 64 frames per radiograph.

2. Displacements on soil specimens can be applied in small steps. When a fixed axis test such as triaxial is used the rotation of the blocks are small as well. Sand particles are different and irregular in shape, which means that no repeating pattern is expected for an image with sufficient resolution to separate the particles.

3. When there is no information about displacements all the voxels in the current image should be searched. However, displacements of the blocks may be estimated based on the external radial and axial displacements in triaxial tests. Therefore, to reduce the processing time a subvolume around a particular block from target is extracted to search for the maximum NCC within that small subvolume. Use of fast NCC algorithms (Lewis 1995; Eaton 2005) reduces the processing 20 times and even more. Finally, use of parallel processing is recommended to further reduce the processing time.

3. An interactive computer code to find 3D displacement fields

An interactive computer code (M-DST) was developed to find the displacement fields by comparing template and target images using the NCC technique in a MATLAB environment (Razavi 2006). M-DST takes advantage of successive data write and read on the hard disk drive to process large volumes. In this way only the necessary portion of the image is loaded from the hard disk drive to the random access memory (RAM). After completing the operations, the results are saved on the hard disk drive and deleted from RAM to provide sufficient free space for the next image portion.

3.1. *Determination of 3D displacement fields using M-DST*

Silica 30-40 sand specimens (US Silica Company) with an average particle diameter of 1.6 mm were prepared (G_s=2.70) in an acrylic cylindrical mold, 27.82 mm in diameter and 150.62 mm in height. They were compacted in five layers by tamping on the sides of the mold. All of the specimens were scanned using x-ray CT at Washington State University High-Resolution X-ray CT Lab and their 3D images obtained by using a current of 0.284 mA, an x-ray beam energy of 160 keV, and 64 frames per radiograph (each radiograph is an average of 64 images).

Several preliminary scans were performed to choose the optimum magnification to attain the best resolution within the constraints of the computer memory and the capability to process the images. Based on repeated trials, it was found that a magnification of 3.1 times was sufficient.

Different tests were used to verify the computer code. In the first, the current image is generated by applying known displacements to the reference image. Then the displacements of the reference image are determined by the program and the results are compared. Figure 1.a shows the displacement fields of a 3D x-ray CT image after a translation of 3, 3, and 5 pixels in X, Y, and Z directions.

(The volume size is $337 \times 337 \times 246$ voxels and the total processing time on a 3.2 GHz dual processor Pentium IV machine is about 17 hours. In this particular problem template size is $35 \times 35 \times 35$ voxels, search radius is 60 pixels and the displacements are determined for 20 voxels intervals with 100% similarity threshold).

The results of M-DST are exactly the same as imposed displacements. To check the program for larger displacements, the 3D x-ray CT image is translated 10, 10, and 20 pixels in X, Y, and Z directions, respectively. The results are shown the Figure 1.b, which are identical to the imposed displacements. (To reduce the processing time, the search radius is reduced to 25 pixels and the displacements are determined for 35 voxels intervals. The other parameters in the program remained unchanged. These changes reduce the processing time significantly to 00:48:38 hours using the same machine.)

In case of rotation of template blocks, M-DST determines rotation pattern correctly, but there is a significant error in displacement values. In this case M-DST could not find any template and target blocks to have a similarity more than 95%. It means that for significant rotations considerable error is expected, though pattern of the displacement fields seems to be valid.

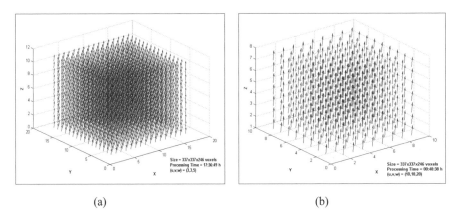

(a)   (b)

Figure 1. (a) *Determined 3D displacement fields for imposed displacements of 3, 3, and 5 pixels in X, Y, and Z directions, respectively (search radius is 60 pixels); (b) determined 3D displacement fields for imposed displacements of 10, 10, and 20 pixels in X, Y, and Z directions, respectively (search radius is 20 pixels).*

4. Conclusions

The method of template matching to find the 2D displacement fields between two 2D digital photographs was extended to find the 3D displacement fields between two 3D x-ray CT images. An interactive computer program (M-DST) was developed to process two 3D CT images, find the displacement fields, and represent it as a vector field or contour plots in three different planes.

The major problems with extension of the correlation method for 3D x-ray CT images are fluctuations of the x-ray beam, limited computer memory to process two 3D x-ray CT images simultaneously, and extremely long computing time. Fluctuation of the x-ray intensity was fixed using large numbers of averaging captured frames to form each digital radiograph. To overcome to the memory problem with loading of two 3D x-ray CT images at the same time, the method of successive read and write on the hard disk drive was applied. Instead of searching the whole voxels of the current image, a search region is defined in the computer code and in this way only the part of the image, which is required for processing, is loaded in memory to avoid memory overflow. To speed up the processing time, the computer code was developed so that user can install it on as many as machines he wants to find the displacement fields for each part individually. The results for each part are put together to have the displacement fields for the whole image.

5. Acknowledgements

The study presented in this paper was sponsored by the National Science Foundation under the grants CMS-0116793 and CMS-0010124 to Washington State University. The x-ray CT system was established from funds contributed by the National Science Foundation, Murdock Charitable Trust Fund, and Washington State University. The authors gratefully acknowledge the support of these agencies.

6. References

Alshibli, K. A., and Al Ramahi, B., "Microscopic evaluation of strain distribution in granular materials during shear", *Journal of Geotechnical and Geoenvironmental Engineering*, vol. 132, no. 1, p. 80-91, 2006.

Clocksin, W. F., Quinta Da Fonseca, J., Withers, P. J., and Torr, P. H. S., "Image processing issues in digital strain mapping", *Proceedings of SPIE-The International Society for Optical Engineering*, vol. 4790, p. 384-395, 2002.

Dennis M. J., "Industrial computed tomography", *Reprinted from Metals Handbook*, vol. 17, p. 358–386, 1989.

Eaton D., "Fast NCC", *http://www.cs.ubc.ca/~deaton/remarks_ncc.html*, 2005.

Haralick, R. M. and Shapiro, L. G., *Computer and Robot Vision*, vol. II, Addison-Wesley Inc., 1993.

Kak, A. C., and Slaney, M., *Principle of Computerized Tomographic Imaging*, Siam, Philadelphia, 2001.

Liu, J. and Iskander, M., "Adaptive cross correlation for imaging displacements in soils", *Journal of Computing in Civil Engineering*, vol. 18, no. 1, p. 46-57, 2004.

Lewis, J. P., "Fast normalized cross-correlation", *Industrial Light & Magic*, http://www.idiom.com/~zilla/Papers/nvisionInterface/nip.html, 1995.

Nemat-Nasser, S. and Okada, N., "Radiographic and microscopic observation of shear bands in granular materials", *Geotechnique*, vol. 51, no. 9, p. 753-765, 2001.

Razavi, M., Characterization of Microstructure and Internal Displacement Field of Sand using X-ray Computed Tomography, PhD Thesis, Washington State University, Pullman, WA, USA, 2006.

Sadek, S., Iskander, M., and Liu, J., "Accuracy of digital image correlation for measuring deformations in transparent media", *Journal of Computing in Civil Engineering*, vol. 17, no. 2, p. 88-96, 2003 .

White, D. J., and Bolton, M. D., "Displacement and strain paths during plane-strain model pile installation in sand", *Geotechnique*, vol. 54, no. 6, p. 375-397, 2004.

Wood, D. M., "Some observations of volumetric instabilities in soils", *International Journal of Solids and Structures* 39, Pergamon, p. 3429-3449, 2002.

Characterization of Shear and Compaction Bands in Sandstone Using X-ray Tomography and 3D Digital Image Correlation

E-M. Charalampidou*'** — S.A. Hall** — S. Stanchits*** — G. Viggiani**— H. Lewis*

* Institute of Petroleum Engineering, Heriot-Watt University
Edinburgh EH14 4AS, Scotland
elma.charalampidou@pet.hw.ac.uk
helen.lewis@pet.hw.ac.uk

** Laboratoire 3S-R- CNRS/Université Joseph Fourier/Grenoble INP
Domaine Universitaire, BP53, 38041 Grenoble, France
hall@geo.hmg.inpg.fr
cino.viggiani@hmg.inpg.fr

*** Deutsches GeoForschungsZentrum GFZ
Telegrafenberg, D 423, D-14473 Potsdam, Germany
stanch@gfz-potsdam.de

ABSTRACT. *In this work we employ x-ray tomography, 3D digital image analysis and 3D-volumetric digital image correlation techniques to characterize localized deformation phenomena in sandstone. The specimens considered have been deformed in triaxial compression under a range of confining pressures (20-190MPa). Shear or compaction bands were observed, at low and higher confinement respectively. X-ray tomography images have been acquired before and after loading (unconfined) at different spatial resolutions (30 and 90 μm voxel size). The combination of both x-ray tomography and 3D DIC provides insights into the geometry and mechanisms of the localized features.*

KEYWORDS: *sandstone, localization, shear bands, compaction bands, x-ray tomography, 3D digital image correlation*

1. Introduction

Porous sandstones generally exhibit strain localization when loaded in triaxial compression. The geometry and mechanisms of the localization features depend on the applied confining pressure; in general, shear bands are observed for low to middle confining pressures while compaction bands occur with higher confinements. Studies of such localization phenomena at the laboratory scale can provide useful insight into the mechanisms and internal structure of shear zones, faults and compaction bands at a reservoir scale. This can thus lead to better understanding of fault-leakage or -sealing and flow-barriers due to compaction features, which is of keen interest for hydrocarbon production and subsurface storage of CO_2.

In the laboratory, conventional measurements, of forces and displacements made at a specimen boundary, describe only a "global" behavior of the studied specimen. Even when local measurements are made, for example displacement measurements via LVDTs mounted on the surface of the specimen, the evolution and propagation of localized deformation are difficult to characterize. Consequently, the development and application of full field techniques, (e.g. ultrasonic tomography (UT), acoustic emissions (AE), x-ray tomography (x-ray CT) or digital image correlation (DIC)), has become increasingly common (see for example the recent review paper by Viggiani and Hall, 2008). The greatest advantage of these non-conventional laboratory techniques is the measurement of fields of some parameters, i.e. ultrasonic velocity (UT), density (x-ray CT) and displacement/strain field (DIC).

In this paper some results obtained by x-ray tomography and 3D-DIC, using both High Resolution (hereafter HR, ~30 μm voxel size) and Low Resolution (hereafter LR, ~90 μm voxel size) x-ray images, are presented. The materials studied are porous sandstones that have been deformed under triaxial compression. The two full field techniques (x-ray CT and 3D-DIC) are first presented followed by a brief description of the experimental program. Subsequently a few selected results are provided for three sandstone specimens that have been deformed at different confining pressures.

2. Full field non-destructive techniques

2.1. X-ray computed tomography

X-ray tomography is a powerful tool to map 3D density variations, and thus 3D structure, in material. This technique has been widely used for the study of geomaterials, including, in a few cases, for the characterization of localized deformation in sandstones (e.g. Louis et al., 2006).

The x-ray tomography images presented in this paper were obtained using the x-ray facility at Laboratoire 3S-R in Grenoble. A number of specimens of Vosges sandstone were scanned (unconfined) before and after triaxial compression tests plus a single specimen of Bentheim sandstone has been scanned after loading. These x-ray scans were either HR (30 x 30 x 30 μm^3 voxel size) or LR (90 x 90 x 90 μm^3 voxel size).

2.2. *Digital Image Correlation*

Digital Image Correlation (DIC) is a method that is being used increasingly in experimental mechanics to measure displacement and deformation fields over surfaces (e.g. Pan *et al.*, 2009) or through a volume (e.g. Bay, 2008). DIC is based on the comparison of two digital images of a specimen at different deformation states. First a set of node points is defined in one image, then, for small subsets of this image defined around each node, a search is made for the equivalent image subset in the second image, taking into account the geometrical transformation between the images (this transformation can include displacements, rotations and strain, although in the current case just rigid translation is assumed). This procedure provides a displacement-vector field, which describes the transformation of one image to the other. In this work, we employ a 3D-volumetric DIC code developed at Laboratoire 3S-R (see Hall *et al.*, 2009, for an overview) on HR and LR x-ray images of the each specimen before and after triaxial compression

3. Experimental work: materials studied and experimental program

The principal material in this study is Vosges sandstone. This rock has a pink color, contains 93% quartz, 5% microcline, 1% kaolinite and 1% mica, with a mean grain size of about 300μm (Bésuelle, 2001). The average porosity of the samples studied is 22%. In addition analysis of a specimen of Bentheim sandstone is provided as a comparison; in this case only post-mortem HR x-ray images are presented and no mechanical data are discussed. The Bentheim sandstone is a more homogenous sandstone of yellow color containing 95% quartz, 3% feldspar and 2% kaolinite with a porosity of about 22% and mean grain size of about 300 μm (Stanchits, 2009).

In the larger project of which this work is a part, we study the mechanical behavior of the Vosges sandstone under triaxial compression for a range of confining pressures (20-190 MPa) and implement a series of experimental techniques (UT, AE, x-rays, DIC) to study the deformation mechanisms and structures. This porous sandstone loaded under triaxial compression develops dilating and compacting shear bands for low (10-30 MPa) and medium (40-60 MPa) confining pressures, respectively (Bésuelle, 1999). At higher confinements, roughly horizontal compaction bands are observed. The experimental campaign included 7 tests with loading under low and middle confinements and 6 with loading under higher confinement. Cylindrical specimens (40mm in diameter and 80 mm in height) were cored perpendicular to the bedding. The specimens have two opposite flattened surfaces throughout their height. This particular geometry is necessary for the UT (not presented in this paper). Moreover, notches were machined on the surface of the specimens to encourage the localization to occur between these two flattened faces. Samples deformed at low and middle confinements have two shifted notches (2.5 mm depth and 2 mm wide) on the flattened surfaces while samples deformed at higher confining pressures have a circumferential notch (4mm depth and 0.8 mm wide) in the middle of their height. A teflon O-ring of about 0.7 mm thickness was used to fill the circumferential notch. All specimens were tested dry. Ultrasonic (P-waves) and x-ray measurements were carried out before and after the majority of the experiments. AE were recorded during a few tests.

4. Results and discussion

In this paper, x-ray images and DIC of two Vosges specimens are presented. VEC4 was loaded under 50 MPa confinement and Ve4 under 130 MPa. Mechanical data and photographs of both specimens after testing are presented in Figure 1. In test VEC4 the loading was stopped well after the peak stress (~0.4% after the peak strain) while for Ve4 loading was stopped before a peak in stress was observed. X-ray tomography images of BE6 Bentheim specimen, which was deformed at 185 MPa confinement, are also presented for comparison with Ve4.

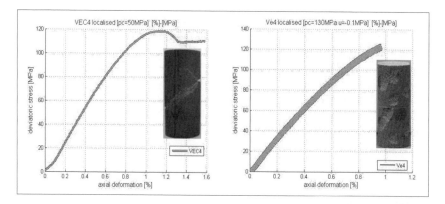

Figure 1. *Stress-strain curves for the triaxial compression tests and photographs of specimens VEC4 deformed in 3S-R and Ve4 deformed in GFZ (the circular marks on the surface of the latter are the traces of transducers used during AE measurement)*

4.1. *Characterization of shear band structure in the Vosges sandstone*

For VEC4 (Vosges) HR x-ray tomography has been used to visualize the shear band structure in 3D. Figure 2 shows a set of vertical slices through the x-ray tomography volume of the specimen. Three bands of darker colors, which indicate higher density, can be seen; these are interpreted as compacting shear bands. From the images it can be inferred that these bands initiated from both notches and traversed the sample to meet near the middle. Another band can be seen heading upwards from the upper notch at a similar angle to the two main bands, but is not so well developed. The bands are not planar and their structure evolves in 3D. The width of the bands ranges from 270 μm to 750 μm (i.e. about 1-2.5 grain diameters - recall that the mean grain size is 300 μm). It is worth noting that this specimen was loaded until after peak stress, so we expect a wider damaged zone around the central part of the shear band. Near the notches cracks are visible (light color indicates lower density); these have a mean width of approximately 60 μm.

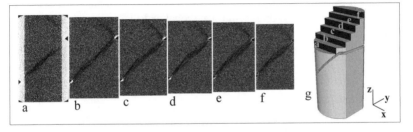

Figure 2. *Images from a to f represent sequence of views (x-z plane) through the image (y axis) showing the 3D structure of the shear band in VEC4 Vosges specimen (g)*

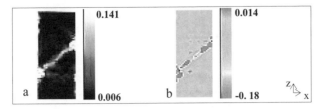

Figure 3. *DIC-derived strain maps from the LR x-ray images of VEC4: a) shear strains, b) volumetric strains. These 2D images represent the median values of all the vertical slices (x-z plane) projected along the y-axis*

Figure 3 presents the shear and volumetric strains derived from the 3D DIC using LR x-ray images. Shear strain inside the band ranges from 4% to 11%. The shear band is compacting reaching even 18% of volume reduction. Some dilation in the zone just outside the band is also seen. Note that the thickness of the shear band is probably over-estimated in the 3D DIC images as the resolution depends on the image subset, over which the correlation is calculated (in this case 15 voxels, i.e. 1.35^3 mm^3) and the analysis grid spacing (in this case 10 voxels or 0.9 mm).

4.2. Characterization of compaction band structure in the Vosges sandstone

The detection of the localized band in Ve4 Vosges specimen, which was deformed at 130 MPa confining pressure, was quite challenging. AE data recorded during the loading indicate that the band was not fully developed. Moreover, under this level of confinement, the expected localized structures are much narrower. Using only raw x-ray tomography images the compaction band indicated by the AE recording could not be resolved. In previous work (Louis *et al.*, 2006) local statistical measures of the gray-level values such as skewness and standard deviation have been demonstrated to be useful in highlighting the complex geometry of compaction bands. For specimen Ve4, we found that the local standard deviation of the image grayscale allowed the detail of the localized features to be seen (here the calculation was made over a small volume centered on each pixel of 7 voxels^3). In this case, the compaction bands appear as zones of decreased standard deviation, which probably indicates a reduction in grain size, due to grain crushing, to below the voxel size and thus a homogenization of the image.

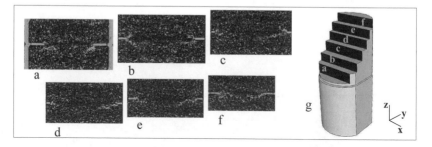

Figure 4. *Sequence of slices cut through the volumes of local standard deviation of the HR x-ray tomography images of Ve4 specimen (g). The slices are cut parallel to the specimen axis in the plane perpendicular to the flattened faces that contain the notches: (a) to (f) are progressively deeper into the volume as in Figure 2.*

Figure 4 shows the local standard deviation images for the Ve4 specimen and the resolution of the structural complexity of the compaction band. Red depicts the low standard deviation values, while black represents less homogenous regions compared to the compaction band. The band initiates and propagates near the notch, but does not evolve to the central part of the specimen (something already shown in AE of this specimen).

Figure 5 shows vertical slices through the 3D-DIC-derived shear- and volumetric-strain volumes. HR x-ray image volumes were used for the DIC in this case. The shear strain in the band ranges from 0.6% to 1.8% and the volumetric strain reaches 2.7% of the volume reduction. Figure 5c shows a thresholded 3D view of the volumetric strain field highlighting the 3D geometry of the compaction band.

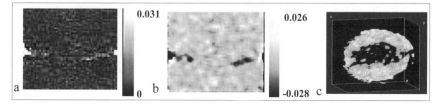

Figure 5. *DIC-derived strain maps from the HR x-ray images of Ve4: (a) shear strains, (b) volumetric strains (c) thresholded 3D view of the volumetric strain field*

4.3. *Visualization of compaction band on Bentheim sandstone*

BE6 Bentheim specimen was deformed at 185MPa under triaxial compression. Figure 6 shows HR x-ray images of a vertical projection near the center of the specimen. The top image shows the raw data focusing on high (white) and low (black) density regions in the histogram. A crack initiating from the right notch is apparent. However, by visualizing only this gray level, it is not easy to capture the compaction band. In the middle image, local standard deviation is shown. The more homogenous parts of the specimen appear in red. In fact, a low standard deviation might indicate either pore space or compacting material. Thus, in the bottom image we show both raw data and local standard deviation. The compaction bands are better defined. Moreover, details on low or high density regions as well as on more or less homogenous regions can be defined.

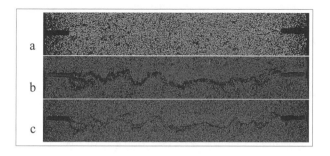

Figure 6. *Compaction band in BE6 specimen; a) raw data, b) local standard deviation, c) raw data and local standard deviation superimposed*

5. Conclusions

X-ray tomography and 3D DIC have been employed to characterize shear and compaction bands in two different sandstones. HR x-ray images of the Vosges samples depict clearly the shear band, especially when the band is well developed. However the local standard deviation of the x-ray tomography data was used for a better visualization of the compaction band. In the Bentheim sandstone compaction bands could be more easily captured, even by the x-ray tomography gray-level images perhaps because this is a more homogeneous sandstone and because, for the sample considered, the band was well evolved (confirmed by AE data), which was not the case for the Vosges specimen. DIC allowed a qualification of strain (shear or volumetric) and shed light on the mechanisms of the localized features.

6. Acknowledgements

The authors would like to thank P. Charrier, C. Rousseau and D. Takano. G. Dresen is gratefully acknowledged for the financial support of the experiments carried out in GFZ. E-M.Charalampidou is supported by the Ali Danesh scholarship from HWU.

7. References

Bay B. K., "Methods and applications of digital volume correlation", *J. Strain Analysis*, vol. 43, 2008, p. 745-760.

Bésuelle P., "Evolution of Strain Localisation with Stress in a Sandstone: Brittle and Semi-Brittle Regimes", *Phys. Chem. Earth*, vol. 26, No. 1-2, 2001, p. 101-106.

Hall S.A., Lenoir N., Viggiani G., Desrues J., Bésuelle P., "Strain localisation in sand under triaxial loading: characterisation by x-ray micro tomography and 3D digital image correlation", *1st Int. Symp. on Computational Geomechanics*, 2009, IC^{2}E, p.239-247.

Louis L., Wong T.-F., Baud P., Tembe S., "Imaging strain localization by X-ray computed tomography: discrete compaction bands in Diemelstadt sandstone", *Journal of Structural Geology*, vol. 28, 2006, p. 762-775.

Pan B., Qian K., Xie H., Asundi A., "Two-dimensional digital image correlation for in-plane displacement and strain measurements: a review", *Meas. Sci.Technol.*, vol. 20, 2009, doi:10.1088/0957-0233/20/6/062001.

Stanchits S., Fortin J., Guenguen Y., Dresen G., "Initiation and propagation of Compaction Bands in Dry and Wet Bentheim Sandstone", *Pure appl. Geophys.*, vol. 166, 2009, p. 843-868.

Viggiani G., Hall S. A., "Full-field measurements, a new tool for laboratory experimental Geomechanics", *Deformation Characteristics of Geomaterials*, IOS Press, Atlanta, USA, 2008, p. 3-26.

Deformation Characteristics of Tire Chips-Sand Mixture in Triaxial Compression Test by Using X-ray CT Scanning

Y. Kikuchi* — T. Hidaka** — T. Sato*** — H. Hazarika****

Port and Airport Research Institute
3-1-1, Nagase, Yokosuka, Kanagawa 239-0826, Japan
kikuchi@pari.go.jp

***Toa Corporation*
1-3, Anzen-cho, Tsurumi-ku, Yokohama 230-0035, Japan
ta_hidaka@toa-const.co.jp

****Kumamoto University*
2-39-1, Kurokami, Kumamoto City, Kumamoto 860-8555, Japan
sato@tech.eng.kumamoto-u.ac.jp

*****Akita Prefectural University*
84-4, Aza Ebinokuchi, Tsuchiya, Yurihonjo City, Akita 015-0055, Japan
hazarika@akita-pu.ac.jp

ABSTRACT. *In this paper, the deformation characteristics of a mixture of tire chips and sand are discussed. First a series of drained triaxial compression tests (CD tests) on a rubber sphere specimen and the tire chips specimen was conducted to observe the deformation behavior of a rubber particle aggregate. From the movement of the rubber sphere traced on computer tomography (CT) images, the basic deformation characteristics of the rubber particle aggregate were considered. Then, a series of CD tests on the tire chips and sand mixture was conducted and the changes in the shape of each specimen before and after compression were shown by CT images. Moreover, the Particle Image Velocimetry (PIV) method was applied to CT images scanned during the test, and the deformation characteristics of the tire chips and sand mixture based on the PIV results were shown. The effect that adding tire chips had on the deformation characteristics of the mixture is described while considering the deformability of the rubber particle aggregate.*

KEYWORDS: *rubber sphere, tire chips and sand mixture, drained triaxial compression test, micro-focus x-ray CT scanner, PIV method*

1. Introduction

Since the 1990s, a great deal of research has been conducted on using shredded scrap tires for backfills of quay wall structures (e.g., Edil *et al.* 2002; Humphrey *et al.* 1998). Recently many research projects on the beneficial functions of using tire chips to mitigate the effects of earthquakes have been conducted. They include research using a model test of earth pressure reduction on quay wall structures (Hazarika *et al.* 2006), a numerical simulation to discover the mechanism of this reduction (Kaneda *et al.* 2007), and a series of undrained cyclic triaxial tests related to the earthquake-proof effect (Hyodo *et al.* 2007). Kaneda *et al.* (2007) presented a study showing that earth pressure on quay wall structures was reduced effectively by having a smaller Young's modulus and Poisson's ratio and this earth pressure reduction could be achieved with tire chips.

The tire chips and sand mixture form a lightweight soil because of the lower particle density of the tire chips. This soil can be used to help reduce the earth pressure on quay wall structures. On the other hand, the mixture may deform easily against a surcharge because of the deformability of the tire chips. These interactions should be observed by visualizing the deformation features of the mixture.

The main objective of this paper is to examine the basic deformation and shear strength features of a mixture of sand and tire chips. First, in order to observe the basic deformation behavior of a specimen made with homogeneous elastic particles, a series of CD tests were conducted on a rubber sphere specimen. Then a series of CD tests on a specimen consisting of a mixture of tire chips and sand were conducted and the characteristics of shear strength and compressibility of the mixture were observed. These investigations on a rubber sphere and on the tire chips and sand mixture were conducted by using micro-focus x-ray CT scanning.

2. Deformation characteristics of rubber particle aggregate

2.1. *Experimental method*

A series of CD tests was conducted to investigate the deformation characteristics of a rubber sphere and tire chips during compressive shearing. The particle density of the rubber spheres was 1.270 g/cm^3, and their particle size was 8.0 mm. This particle size was selected by referring to a previous study (Holtz *et al.* 1956). The particle density of the tire chips was 1.150 g/cm^3, their mean particle size was 2.0 mm, and the maximum and minimum void ratios were 1.632 and 1.091. The rubber spheres and tire chips were washed with a detergent to remove any impurities adhering to their surfaces and then dried. The diameter and height of both specimens were 5 cm and 10 cm, respectively. The target relative density of the tire chip specimen (TC) was 100%. The number of rubber sphere particles was controlled to

405 when preparing the rubber sphere aggregate (RS). The consolidation pressures used were 30 and 100 kN/m². The void ratios of the specimen after consolidation were 0.4–0.5 and 0.2–0.3 at consolidation pressures of 30 and 100 kN/m², respectively in both TC and RS, although the accuracy of the measured void ratios was low because of membrane penetration. The CD tests were conducted with a compression strain velocity of 1.0%/min.

The settings of the micro-focus x-ray CT scanner were as follows: the x-ray power was set to 200 kV, 300 μA; x-ray imaging was used in cone CT (3D-CT) mode; the scanner took 200 images at a time with a slice thickness of 0.5 mm; one slice was a 512 × 512 matrix with a pixel size of 0.12 × 0.12 mm; and the three dimensions of the reconstructed images were 61.44 × 61.44 × 100 mm. The test apparatus was kept in an x-ray shield box during the test and scanning.

2.2. Experimental results

Figure 1 shows the deviator stress-volumetric strain-axial strain relationships of TC and RS. In the cases of RS, these curves followed the same processes during the loading in spite of their cyclic loading between an axial strain of 0% to 10%. The deviator stress of RS was slightly higher than that of TC at the same axial strain at each consolidation pressure. Except in the case of RS consolidated at 30 kN/m², there were no maximum deviator stresses observed until an axial strain of 23%. From these results, the deformation characteristics of TC and RS were found to be originally similar. The reason for this similarity is that the particle deformability between tire chips and rubber spheres is similar.

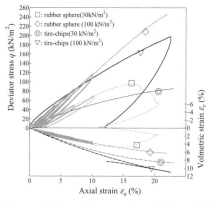

Figure 1. *Triaxial test results of TC and RS*

Figure 2 shows the results of the traced movement of each particle in order to observe the deformation mode change of RS during the CD test. The monotonous loading was conducted at a consolidated pressure of 30 kN/m². Particle displacement vectors during axial strain from 0% to 15% are shown in (a), and those from 15% to 18% are shown in (b). As shown in (a), most particles moved almost vertically up to an axial strain of 15% and the displacement of the particles was large in the upper particles and small in the lower particles. As shown in (b), particles in the upper part moved not only vertically but also horizontally, even though the particles in the lower part seldom moved during axial strains from 15% to 18%. During this observation, a sliding plane was seen, as shown in (b).

This shear deformation produced the deviator stress reduction and dilative volume strain as shown in Figure 1. The dilative deformation was observed only in RS consolidated at 30 kN/m² as shown in Figure 1. The similarity of the deformation characteristics between RS and TC described above was because of the one-dimensional vertical compression of the particles themselves as shown in Figure 2 (a). Therefore, it is considered that the main deformation mode of tire chips is the compression mode in triaxial compression and that the shear deformation mode is seldom observed.

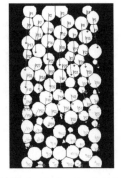
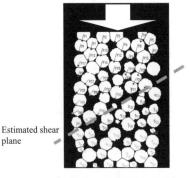

Estimated shear plane

(a) Displacement vectors at axial strain of between 0% and 15%

(b) Displacement vectors at axial strain of between 15% and 18%

Figure 2. *Displacement vectors during each axial strain of rubber sphere particle*

3. Deformation characteristics of tire chips and sand mixture

3.1. *Experimental method*

Tire chips (ρ_s=1.150 g/cm³, e_{max}=1.556, e_{min}=1.126, D_{max}=2.0 mm), Sohma sand #4 (ρ_s=2.644 g/cm³, e_{max}=0.970, e_{min}=0.634, D_{50}=0.77 mm, D_{max}=2.0 mm), and a mixture of these two materials were used in this series of experiments. The tire chips and Sohma sand were mixed in various proportions. Specifically, the mixing ratios of sand to tire chips by volume were set at 10:0, 7:3, 5:5, 3:7, 1:9 and 0:10. Here, sand fraction (*sf*) indicates the proportion by volume occupied by Sohma sand in the tire chips-sand mixture. Thus, when *sf* =1.0 this indicates a sample consisting of sand only.

The specimen used for the CD test was prepared by moist tamping. The tire chips and Sohma sand were mixed at the prescribed mixing ratio. Water was added to the mixture to obtain an initial water content of w=10%. The diameter and height of the specimens were 5 cm and 10 cm, respectively. The specimen was prepared by arranging the mixture in five layers, with each layer compacted a prescribed number of times by dropping a rammer from a prescribed height. This was done to control the compaction energy so that it reached 50% of the relative density in the specimen of *sf*=1.0.

The same amount of compaction energy was applied to each specimen with a different sand fraction. The specimen was saturated and consolidated at a confining pressure of 50 kN/m^2 with a back pressure of 100 kN/m^2. The CD test was conducted with a strain velocity of 1.0%/min. CT scanning was conducted at axial strain increments of 2.0% each in all cases.

The settings of the CT scanner different from 2. were as follows: the voltage was 180 kV; the scanner obtained a picture by taking 500 images at a time; the slice thickness was 0.2 mm; the pixel size was 0.15 × 0.15 mm; and the three dimensions of the images were 76.8 × 76.8 × 100 mm.

3.2. *Experimental results*

Figure 3 (on the next page) shows the relationships among deviator stress, volumetric strain and axial strain during the triaxial tests. The solid lines show the relationship between deviator stress and axial strain. The broken lines show the relationship between volumetric strain and axial strain.

From the stress-strain curves, it can be seen that as the sand fraction increases and the stress increment ratio becomes larger, the strain at the maximum deviator stress becomes smaller. Especially, in the cases of *sf*=0.1 and *sf*=0.0, there was no maximum deviator stress observed until the axial strain reached 20%. Looking at the maximum deviator stress, there was no large difference observed except at *sf*=0.1 and *sf*=0.0. In addition, in terms of compressibility it was observed that the dilative tendency increased as the sand fraction increased.

Figure 4 (on the next page) shows how the internal friction angle ϕ_0 and Young's modulus E_{50} change with different sand fractions. When the sf was larger than 0.3, a small difference was observed in ϕ_0, although a large reduction of ϕ_0 was observed when the sf was less than 0.3. Furthermore, a large difference in E_{50} was observed between an sf of 1.0 and 0.7, but the difference in E_{50} values between an sf of 0.7 and 0.0 was small.

This difference in change in ϕ_0 and E_{50} arose because of the different deformation characteristics of the mixture. E_{50} is mainly affected by the early stage of compression and at this stage, the deformation characteristics are mainly affected by the compressibility of the tire chips. On the other hand, ϕ_0 is affected by the final stage of compression and the final shear strength is mainly affected by the shear resistance of the sand.

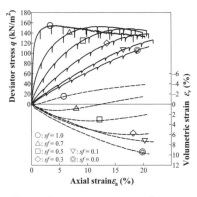

Figure 3. *Deviator stress - volumetric strain - axial strain relationship of tire chips and sand mixture*

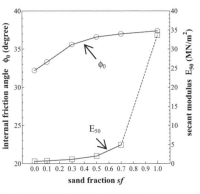

Figure 4. *Change in internal friction angle and Young's modulus with changes in sand fraction*

Figure 5 shows the deformation of the specimens before and after compression. This figure shows CT images of the vertical cross-sections on a central axis of the specimen. In the cases of sf=0.7, 0.5, 0.3 and 0.1, the images include a dark part and a lighter part.

The dark parts are tire chips and the lighter parts are sand particles. When the specimen was compressed to an axial strain of 20%, the specimen in which sf=1.0 deformed into a barrel shape. This kind of horizontal deformation decreases with a decrease in sand fraction and if sf =0.1, very little horizontal deformation is observed.

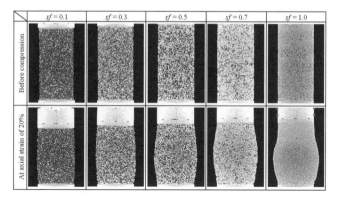

Figure 5. *X-ray CT images of vertical cross-sections on a central axis before and after compression*

Figure 6 shows the distributions of shear strain and volumetric strain at an axial strain of 20%. The distributions were calculated from the displacement vector analyzed by the PIV method. Shear strain distributions are shown in the upper row, where it can be seen that the shear strain increases as the darkness increases, with almost no shear strain seen in the white part. In the case of sf =1.0, a clear shear band was observed at the middle height of the specimen. In this case, no kind of shear zone was clearly observed in the CT image as shown in Figure 5. However, a shear zone was observed by applying the PIV method as shown in Figure 6.

This kind of localization of shear strain became less distinct with decreasing sand fraction. As shown in the cases of sf =0.3 and sf =0.1, the distribution of shear strain is almost homogeneous and the shear strain is smaller. Volumetric strain distributions are shown in the bottom row, where the color of the initial state is light gray as shown in the part of the wedge in the case of sf =1.0. The darker part is more compressed and the white part is more dilated. In the case of sf =1.0, the specimen is remarkably compressed at the middle height area, and the dilated part is seen on the boundary of the shear zone and active zone. With decreasing sand fraction, the compressed area increases to the upper and lower parts and the dilated area diminishes.

The tendency for the dilated area to diminish with decreasing sand fraction is expressed well although the accuracy of the dilated area is not exactly high in comparison with the measured volumetric strain. When the active zones are compared with each other in the cases of sf =1.0 and sf =0.7, the wedge of sf =0.7 is compressed. It is considered that shear deformation of the mixture occurs after the specimen is compressed.

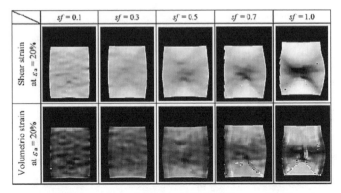

Figure 6. *Shear strain distribution and volumetric distribution on vertical cross-section on central axis derived from PIV analysis*

Based on the results mentioned above, the deformation characteristics of the tire chips and sand mixture are different from those of sand alone. When tire chips are mixed with sand, the Young's modulus of the mixture is smaller than the Young's modulus of sand alone. The shear band spreads out and the shear strain becomes smaller with decreasing sand fraction. The tendency of the volume change shifts from dilative to compressive as the compressed area increases. These results show that the deformation characteristics of the mixture are controlled by the compressibility of the tire chips in the mixture as shown in Figure 2(a), more than the shear properties. However, the shear strength and shear resistance of the mixture are quite similar to sand and are affected at the final stage of compression.

4. Conclusions

In this paper, the deformation characteristics of a mixture of tire chips and sand were discussed. Initially, the deformation characteristics of the tire chips were observed independently from the mixture. Based on these observations, it was understood that the main deformation mode of the tire chips is the compression mode during triaxial compression, while the shear deformation mode is seldom observed.

The deformation characteristics of the tire chips and sand mixture are different from those of sand alone. The Young's modulus of the mixture is very small, the generated shear strain is smaller than that observed with sand alone, and the localization of shear strain diminishes with decreasing sand fraction. The tendency of the volume change shifts from dilative to compressive as the compressed area is increased. However, ultimate shear resistance seldom changes with decreasing sand fraction. This kind of change in characteristics is due to the deformation characteristics of the tire chips, as was shown.

5. References

Edil, T.B., "Mechanical properties and mass behavior of shredded tire-soil mixtures", *Proc. of the International Workshop on Lightweight Geo-Materials, JGS*, pp. 17-32, 2002.

Humphrey, D.N., Whetten, N., Weaver, J., Recker, K., Cosgrove, T.A., "Tire TDA as lightweight fill for embankments and retaining walls", *Proc. of Conference on Recycled Materials in Geotechnical Application, ASCE*, pp. 51-65, 1998.

Hazarika, H., Kohama, E., Suzuki, H., Sugano, T., "Enhancement of earthquake resistance of structures using tire chips as compressible inclusion", *Report of the Port and Airport Research Institute*, Vol. 45, No. 1, 2006.

Holz, W.G., Gibbs, H. J., "Triaxial shear tests on pervious gravelly soils", *ASCE*, vol. 82, no. SM 1, pp. 1-22, 1956.

Hyodo, M., Yamada, S., Orense, R.P., Yamada, S., "Undrained cyclic shear properties of tire chip-sand mixtures", *Scrap Tire Derived Geomaterials*, Taylor & Francis, pp. 187-196, 2007.

Kaneda, K., Hazarika, H., Yamazaki, H., "The numerical simulation of earth pressure reduction using tire chips in backfill", *Scrap Tire Derived Geomaterials*, Taylor & Francis, pp. 245-251, 2007.

Strain Field Measurements in Sand under Triaxial Compression Using X-ray CT Data and Digital Image Correlation

Y. Watanabe* — N. Lenoir** — S. A. Hall*** — J. Otani*

*X-Earth Center, Kumamoto University
2-39-1, Kurokami, Kumamoto City
Kumamoto 860-8550, Japan
089d9408@st.kumamoto-u.ac.jp

**Civil and Environmental Engineering
Northwestern University
Evanston, IL, USA
n-lenoir@northwestern.edu

***Laboratoire 3S-R
CNRS/Grenoble University
Grenoble, France
hall@geo.hmg.inpg.fr

ABSTRACT. *This paper presents the results of an experimental study on strain localization in soil. The purpose is to characterize in space and time the internal strains and strain localization patterning in a sand undergoing triaxial compression at different confinement pressures. In-situ triaxial tests were conducted on Yamazuna sand at the X-Earth Center (Japan) at different confinement pressures. Complete 3D images of the specimen were obtained at several loading stages throughout the test and analyzed by 3D-volumetric digital image correlation (3D-DIC) in order to obtain 3D incremental displacement and strain fields. Based on the results of XRCT (x-ray computed tomography)/3D-DIC combination, the deformation process and especially the strain localization are quantitatively characterized in space and time under triaxial compression conditions.*

KEYWORDS: *strain localization, sand, in-situ triaxial test, digital image correlation*

1. Introduction

In geotechnical engineering, characterization of the deformation behavior and failure of soils is essential. In particular, for the failure of soils, strain localization is a key issue. In laboratory mechanical testing this can only be investigated through full-field measurements of the displacements and strain in a soil specimen. There are some research activities, which have investigated 3D strain field measurements in granular media including particle displacements (e.g., Yamamoto and Otani, 2001; Nielsen *et al.*, 2003; Hall *et al.*, 2010). In this work, the objective is to characterize the displacement and strain field evolution in a soil under triaxial compression with *in-situ* (i.e. with x-ray scanning during the loading test) x-ray computed tomography (CT) and Digital Image correlation (DIC). The experimental data analyzed in this work concern a sand of wide grading, for which previously the movements of the larger soil particles have been traced in the CT images (Watanabe *et al.*, 2008). Based on the results of this particle tracking, the 3D displacement fields were obtained for various increments of deformation of the soil specimen and the associated strain fields were calculated using the finite element method. However, the highlighted behavior seems to not truly represent the behavior of the soil because only a restricted number of particles were traced in the CT images.

This paper is concerned with the characterization, using 3D-volumetric DIC, of internal strains and strain localization patterning in a sand undergoing triaxial compression at different confinement pressures. The full 3D fields of displacement and strain are evaluated using x-ray CT and 3D-volumetric DIC. Complete 3D images of the specimens have been recorded at several stages throughout the tests and analyzed by DIC in order to obtain 3D incremental displacement and strain fields. Based on the results of the combined x-ray CT and DIC, the deformation process and especially the strain localization are quantitatively characterized in space and time under different triaxial compression conditions.

2. Summary of test procedure

In-situ triaxial compression tests under drained condition were performed on Yamazuna sand at the X-Earth Center (University of Kumamoto, Japan). The specimen was scanned during the test at different levels of loading in order to obtain full 3D images during the process of compression. Figure 1 shows the grain size distribution of this sand in which the minimum particle size was 0.001 mm and the maximum size was 10 mm with a D_{50} of about 0.5 mm. Note that the grain size distribution was obtained by sieve and sedimentation analysis for particle sizes over and under 75 μm respectively. The specimen size was 50 mm in diameter and 100 mm in height with a dry density of 1.63 t/m^3 and D_r=90%.

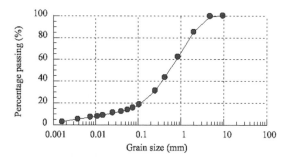

Figure 1. *Grain size distribution*

The axial loading rate was 0.3%/min and confining pressures of 50 kPa and 150 kPa were applied in the two tests presented. The scanner used was a Toshiba industrial scanner (see Otani *et al.*, 2002 for details on the scanner and the triaxial test apparatus). For the CT scanning, the voltage used was 150 kV and the electric current was 4 mA. The spatial resolution, i.e. the size of the CT image voxels was $0.073 \times 0.073 \times 0.3$ mm^3. The specimen was scanned in continuous subsequent sections from bottom to top in order to obtain a 3D image of the whole specimen.

3. Image analysis

DIC techniques have been used increasingly over the last 20 years in studies of the mechanics of a diverse range of materials, including recently geomaterials (Viggiani and Hall, 2008). DIC is essentially a mathematical tool for assessing the spatial transformation between images. The DIC analysis presented in this paper was carried out using the TomoWarp code developed at Laboratoire 3S-R (see Hall *et al.*, 2009 for details). This code is a 3D (volumetric) DIC code providing a 3D volume of 3D displacement vectors between *in-situ* acquired CT images. Tomowarp follows the same basic steps as most DIC procedures for strain analysis:

1) definition of nodes distributed over the first image;

2) definition of a region centered on each node (the correlation window);

3) calculation of a correlation coefficient for all 3D displacements of the correlation window within an area (the search window) around the target node in the second image;

4) definition of the discrete displacement (integer number of pixels), given by the displacement with the best correlation;

5) sub-pixel refinement (because the displacements are rarely integer numbers of pixels);

6) calculation of the strains based on the derived displacements and a continuum assumption.

4. Results and discussion

4.1. *Test results*

Figure 2 shows the force-displacement responses from both triaxial compression tests presented here. The annotations on the curve (Initial, A, B, C and D for the 50 kPa test and Initial, 1, 2, 3 and 4 for the 150 kPa test) indicate the moments of the CT scans. Both curves show a similar behavior. First, there is a roughly linear increase that is followed by a curvature to the peak force at around 7%. Then, the force decreases until the end of the test where there is the beginning of a plateau. As expected, the 150 kPa specimen presents a stiffer behavior with higher values of force. Note that there are some stress relaxation phases due to stopping the loading during CT scanning. Most of the relaxation phase occurs almost immediately after the loading is stopped and not over the whole scanning period.

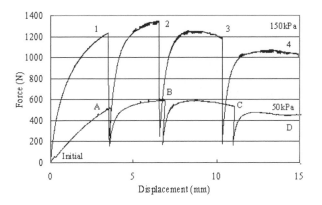

Figure 2. *Force-displacement response*

Figure 3, on the next page, shows vertical slices of the middle part of the specimen at each scanning step. These slices are roughly perpendicular to the final plane of localization obtained at the end of the tests. The images from the 50 kPa and 150 kPa tests are presented in the top and bottom part of the figure respectively. In these grey scale images, black represents the air and white represents high-density material. In these images, the spatial variation in the soil density is clear and, due to the wide range of grain sizes and the given spatial resolution, the specimens appear to be a soil with large grains embedded in a matrix of fine grains (smaller than the resolution). Both series of vertical sections along the test show the shortening of the specimen due to the axial loading from the top and the well-known barreling effect caused by the friction at the top and bottom boundaries. For the 50 kPa test, some areas of low density appear at step B (i.e. at the stress peak) in the upper part of the specimen, mainly in the middle and the right and left top corners.

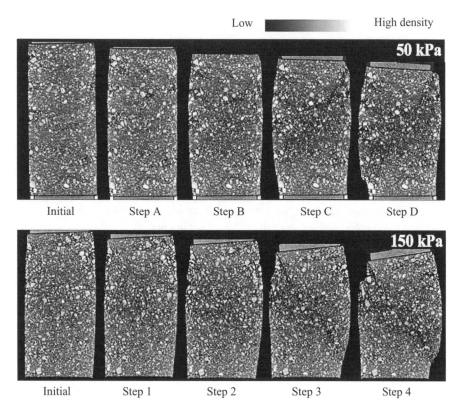

Figure 3. *Vertical CT slices through the loading for the 50 kPa test (top) and the 150 kPa (bottom)*

At the next step C, the previous low-density areas in the middle and the right top corner develops and forms one localized low-density band inclined from the top-right to the bottom-left. At the end of the test (i.e. step D) this band is more evident. For the 150 kPa test, the scenario is roughly the same with a low-density zone appearing at the peak stress and developing into an inclined low-density band from the left top to the right bottom. For both tests, the CT images clearly show that only after the peak stress that the strain is localized in a dilative inclined band.

4.2. Digital image correlation results

The CT images were analyzed using 3D-volumetric DIC in order to obtain the full incremental strain field for each load step. Figure 4, on the next page, shows the distribution of the incremental shear strain at a vertical cross section for the 50 kPa (top) and 150 kPa tests (bottom). The section is roughly located at the same place as the vertical slices shown in Figure 3. The white represents a strain greater than 0.3 and black represents no strain. Note that the high strain zones at the boundaries are artefacts of the DIC. For the 50 kPa test, a wide inclined zone of shearing from the left top to the right bottom appears at the increment Initial-A. At increment A-B, just before the peak stress, the shear strain is localized in two crossing bands inclined from the top to the bottom corners. In increment B-C, just after the peak, the shear strain increases in the band inclined from the right corner and decreases in the other one. At the final increment C-D, the shear is only localized in the band from the top-right corner, which corresponds to the band previously observed in the CT images. For the 150 kPa test, the scenario is roughly the same except that only one inclined shear band evolves through the test. The shear strain is localized in a wide, diffuse, inclined zone from the top-left to the bottom-right corners at the increment Initial-1. The zone starts to narrow before the peak (increment 1-2) and even more after with an increase of the shear strain inside the band with an increase in axial loading (increments 2-3 and 3-4). These results clearly show the evolution of the shear strain from two inclined shear bands before the stress peak into a single band after the peak for both the 50 kPa test and for the 150 kPa test. Moreover, it can be seen that the band starts as a wide, diffuse zone and converges to a narrow one in which the shear strain increases with an increase in axial loading.

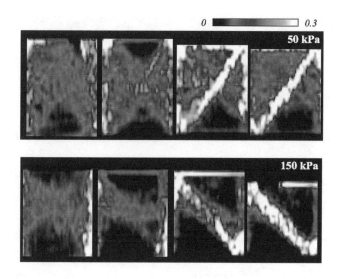

Figure 4. *Distribution of incremental maximum shear strain from DIC at a vertical cross section for the 50kPa test (top) and the 150kPa (bottom)*

Figure 5 shows the distribution of the incremental volumetric strain at the same vertical cross section position as in Figure 3 for the 50 kPa (top) and 150 kPa (bottom) tests. The white represents a volumetric strain of -0.4 and less (i.e. dilation with the soil mechanics convention) and black represents a compressive strain greater than 0.4 (i.e. compression). Note that the horizontal bands might be an artefact or could be the result of a non-uniform sample deposition. For the 50 kPa test, the volumetric strain indicates dilation as the loading increases. Before the stress peak, in step A-B, two dilating bands have formed in the specimen. After step A-B, it is observed that zones of compression appear in the shear band. At step C-D, zones of dilation and compression exist through the whole of the band. For the 150 kPa test, at steps Initial-1 and 1-2, dilation areas appear in the middle of the specimen. After stress peak, it is observed that compression and dilation areas appear from the top-left to the bottom-right. At step 3-4, dilation and compression occur throughout the shear band as shown in the case of 50 kPa. The results from DIC clearly demonstrate that XRCT is not enough by itself to fully characterize the shear localization process. For instance, the shear band is only detectable in this study after the peak stress once the local volumetric is large enough to be 'seen' with XRCT. The combination XRCT/DIC is necessary and is a powerful method to characterize in detail, both qualitatively and quantitatively, the process of shear localization in space and time.

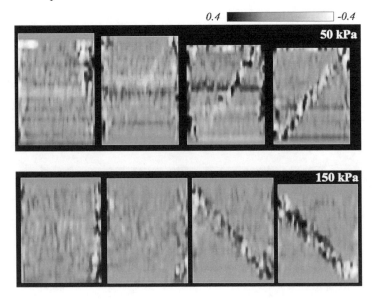

Figure 5. *Distribution of incremental volumetric strain from DIC at a vertical cross section for the 50kPa test (top) and the 150kPa (bottom)*

5. Conclusions

CT images of a sand under triaxial compression at two different confinement pressures were obtained and analyzed by DIC in order to characterize the strain localization process. Three dimensional incremental strain fields were obtained. Thus, the shear localization process was characterized in detail in space and time. In particular, it has been shown that the shear band starts before the stress peak as a wide, diffuse zone and then narrows into a concentrated shear zone after the peak. Moreover, a certain degree of structure has been observed in the band with alternating of compactive and dilative zones. However, it should be noted that only two tests were analyzed in this paper. More tests in different conditions, such as a wider range of confinement pressure or loading rate, have to be conducted to fully characterize the strain localization in this sand.

6. References

Hall S.A., Lenoir N., Viggiani, G., Desrues J., Bésuelle, P., "Strain localisation in sand under triaxial loading: characterisation by x-ray micro tomography and 3D digital image correlation", *Proc.of Int. Symp. on Computational Geomechanics COMGeo09*, 2009.

Hall S.A., Bornert M., Desrues J., Pannier Y., Lenoir N., Viggiani, G., Bésuelle, P., "Discrete and Continuum analysis of localised deformation in sand using X-ray micro CT and Volumetric Digital Image Correlation", accepted, *Géotechnique*, 2010.

Nielsen, S.F., Poulsen, H.F., Beckmann, F. Thorning, C. and Wert, J.A. "Measurements of plastic displacement gradient components in three dimensions using marker particles and synchrotron X-ray absorption microtomography", vol.51, No.8, pp.2407–2415, *Acta Materiala*, 2003.

Viggiani, G. and Hall, S.A., "Full-field measurements, a new tool for laboratory experimental geomechanics", pp.3-26, *Fourth Symposium on Deformation Characteristics of Geomaterials*, IOS press, 2008.

Watanabe, Y., Otani, J., Lenoir, N., Takano, D. and Mukunoki, T. "Visualisation of Strain Field in Sand under Triaxial Compression Using X-ray CT", vol.56, pp.119-124, *Theoretical and Applied Mechanics Japan*, 2008.

Yamamoto, K. and Otani, J., "Microscopics observation on progressive failure on reinforced foundations", vol.41, No.1, pp.25–37, *Soils and Foundations*, 2001.

Latest Developments in 3D Analysis of Geomaterials by Morpho+

V. Cnudde*,** — J. Vlassenbroeck** — Y. De Witte** — L. Brabant** — M. N. Boone** — J. Dewanckele*,** — L. Van Hoorebeke** — P. Jacobs*,**

* Department of Geology and Soil Science, Ghent University
Krijgslaan 281/S8, B-9000 Ghent, Belgium
veerle.cnudde@ugent.be
jan.dewanckele@ugent.be
patric.jacobs@ugent.be

** Department of Subatomic and Radiation Physics, Ghent University
Proeftuinstraat 86, B-9000 Ghent, Belgium
jelle.vlassenbroeck@ugent.be
yoni.dewitte@ugent.be
loes.brabant@ugent.be
matthieu.boone@ugent.be
luc.vanhoorebeke@ugent.be

ABSTRACT. At the Center for X-ray Tomography at Ghent University (Belgium) (www.ugct.ugent.be), besides hardware development for high-resolution x-ray CT scanners, a lot of progress is being made in the field of 3D analysis of the scanned samples. Morpho+ is a flexible 3D analysis software which provides the necessary petrophysical parameters of the scanned samples in 3D. Although Morpho+ was originally designed to provide any kind of 3D parameter, it contains some specific features especially designed for the analysis of geomaterial properties like porosity, partial porosity, pore-size distribution, grain size, grain orientation and surface determination. Additionally, the results of the 3D analysis can be visualized which enables us to understand and interpret the analysis results in a straightforward way. The complementarities between high-quality x-ray CT images and flexible 3D software are opening up new gateways in the study of geomaterials.

KEYWORDS: 3D analysis, x-ray CT, geomaterials, Morpho+, high-resolution

1. Introduction

At the Center for X-ray Tomography at Ghent University in Belgium (UGCT, www.ugct.ugent.be) there is a continuous drive to perform x-ray tomography scans at the highest quality and highest resolution with laboratory equipment. The combination of x-ray CT and 3D visualization is a powerful technique which allows us to look inside the samples in a non-destructive way. Although 3D visualization is a powerful tool to investigate the sample after reconstruction, it has its limitations. While large and apparent structures can be rendered and their shapes and relative sizes can be assessed qualitatively, a lot of information remains hidden. To extract quantitative information, the volume has to be analyzed using appropriate computer algorithms. Since a complete volume of linear attenuation coefficients can be obtained from x-ray computed tomography, the possibilities to analyze the resulting data are extensive. While a single cross-section can already reveal a lot of useful information, the true power of x-ray tomography lies within the capability to extract quantitative information about internal and external three-dimensional structures. Therefore Morpho+, a flexible 3D analysis software package, was developed in-house, in order to provide any kind of 3D parameter for any kind of scanned material.

2. Morpho+

Morpho+ is a new software package developed at the UGCT (Belgium) which provides 3D data of CT scans like porosity, pore-size distribution, grain orientation, sphericity, etc. The development of Morpho+ was initiated by the observation that the in-house developed Matlab analysis package μCTanalySIS (Cnudde *et al.*, 2004) showed several shortcomings like the lacking of user interface and advanced level of functionality. Although Morpho+ builds on some of the concepts used in μCTanalySIS, it combines this with more performance, a more extensive set of analysis tools and an interactive and intuitive user interface. Due to the versatility of the applications at the UGCT, full control over the analysis process and a thorough understanding of the underlying algorithms was an essential requirement for this software. Additionally, the possibility to implement custom algorithms for specific applications was a prerequisite. These conditions had to be combined with a high computational performance and memory efficient code. In order to obtain the necessary user-friendly interface, performance and coding flexibility, Morpho+ was programmed in C++ using the Qt® application framework.

The typical work-flow for the 3D analysis in Morpho+ will be illustrated on the dataset of a North Sea reservoir sandstone. This sandstone was scanned with the flexible micro CT set-up from UGCT. After scanning and reconstruction of the images, the different steps possible in Morpho+ in order to obtain 3D data are described.

2.1. *Scanning conditions*

The North Sea reservoir sandstone sample was scanned with the flexible micro CT scanner of UGCT. The transmission head (Feinfocus, FXE-160.51) was selected with a tube voltage of 110 kV and a tube current of 73 µA. The Varian PaxScan 2520V (1880 x 1496 pixels) was used as the detector. 1,200 projections were taken over 360°, with an exposure time of 300 ms per frame and 5 frame averages were taken per projection. This resulted in a total scan time of around 1 hour. The highest spatial resolution of this system, depending on the sample size, the spot size of the x-ray source and the resolution of the detector, is around 1 µm. For this set-up, with a geometrical magnification of 23, a resolution of 3.97 µm was obtained in all directions. After reconstruction of the projections with the software package Octopus (Vlassenbroeck *et al.,* 2007), a series of horizontal cross-sections (Figure 1) was obtained, ready to be analyzed.

Figure 1. *Reconstructed cross-section through a North Sea reservoir sandstone*

2.2. *3D analysis by Morpho+*

2D reconstructed images all contain information which is shown in a grayscale representation. This gray sale is used for further 3D analysis in Morpho+. However, a grayscale volume is only a discrete representation of a real object. This poses several limitations, like the fact that features smaller than the voxel size cannot be distinguished. Although only cross-sections of the analyzed volume are shown in the following paragraphs, all analysis steps in Morpho+ are performed in 3D.

2.2.1. *Morpho+ analysis steps*

2.2.1.1. Volume of interest selection

When performing 3D analysis, first the volume of interest (VOI) needs to be selected in order to omit irrelevant data which can give wrong analysis results. For example, when analyzing the porosity distribution in a geological sample, the air outside the sample should not be included in the analysis. For cylindrical or cuboid samples, a VOI can be selected by defining a circular or rectangular region in one cross-section and by propagating this selection over a number of cross-sections in the stack. Additionally, Morpho+ allows rotating a cylindrical or cuboid VOI to compensate for sample tilt. Irregular shaped samples need a more intelligent approach for the selection of a VOI. In Morpho+ this is possible by using segmentation and binary operations in order to select the correct VOI.

2.2.1.2. Segmentation and filtering techniques

Segmentation algorithms are aimed at extracting structural features from the volume, like the pore network or the grains inside the sandstone. To facilitate the segmentation several noise filters (median, Gaussian and bilateral filter) are available in Morpho+, where each filter has its own unique properties in terms of noise reduction and preservation of image sharpness. The median filter is a non-linear filter, where each voxel value is replaced by the median of its neighboring voxels. This filter can reduce the spread and overlap of the distributions of the gray values of the different components inside an image, which makes it easier to segment the different components since it preserves straight edges, although sharp corners are rounded. The Gaussian filter is a linear filter, since each voxel of the volume is replaced by a linear combination of voxels within a kernel of size N x N x N centered around the voxel. This filter results however in a blurring effect around edges. The bilateral filter is an extension of the Gaussian filter, where the actual grayvalues of the neighboring voxels are taken into account. When a material voxel close to an edge is evaluated in the Gaussian filter, the neighboring voxels containing the same material are processed in the same way as the voxels corresponding to air. The bilateral filter (Tomasi & Manduchi, 1998) solves this problem by adapting the multiplication factors of the filter kernel based on the actual grayvalues of the neighboring voxels.

After the filtering operations, the data is thresholded to separate the material of interest from the background. During this operation a binary volume is created and voxels are categorized as foreground voxels, when their grayscale value lies within a certain interval, and as background voxels, when they are outside of the interval. The threshold level can be calculated automatically in Morpho+ by using Otsu's method which assumes the volume is composed of two components, where each component shows a certain distribution of grayvalues in the histogram, and both distributions have an overlapping region (Otsu, 1979). When applied to real data,

Otsu's method and other automatic thresholding techniques often fail, especially when the partial volume effect (the grayvalue of a voxel corresponds to the average linear attenuation coefficient of the different materials inside the voxel) play a significant role. Besides automatic thresholding several classes of manual thresholding techniques exist in Morpho+ like single thresholding, which is sensitive to residual image noise, and dual thresholding, which uses two separate threshold values (or intervals) to reduce the sensitivity to residual image noise. In the scanned sandstone, besides the dense inclusions, the grains or the pore structure can easily be segmented by using the dual thresholding technique.

Several binary operations, including opening, closing, removal of isolated foreground/background voxels and the filling of holes are implemented in Morpho+ which can help to remove or reduce the still remaining noise after thresholding. For our sample this was not necessary however.

2.2.1.3. Labeling

The next step in the analysis process is the labeling. The binary volume is therefore divided into several connected components called objects, where each object is assigned a unique value (color code). This object labeling is used to detect each connected component of foreground voxels in the binary image and to assign a unique label to all voxels of that object. In Morpho+, a series of algorithms is used to detect the connected components (Roselfeld & Pfaltz, 1966; Knuth, 1997). The labeling algorithm results in a 3D array of integer labels and a list of objects is extracted from this array. Each object can then be analyzed separately, which adds a significant flexibility to the analysis process. Each object is characterized by its bounding box, the position of its bounding box in the original volume and the object voxels inside the bounding box. After the extraction of the list of objects, several operations are possible. Before proceeding with other analysis steps, it is often interesting to apply the fill holes algorithm to the individual objects.

2.2.1.4. Distance transform

After the labeling algorithm, the Euclidean distance transform (Guan & Ma, 1998; Delerue, 2001) is applied which serves several purposes in Morpho+. It can be used to separate objects which are initially connected, combined with the watershed segmentation, based on the algorithm by Vincent et al. (1991). Also, the maximum opening of an object, defined as the diameter of the maximum inscribed sphere which fits inside the object, can be extracted from its distance transform.

2.2.1.5. Quantitative information

After obtaining the distance transform, several parameters can be determined quantitatively in Morpho+. The total porosity can be obtained by dividing the total number of foreground voxels by the total number of voxels in the VOI. For the total analyzed volume of the scanned sandstone a total porosity of 13% was found. Since the total volume can be subdivided in several blocks in any direction (X, Y or Z) it is possible to evaluate the porosity per block. This way the porosity distribution in the sample can be extracted. The determination of the percentage of open and closed pores can also be distinguished. The open porosity is defined as the porosity of all border objects, while the closed porosity is composed of the isolated objects. Additionally, it is possible to extract information from each object. This data can be exported for each object and/or distributions can be extracted. The maximum opening and the total volume of each object can be determined. If we construct an equivalent sphere with the same total volume, the corresponding diameter is defined as the equivalent diameter. Based on the maximum opening and equivalent diameter, the sphericity S is defined as the ratio of both. This parameter gives a rough approximation of the shape of objects, since it expresses how much an object resembles a sphere (S = 1). Since Morpho+ can determine the orientation of the objects of interest, such as the grains inside the sandstone, it is possible to plot a stereoplot derived from the 3D analysis of the grains. Additionally, the Euler number, which expresses the multi connectivity of a volume and can be used to compare the connectivity of different (similar) samples, and the fractal dimension, determining how efficiently an object fills the space under consideration, as well as surface extraction, the determination of connections, and skeletonization are also implemented in Morpho+ (Brabant, 2009).

In the sandstone sample the Euler number of the pore network was -68929. A single value of the Euler number is difficult to interpret and these values are more relevant when comparing very similar rocks. Morpho+ also makes it possible to count the number of objects that are not connected to each other, so in this case the number of isolated pore networks could be obtained. In the analyzed sandstone sample 61,038 isolated pore networks were identified. It is possible to calculate the number of tunnels through the pore network. There were 140,510 tunnels identified through the pore network of the analyzed sample, which illustrates that it has a high multi-connectedness. The fractal dimension of the grains of the sandstone sample is 2.769. The surface of the separate objects or the complete volume can also be extracted by applying the marching cubes algorithm. The resulting 3D mesh can be saved as an STL file and can be used in other software packages for visualization and analysis purposes (for example, finite element simulations). The result is shown for the grains of the sandstone sample (Figure 2).

It is also possible to calculate the surface based on the marching cube algorithm. The pore network within a volume of height 3.08 mm, depth 2.01 mm and width 2.33 mm has a surface of 369 mm² for the scanned sample.

Figure 2. *The visualization of grains of a sandstone sample. The outside of the grains is colored in red, the inside in green and the dividing planes are colored in yellow (see CD of this volume for color version)*

2.3. *Data visualization*

The processed volume can be visualized after each step in the analysis process. Since all analysis algorithms operate in three dimensions, cross-sections of the volume according to the different principal planes can be visualized. Each object can be relabeled (color coded) based on a certain parameter (Figure 3), like for example the equivalent diameter where the blue objects have a small equivalent diameter and the red objects are the largest.

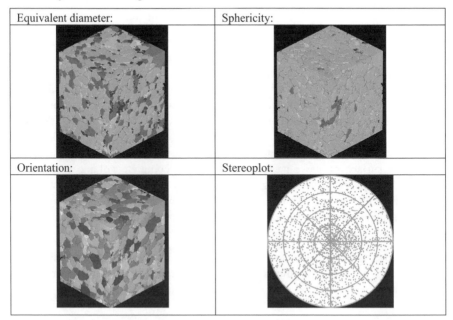

Figure 3. *3D renderings of the color-labeled images after 3D analysis with Morpho+ (see CD of this volume for color version)*

Figure 3 illustrates the different possibilities to process the object data: it shows the distribution of the equivalent diameter, sphericity, the orientation and the stereoplot of the analyzed quartz grains inside the sandstone.

3. Conclusions

Morpho+ contains some specific features especially designed for the analysis of geomaterial properties like porosity, partial porosity, pore-size distribution, grain size, grain orientation, sphericity, orientation, Euler number, fractal dimension and surface determination. Additionally, the results of the 3D analysis can be visualized which enables to understand and interpret the analysis results in a straightforward way. The complementarities between high-quality x-ray CT images and flexible 3D software are opening up new gateways in the study of geomaterials.

4. Acknowledgements

The Fund for Scientific Research—Flanders (FWO) is gratefully acknowledged for the post-doc grant to V. Cnudde. The Institute for the Promotion of Innovation by Science and Technology in Flanders, Belgium is acknowledged for the PhD grant of J. Dewanckele.

5. References

Cnudde, V., Cnudde, J.P., Dupuis, C., Jacobs, P. "X-ray micro-CT used for the localization of water repellents and consolidants inside natural building stones", *Materials Characterization*, vol. 53 no. 2-4, 2004, p. 259 – 271.

Tomasi, C., Manduchi, R., "Bilateral Filtering for Gray and Color Images. Computer Vision", *IEEE International Conference*, 0:839, 1998.

Otsu, N. "A threshold selection method from gray-level histograms". *IEEE Transactions on Systems, Man and Cybernetics*, vol. 9 no. 1, 1979, p. 62–66.

Rosenfeld, A., Pfaltz, J.L. "Sequential Operations in Digital Picture Processing", *J. ACM*, vol. 13 no. 4, p. 471–494, 1966.

Knuth, D.E. *Art of Computer Programming, Volume 1: Fundamental Algorithms (3rd Edition).* Addison-Wesley Professional, July 1997.

Guan, W., Ma, S. "A List-Processing Approach to Compute Voronoi Diagrams and the Euclidean Distance Transform", *IEEE Transactions on Pattern Analysis and Machine Intelligence,* vol. 20 no. 7, 1998, 757–761.

Delerue, J.F. Segmentation 3D, application l'extraction de réseaux de pores et la caractérisation hydrodynamique des sols, PhD thesis, Paris XI Orsay, 2001.

Vincent, L., Soille, P. "Watersheds in digital spaces: an efficient algorithm based on immersion simulations", *IEEE Transactions on Pattern Analysis and Machine Intelligence*, vol. 13 no. 6, 1991, p. 583–598.

Brabant, L. Geavanceerde algoritmes voor 3D-analyse van micro-CT data, Master's thesis, Ghent University, Belgium, 2009.

Vlassenbroeck, J., Dierick, M., Masschaele, B., Cnudde, V., Van Hoorebeke, L., Jacobs, P. "Software tools for quantification of X-ray microtomography at the UGCT", *Nuclear Instruments and Methods in Physics Research Section A: Accelerators, Spectrometers, Detectors and Associated Equipment*, 2007, 580(1):442-445.

Quantifying Particle Shape in 3D

Experimental and Mathematical Considerations

E. J. Garboczi

100 Bureau Drive Stop 8615
National Institute of Standards and Technology
Gaithersburg, Maryland 20899-8615 USA
edward.garboczi@nist.gov

ABSTRACT. Quantifying the shape of particles in three dimensions (3D) is important in particle technology. In concrete, the 3D shape of particles like sand, gravel and cement is of great interest for applications including suspension rheology, mechanical properties and realistic microstructure models. When particles are classified as star-shaped, a weaker condition than convexity, a combination of x-ray computed tomography (CT) and spherical harmonic series analysis can quantitatively describe their 3D shape. Since this analysis results in an analytic function for the particle's surface, one can perform almost any kind of volume or surface integral and so compute many geometrical properties. This paper reviews how 3D particle shape can be measured and analyzed and gives examples for the classes of particles found in construction. Some data will also be given on how particle shape can influence particle size measurement.

KEYWORDS: particle, shape, 3D, spherical harmonics, x-ray tomography, concrete

1. Introduction to particle shape in 3D

The geometry of particles is of interest to researchers in many fields. For regular objects like spheres, cubes, ellipsoids and rectangular boxes, relationships between their dimensions and their shape parameters (e.g. volume, surface area, moment of inertia tensor) have been known for millennia. For a cube or a sphere, a single dimension totally characterizes all properties of the particle. This is not the case for random particles, especially in 3D. It is not hard to define various dimensions for irregular objects – for example, the diameter of the sphere with volume equal to the particles (ESD). However, one length parameter alone cannot define the size and shape of a randomly shaped particle.

The shape of a particle can be important in several ways. In composites, where distinct particles (usually called inclusions) are embedded in a homogeneous phase (usually called the matrix), the particle shape of the inclusion phase can strongly influence the composite properties (Douglas *et al.* 1995). The percolation properties of the inclusions, as the inclusion packing density becomes higher, also depend on shape (Garboczi *et al.*, 1995). The shape of the solid particles that are suspended in a fluid controls the effective suspension viscosity at a given packing density (Douglas *et al.*, 1995). To accurately model processes where a solid is built up from collections of particles, like in cement hydration in concrete, realistic 3D particle models are required (Bullard *et al.*, 2006). Sometimes, as in the case of Portland cement, particle shape can be linked to particle mineralogy when particles are multi-phase and are prepared by mechanical grinding (Erdoğan *et al.*, 2009).

In general, irregular bodies may or may not be *convex*: a body in which a straight line joining (any) two points in the interior or on the boundary of the body always remains within the body. Most construction particles have some concavities on their surface – especially if they have been crushed – and hence are not convex. A weaker criterion uses the concept of "star shaped". A star-shaped object requires that all lines that connect the center of volume to any point inside the body (or on its surface) lie entirely within the body. Almost all particles used in the construction industry are star-shaped (Garboczi 2002), as are most geologically-formed particles.

2. Measuring particle shape – experiment and mathematical analysis

The x-ray CT results presented in this paper were collected on two instruments: (1) beamline X2B, at the National Synchrotron Light Source (NSLS), Brookhaven National Laboratory, and (2) on a benchtop instrument located at the National Institute of Standards and Technology (NIST). Pixel sizes down to slightly less than one micrometer were used. Small particles were embedded in epoxy in a cylindrical polymer mold and then scanned. Large particles were mechanically held singly or in some kind of packing so that the particles did not touch each other. The reconstructed image slices were stacked into 3D microstructures consisting of gray scale images of particles embedded in a matrix. A suite of image analysis and error correction software was used to create a clean image of white, separated particles in a black matrix. Custom software was used to extract the particle voxels in 3D (Erdoğan *et al.*, 2006).

The 3D voxel data for a single particle is used to generate the function $r(\theta,\phi)$, which is the distance from the center of volume to the surface in the direction given by the spherical polar angles (θ,ϕ). Spherical harmonic (SH) functions are then used to create a smooth approximation, within the uncertainty of the original x-ray CT image, to the function $r(\theta,\phi)$, using the following expansion for star-shaped particles (Arken, 1970) :

$$r(\theta,\phi) = \sum_{n=0}^{\infty} \sum_{m=-n}^{n} a_{nm} Y_{nm}(\theta,\phi) \qquad [1]$$

The star-shaped criterion must be met for this expansion to be valid (Garboczi, 2002). This analytic approximation accurately represents a random-shaped particle and can be used to compute any geometric quantity of the particle like volume, surface area, or moment of inertia (Garboczi, 2002).

A useful set of three orthogonal dimensions for an irregular object, which probably cannot be defined by integrals, is found by directly measuring three dimensions, called length, L, width, W, and thickness, T, defined the following way. One measures the longest line that connects any two points within the body and defines this as the "length" of the body L; the longest such line that is orthogonal to L is called the width W. A similar procedure yields T, which must be orthogonal to both L and T (ASTM D4791). The SH mathematical approximation of the particle can be used in a simple algorithm to find approximations for L, W, and T simply by searching for pairs of surface points that satisfy the length and direction criteria. Often, L = L/T and W = W/T are used and referred to as the normalized dimensions.

3. Examples of particle shape analysis

3.1. *Simulated lunar soil*

A sample of lunar soil simulant JSC-1A[1] (Metzger, 2008) was wet screened with water into four size fractions: (1) particles retained on a 300 μm ASTM screen (+300), (2) those passing the 300 μm screen but retained on a 75 μm screen (75-300), (3) those passing the 75 μm screen but retained on a 38 μm screen (38-75), and (4) those passing the 38 μm screen but retained on a 20 μm screen (20-38).

---

1. Material supplied by P.T. Metzger, Kennedy Space Center, Florida, 2008.

Samples of all four size fractions were embedded in epoxy and formed into cylindrical specimens. A projection algorithm was used to compute the apparent images as would be seen by a 2D optical scanning instrument, which are used commonly to attempt to measure particle shape. Approximately 130,000 particles were measured and analyzed (Metzger *et al.*, 2009). Figure 1 shows two of these particles. The 3D L, W and T dimensions, the moment of inertia tensor, as well as the mean curvature, were computed for each particle, along with many 2D parameters for purposes of comparison with true 3D values.

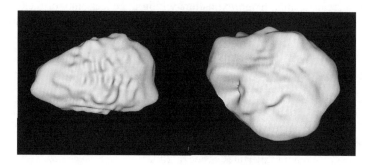

Figure 1. *Virtual Reality Modeling Language (VRML) images of simulated lunar soil particles from two size classes - (left) 20-38 (right) 75-300 (images not to scale)*

3.2. *Portland cement*

Portland cement powder is prepared by grinding 10 mm to 20 mm particles called clinkers down to micrometer size particles. Since the clinkers are multi-phase at the micrometer length scale, so is the resulting powder. The different phases are known to have different hardnesses, so particle phase composition could be related to particle shape. In a study examining eight different Portland cements (Erdoğan *et al.*, 2009), parallel x-ray diffraction studies of particle mineralogy and x-ray CT/SH studies of particle shape found a strong correlation between mineralogy and shape. x-ray CT has also been used to examine the 3D shape of particles in the 20 μm to 60 μm size range (the upper size range for cement), and focused ion beam nanotomography (FIB) has been used to examine the 3D shape of cement particles found in the 0.4 μm to 2.0 μm size range (about the smallest size range for cement) for the same cement (Holzer *et al.*, 2009). Both size ranges were defined by sieving. By comparing various kinds of computed particle shape data for each size class, the conclusion was made that, within experimental uncertainty, the smaller size class particles tended to be marginally more prolate than the larger size class. Figure 2 shows the probability, in terms of number fraction, vs. length and width distributions for both size classes of particles.

3.3. *Gravel*

Twelve reference rocks, all between 12 mm and 19 mm in size, were extensively studied via x-ray CT and SH analysis (Taylor *et al.*, 2006). Their volumes were measured using Archimedes' technique, and compared very closely (within 1% to 2%) with the CT/SH technique. Their L, W and T values were also measured with a digital caliper and agreed within experimental uncertainty with the CT/SH values. Several hundred microfine rocks, which passed through a 75 μm sieve but were retained on a 20 μm sieve, were also studied. Several new equivalent shapes, including rectangular boxes and tri-axial ellipsoids, were defined using either the three principal moments of inertia or the L, W and T parameters. Several of these new equivalent shapes proved to be good predictors of particle volume and surface area (as measured at the x-ray CT pixel size).

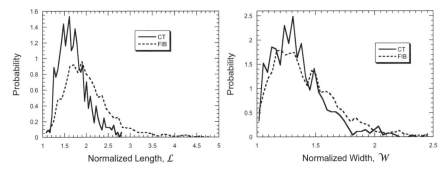

Figure 2. *Histograms for probability (in terms of number fraction)
vs. length and width distributions for both (CT, FIB) size classes of cement particles*

3.4. *Sand*

Several ASTM standards for cement and concrete specify the use of "Ottawa sand", sand that comes from the historic deposits located at Ottawa, Illinois, which have been mined for well over 100 years and are used in laboratories all around North America as the desired sand for many standard tests. For the approximately 2,000 Ottawa sand particles studied in the x-ray CT work, the average value of L = 1.78 ± 0.38 and the average value of W = 1.40 ± 0.28. The uncertainty represents one standard deviation for the average taken over all the particles. Therefore, the average length-to-width ratio of the Ottawa sand investigated was about 1.29. The particles were well-rounded but definitely not spherical. The principal moment of inertia tensor was computed for each particle, and the dimensionless parameter I was defined as the ratio of the maximum principal moment to the minimum. The value of I seemed to vary inversely (but mildly) with particle size, as defined by sieve size.

3.5. *Laser diffraction vs. x-ray CT*

Laser diffraction particle size analyzers are commonly used for measuring particle size distributions of powders. For a sphere, there is one length that matters – the diameter. However, for a randomly shaped particle, specifying one length does not determine all geometric properties of the particle – including light scattering by the particle. It was of interest, therefore, to look at laser diffraction particle size results and compare the use of the equivalent spherical diameter as determined by light scattering to the use of other dimensions of the particle to characterize the particle "size." Particles for the light scattering and the x-ray CT experiments were drawn from the same source, with the same kind of sieving. Figure 3 shows the results for four different choices of "size" compared to the laser diffraction "size": volume equivalent spherical diameter (ESD), and L, W, and T. It appears that using L as the "size" of the particles agrees best with the laser diffraction results (Erdoğan *et al.*, 2007). The label "PF 38-75" indicates the kind of particle and the size class (38 μm to 75 μm as determined by sieving).

3.6. *Generating new virtual particles statistically similar to a real particle dataset*

Can particles be artificially generated (virtual particles) that are like a given set of real particles in a statistical sense? We have explored this question, first using a dataset of 128 real rocks to statistically generate the SH coefficients for new rocks (Grigoriu *et al.*, 2006). This size dataset was found to be rather small, and so, once the algorithm was converted to parallel mode, much larger datasets of real particles were used (on the order of thousands and tens of thousand). Much better agreement between the shape properties of the virtual particles and the real particles from which they were generated were found (Liu *et al.*, 2009), lending support to the use of this procedure.

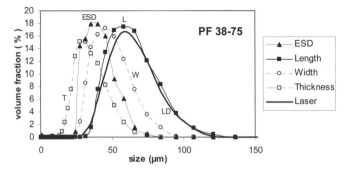

Figure 3. *Histograms for the probability, in terms of volume fraction, vs. "size" for five different measures of particle*

4. Future research

The main research problem at present is not in acquiring and analyzing 3D particle images – these techniques are now well-established and have proven to be very useful – but lies in how to interpret and use the vast array of shape-related data that can now be generated. A precise analytical mathematical representation has been demonstrated for star-shaped particles; enabling practically any surface or volume integral in 3D or 2D (via projection) to be performed and therefore almost any shape-related parameter to be calculated. How does one then use, for example, the principal moments of inertia, or the length, width, and thickness to classify shape? There are very many equivalent shapes that can be generated using these new data – which are the best or most useful? Some attempts to answer these questions have been briefly reviewed in this paper – subsequent progress in the area of particle shape analysis will be made by continuing to address these basic questions.

5. References

Arfken G., *Mathematical Methods for Physicists*, Academic Press, New York, 1970.

Bullard J.W., Garboczi E.J., "A model investigation of the influence of particle shape on portland cement hydration", *Cement and Concrete Research*, vol. 36, 2006, p. 1007-1015.

Douglas J.F., Garboczi E.J., "Intrinsic viscosity and polarizability of particles having a wide range of shapes", *Advances in Chemical Physics*, vol. 91, 1995, p. 85-153.

Erdoğan S.T., Quiroga P.N., Fowler D.W., Saleh H.A., Livingston R.A., Garboczi E.J., Ketcham P.M., Hagedorn J.G., Satterfield S.G., "Three-dimensional shape analysis of coarse aggregates: New techniques for and preliminary results on several different coarse aggregates and reference rocks", *Cement and Concrete Research*, vol. 36, 2006, p. 1619-1627.

Erdoğan S.T., Fowler D.W., Garboczi E.J., "Shape and size of microfine aggregates: X-ray microcomputed tomography vs. laser diffraction", *Powder Technology*, vol. 177, 2007, p. 53-63.

Erdoğan S.T., Nie X., Stutzman P.E., Garboczi E.J., "Micrometer-scale 3-D imaging of eight cements: Particle shape, cement chemistry, and the effect of particle shape on laser diffraction size analysis", submitted to *Cement and Concrete Research*, 2009.

Garboczi E.J., Snyder K.A., Douglas J.F., Thorpe M.F., "Geometrical percolation threshold of overlapping ellipsoids", *Physical Review E*, vol. 52, 1995, p. 819-828.

Garboczi E.J., "Three-dimensional mathematical analysis of particle shape using x-ray tomography and spherical harmonics: Application to aggregates used in concrete", *Cement and Concrete Research*, vol. 32, 2002, p. 1621-1638.

Grigoriu M., Garboczi E.J., Kafali C., "Spherical harmonic-based random fields for aggregates used in concrete", *Powder Technology*, vol. 166, 2006, 123-138.

Holzer L., Flatt R., Erdoğan S.T., Bullard J.W., Nie X., Garboczi E.J., "Shape comparison of a cement between nano-focused-ion-beam and X-ray microcomputed tomography", to be submitted to *J. Amer. Ceram. Soc.*, 2009.

X. Liu, Y. Lu, M. Grigoriu, and E.J. Garboczi, "Spherical harmonic-based random fields for aggregates used in concrete: Further investigation and application", in preparation (2009).

Metzger P.T., Garboczi E.J., "Three dimensional shape analysis of simulated lunar soil particles" (in preparation, 2009).

Taylor M.A., Garboczi E.J., Erdoğan S.T., Fowler D.W., "Some properties of irregular particles in 3-D", *Powder Technology*, vol. 162, 2006, p. 1-15.

3D Aggregate Evaluation Using Laser and X-ray Scanning

L. Wang* — C. Druta** — Y. Zhou*** — C. Harris***

* The Via Department of Civil and Environmental Engineering
Virginia Polytechnic Institute and State University
301N Patton Hall, Blacksburg, VA 24061, (540) 231-5262
wangl@vt.edu

** Virginia Tech Transportation Institute (VTTI)
Center for Sustainable Transportation Infrastructure
3500 Research Transportation Plaza
Blacksburg, VA 24061, (540) 231-1056
cdruta1@vt.edu

*** The Via Department of Civil and Environmental Engineering
Virginia Polytechnic Institute and State University
200 Patton Hall, Blacksburg, VA 24061
(540) 231-6527; (540) 231-1021
zhouy@vt.edu; chharri3@vt.edu

ABSTRACT. In this paper, two scanning systems, an x-ray-based and a laser-based system, have been evaluated in terms of its capability in accurately acquiring three-dimensional (3D) surface data and their potential in evaluating fine and coarse aggregates. The dimension measurements, surface area and volume quantification capability of the two systems was evaluated through comparing these measurements with caliper measurements, theoretical calculations and other proved methods. The evaluations indicate that both the x-ray scanning method and laser scanning method can yield very high resolution up to 6 to 60 μm in determining the dimensions of aggregates, allowing for accurate characterization of the derived characteristics of aggregate shape, angularity and texture in true 3D.

KEYWORDS: aggregate characterization, laser system, x-ray system, 3D evaluation, shape, angularity, texture of aggregates

1. Introduction

Aggregates are an important component in asphalt concrete, cement concrete, granular base and treated base. Their characteristics including shape, angularity, surface texture and surface area significantly affect the properties of mixtures (Masad and Button 2000; Rao *et al.*, 2001; Garboczi *et al.*, 2002; Wang *et al.*, 2004a). Historically, tremendous efforts have been made to quantify these characteristics both directly and indirectly and correlate them to performance. Research using 3D mapping for aggregate evaluation is very limited and so far, only a few published results are available (Kim *et al.*, 2002; Wang *et al.*, 2003; Wang *et al.*, 2004b). Most published results have significant limitations in the true 3D sense: mapping (reconstructing) the 3D coordinates of the surface points of an aggregate particle. Most methods only map partial surfaces of the particle. The methods using laser scanning and CT differ in their mechanisms of acquiring the 3D data. The presented work focused on the evaluation of the accuracy of laser scanning and x-ray tomography scanning in measuring the dimension, surface area and volume.

2. Laser and x-ray scanning systems for aggregate evaluations

Imaging systems using laser and x-ray are frequently used in mapping three-dimensional (3D) surfaces and volumes. Their primary functions reside in acquiring images of the objects of interest and subsequently processing them with the help of analysis tools. Analysis tools are computer programs used to analyze the acquired images in order to obtain the desired information, such as dimensions, shape, texture or angularity. There are various imaging systems in the laser and x-ray scanning categories. The systems used in this study are briefly described in the following sections.

2.1. *Surveyor 3D laser scanner*

The laser scanner used in this study utilizes a line-range laser probe (RPS 120), shown in Figure 1, for profile measurement. The probe contains a laser diode that passively spreads the laser into a line and two charge coupled device (CCD) arrays that acquire the inputs for surface point reconstruction. Both CCDs observe the same area known as the field of view (FOV) of the laser. Solid red line (Figure 1, b) shows a calibrated laser (see CD of this volume for color figures).

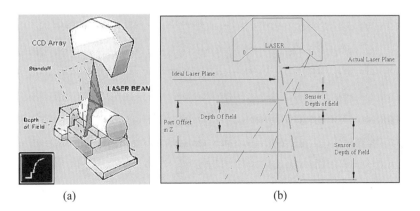

Figure 1. *a) RPS Line range probe and b) probe correction*

A precise method of 3D data acquisition, laser triangulation, is an active stereoscopic technique where the distance of the object is computed by means of a directional light source and a video camera. A laser beam is deflected from a mirror onto a scanning object that scatters the light, which is then collected by a video camera located at a known triangulation distance from the laser. Using trigonometry, the 3D spatial coordinates of a surface point are calculated.

2.1.1. *Image analysis and volumetric reconstruction*

Surveyor laser scanners use 3D point cloud conversion to reconstruct scanned objects into accurate surface files. The process of 3D digitization basically consists of a sensing phase followed by a reconstruction phase. The sensing phase collects or captures the raw data that generate the initial geometry data, usually as a 2D boundary object, or a 3D point cloud.

A surveyor laser scanner uses rapid profile sensors (RPS) to acquire and analyze 2D data. Also, 3D digital models of laser scanned objects can be created from 3D digital coordinate point clouds using specialized computer software.

2.2. *Skyscan 1174 x-ray micro-CT scanner*

The Skyscan 1174 micro-CT scanner uses an x-ray source with adjustable voltage and a range of filters for versatile adaptation to different object densities. Also, a variable magnification (6-30 μm pixel size) is combined with object positioning for easy selection of the part of an object to be scanned. The computed tomography process involves a micro-focus x-ray source which illuminates the object, a planar x-ray detector that collects projection images, and a sample manipulator that rotates and translates a sample. Based on hundreds of angular views acquired while the object rotates, a computer synthesizes a stack of virtual cross-sections or slices through the object. By selecting various volumes of interest, it is possible to measure 3D morphometric parameters and create realistic visual models for virtual travel within the object.

2.2.1. *Image analysis and volumetric reconstruction*

The Skyscan 1174 micro-CT scanner analyzes the tomographic datasets for 2D and 3D morphometry and densitometry. SkyScan's volumetric reconstruction software uses a set of acquired angular projections to create a set of cross-section slices through the object. 3D visual models from scanned datasets can be obtained from selected region-of-interest (ROI) and volume-of-interest (VOI) of an object.

3. Data analysis and results

3.1. *Aggregate dimensioning*

The aggregates used in this project were crushed limestone and granite. Two sets of aggregates, each containing nine particles, ranging from 1 ½ in (37.5 mm) to No. 16 (1.18 mm) were measured using three means: electronic calipers with a precision of two decimals, SkyScan 1174 CT Analyzer software and RPS-120 Geomagic software.

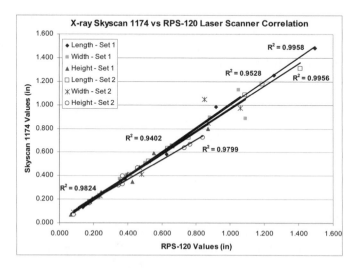

Figure 2. *Correlation between Skyscan 1174 measurements and RPS-120 measurements*

Three dimensions were measured using the above means: length (L) – representing the longest dimension of the particle; width (W) – representing the second longest dimension of the particle normal to the length direction; and height (H) – representing the shortest size of the particle perpendicular to the plane formed by the length and the width. Figure 2 above shows the correlations between Skyscan 1174 analyzer method and the RPS-120 laser method, using linear regression.

3.2. Aggregate surface area and volume evaluation

Using the laser triangulation technique (Figure 3, a), point coordinates can be acquired and then be transformed into a single coordinate system. By selecting the model area field in the program the total surface area of the scanned object is computed. Similarly, by selecting all points or polygons that lie within a user-specified bounding box the total volume of a scanned object is calculated (Figure 3, b).

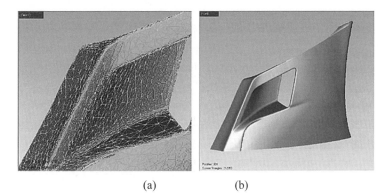

(a)   (b)

Figure 3. *Surface area (a) and volume calculation (b) examples*

Morphometric parameters – object surface and volume – are calculated by CT-Analyser either in 2D from cross-section images or region-of-interest (ROI), individually or integrated over a volume-of-interest (VOI). 3D analysis uses the 2D analysis parameters to create surface area of interest and volume of interest analyses. Data of the surface area and volume of the two sets of aggregates acquired with the laser and x-ray CT software programs indicated that the two methods used to compute (a) the surface areas and (b) the volumes were strongly correlated (Figure 4). This indicates that these two methods give very accurate results and both systems can be employed successfully for evaluating mineral aggregates. Also, surface areas and volumes for four regular geometrical objects, a rectangular prism, a cylinder, a sphere, and a pyramid were determined using both systems. For this particular case, the surface areas and volumes obtained using scanners' software were roughly 4% away of those calculated using mathematical formulas.

Surface area to volume ratios were calculated for both sets of aggregates and for spheres having the same volumes as the aggregates and the results showed good correlations between the specific surface areas of aggregates versus the specific surfaces of spheres with the same volumes. However, slopes for the two aggregate sets were slightly different, indicating minor different morphological characteristics of the two sets of aggregates.

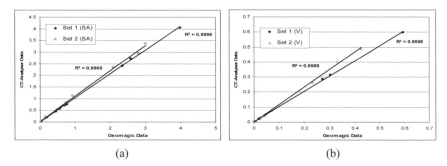

(a)   (b)

Figure 4. *Correlation between CT-Analyser and Geomagic in evaluating the surface area (a) and the volume (b) of aggregate particles*

Similarly, strong correlations were obtained between the masses and volumes of the two aggregate sets computed using software from the two scanning devices. The linear relationship obtained indicates that volume measurements using both scanning methods are accurate, assuming an aggregate has a constant density.

4. Conclusions

This study presents a comparison between two different methods, x-ray computed tomography and laser scanning reconstruction – for evaluating fine and coarse aggregate dimensions, surface areas and volumes. Based on the data obtained it can be concluded that the two scanning systems can successfully be used in evaluating mineral aggregates. Both dimensioning computer programs agreed well with the calipers results. Data from the Skyscan 1174 method were slightly larger than those of the RPS-120 laser meaning that the CT Analyzer software is more accurate in dimensioning samples than Geomagic software. Also, good correlations between the Skyscan 1174 CT Analyzer and RPS-120 Geomagic dimensioning values and surface areas and volumes of aggregates were obtained. These evaluations indicated that both scanning methods can be used to assess the derived aggregate characteristics such as shape, angularity and texture.

5. References

Edil, Garboczi, E.J., "Three-dimensional Mathematical Analysis of Particle Shape Using X-ray Tomography and Spherical Harmonics: Application to Aggregates Used in Concrete", *Cement and Concrete Research*, vol. 32, no.10, 2002, pp. 1621-1638.

Kim, H.K., Haas C.T., Rauch A.F., and Browne C., "Wavelet-based 3D Descriptors of Aggregate Particles", presented at *81th Annual Meeting of the Transportation Research Board*, Washington, D.C., CD-ROM 2002.

Masad, E. and Button J.W., "Unified Imaging Approach for Measuring Aggregate Angularity and Texture", *Journal of Computer-Aided Civil and Infrastructure Engineering*, vol. 15, no. 4, 2000, pp. 273-280.

Rao, C., Tutumluer E., and Stefanski J. A., "Coarse Aggregate Shape and Size Properties Using a New Image Analyzer", *Journal of Testing and Evaluation*, JTEVA, vol. 29, no. 5, Sept. 2001, pp. 461-471.

Wang, L.B. and Frost J.D., "Quantification of Aggregate Specific Surface Area Using X-ray Tomography Imaging", In *ASCE Geotechnical Special Publication*, E. Tutumluer, Y. Najjar, and E. Masad (eds.), GSP No.123, 2003.

Wang, L.B., Paul H. S., Harman T. and D'Angelo J., "Characterization of Aggregates and Asphalt Concrete using X-ray Tomography", *Journal of the Association of Asphalt Paving Technologists*, vol. 73, 2004a, pp. 467-500.

Wang, L.B., Wang X., Mohammad L., and Abadie C., "A Unified Method to Quantify Aggregate Shape Angularity and Texture Using Fourier Analysis", *ASCE Journal of Construction Materials*, vol.17, vo.5, 2004b, pp. 498-504.

T.B., "Mechanical Properties and Mass Behavior of Shredded Tire-Soil mixtures", *Proc. of the International Workshop on Lightweight Geo-Materials, JGS*, pp. 17-32, 2002.

Humphrey, D.N., Whetten, N., Weaver, J., Recker, K., Cosgrove, T.A., "Tire TDA as Lightweight Fill for Embankments and Retaining Walls", *Proc. of Conference on Recycled Materials in Geotechnical Application, ASCE*, pp. 51-65, 1998.

Hazarika, H., Kohama, E., Suzuki, H., Sugano, T., "Enhancement of Earthquake Resistance of Structures using Tire Chips as Compressible Inclusion", *Report of the Port and Airport Research Institute*, vol. 45, no. 1, 2006.

Holz, W.G., Gibbs, H. J., "Triaxial shear tests on pervious gravelly soils", *ASCE*, vol. 82, no. SM 1, pp. 1-22, 1956.

Hyodo, M., Yamada, S., Orense, R.P., Yamada, S., "Undrained cyclic shear properties of tire chip-sand mixtures", *Scrap Tire Derived Geomaterials*, Taylor & Francis, pp. 187-196, 2007.

Kaneda, K., Hazarika, H., Yamazaki, H., "The numerical simulation of earth pressure reduction using tire chips in backfill", *Scrap Tire Derived Geomaterials*, Taylor & Francis, pp. 245-251, 2007.

Computation of Aggregate Contact Points, Orientation and Segregation in Asphalt Specimens Using their X-ray CT Images

M. Kutay

Civil and Environmental Engineering
Michigan State University
3554 Engineering Building
East Lansing, MI, USA
kutay@msu.edu

ABSTRACT. *In this paper, image processing and analysis methods are presented to extract the individual aggregate properties from the x-ray CT images of asphalt specimens. First, a technique was presented (and validated) to segment the clustered aggregates due to elevated pixel intensities near the aggregate-to-aggregate contacts in the image. Once the aggregates were segmented, aggregate-to-aggregate contact points, 3D orientation and segregation of aggregates were analyzed for asphalt specimens compacted at different levels.*

KEYWORDS: *x-ray CT, aggregate, asphalt mixture, image processing*

1. Introduction

Internal structure characteristics related to the aggregate skeleton significantly influence the long-term performance and sustainability of asphalt pavements. Packing and compaction of aggregates within asphalt mixtures can be very important for the overall performance of the system. However, in asphalt mixtures, direct quantification of internal packing characteristics of aggregates in 3D has been very limited. One method of direct quantification of 3D microstructures is through imaging techniques such as x-ray Computed Tomography (CT). Previous studies have utilized x-ray CT to quantify the air void distribution (Masad *et al.* 2002; Tashman *et al.*, 2002), aggregate homogeneity (Azari *et al.*, 2005), and pore-scale fluid flow modeling of asphalt mixtures (Kutay *et al.* 2007a; Kutay and Aydilek 2007b). However, the compacted characteristics of individual aggregates such as location, orientation, and aggregate-to-aggregate contact points have rarely been successfully quantified. One of the main challenges in processing and analyzing the x-ray CT images was the segmentation of aggregates that are in close proximity to other aggregates, which causes clustering of aggregates after thresholding and labeling operations (Figure 1). In this study, an image processing procedure was followed to overcome this shortcoming. In addition, packing/compaction characteristics of aggregates including the aggregate-to-aggregate contact points, 3D orientation angles and segregation were studied for asphalt mixtures with different compaction levels.

2. Materials and characteristics of the x-ray CT equipment

In order to illustrate the aggregate packing characteristics computed by the image analysis algorithms presented in this paper, two asphalt mixture specimens compacted at 30 (low compaction level, labeled N30) and 160 (high compaction level, labeled N160) number of gyrations were utilized. Both specimens were of the same mix design and were compacted using the Superpave Gyratory Compactor. The specimens were scanned at the x-ray CT equipment available at the Federal Highway Administration (FHWA) Turner-Fairbank Highway Research Center (TFHRC). The x-ray CT device had a 420 keV continuous x-ray source and a linear array detector of 512 channels. The diameters of the specimens were 150 mm. The minimum aggregate size that can be measured using the x-ray CT technique is highly dependent on the spatial resolution of the images. In this study, the resolution of the resulting images, corresponding to 150 mm diameter specimens, were approximately 0.3 mm/pixel in the horizontal direction and 0.8 mm/pixel in the vertical direction. Assuming that approximately 8-10 voxels are needed to be able to segment an aggregate, the minimum size of the aggregate that can be measured was of about 3 mm. This size can be reduced if smaller size specimens are scanned.

3. Image processing procedure to segment aggregates

In dense asphalt specimens, where the aggregates are in close proximity to each other, the elevated pixel intensities near aggregate-to-aggregate contacts result in multiple aggregates being incorrectly considered and labeled as a single particle. As seen in Figure 1b and 1c, after thresholding and labeling operations, the regions 11 and 13 are actually composed of two or more aggregates, yet labeled as a single entity when the traditional algorithm is applied.

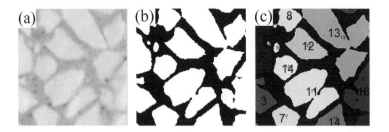

Figure 1. *Illustration of traditional binary thresholding and labeling;*
(a) original grayscale image, (b) thresholded binary image, and (c) labeled image

In order to eliminate the problem of clustering of aggregates, several different image processing methods were investigated. The set of image processing steps that were found to be successful in separating the aggregates is illustrated in Figure 2. In summary, the steps are as follows:

1. Filter the image using a Gaussian filter to eliminate most of the noise (Figure 2b). The size and standard deviation of the Gaussian filter can be varied based on the quality of the image or the material composition of the aggregates.

2. Apply the H-maxima transform to suppress all the maxima in the grayscale image whose height is more than a selected scalar value. The goal of this step is to eliminate the variation in pixel intensity of the aggregates so that they have uniform gray value. H-maxima transform allows this without changing the intensities of the darker (i.e. lower intensity) regions (Figure 2c). This step is very important for a successful watershed transformation.

3. Invert the image and perform watershed transformation where the image is divided into unique watershed regions and their boundaries are determined (Figure 2d).

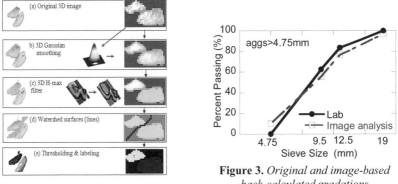

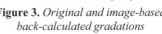

Figure 3. *Original and image-based back-calculated gradations*

Figure 2. *Image processing steps for separation of aggregates*

These steps were implemented in Matlab® and utilized several sub-functions available in the Image Processing Toolbox. The algorithm was validated by calculating the aggregate gradation from the x-ray CT images of a specially prepared asphalt mixture whose aggregate gradation is known. As shown in Figure 3 the gradation of the specimen measured in the lab matches reasonably well with the gradation calculated from the analysis of x-ray CT images. It should be noted that, during image analysis, the individual aggregates were classified into each sieve size using their sphere-equivalent diameter, i.e. $D_{eq} = 2(3V_{ag}/4\pi)^{1/3}$, where V_{ag} is the volume of the aggregate.

4. Calculation of aggregate-to-aggregate contact points

Contact points were calculated using the surface voxels of each aggregate. Surface voxels were isolated from the rest of the voxels using the following rule: a voxel is part of the surface if it is non-zero and it is connected to at least one zero-valued pixel. Then, the minimum distance between the surfaces of neighboring aggregates was calculated (Figure 4). If this distance was less than a pre-selected surface distance threshold (*SDT*) value, the aggregates were assumed to be in-contact.

An important component of the contact point calculation is identifying neighboring aggregates. This step is crucial for the computational efficiency of the algorithm. Otherwise, the surface voxels of all aggregates will be searched, which would increase the computation time significantly. Neighboring aggregates were detected using their centroids (x_c, y_c, z_c), equivalent diameters (D_{eq}), and the *SDT* value. Before the contact point search, the distances between the centroids of all aggregates were computed. If the distance between any two aggregate was less than the sum of the equivalent diameters of the aggregate pair plus the *SDT* value, they were considered "neighbors".

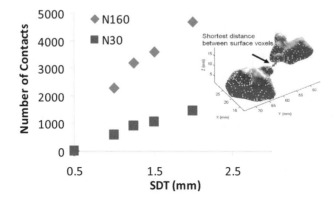

Figure 4. *Number of contact points as a function of SDT*

Figure 4 shows the variation of number of contact points as a function of the *SDT* values. It can be observed from Figure 4 that the number of contact points varies greatly with the selected Surface Distance Threshold (*SDT*) value. In addition, the number of contact points is significantly higher for N160 (high compaction) than for N30 (low compaction) at all *SDT* values. This was expected since the aggregates get close to each-other with additional compactive effort. It should be noted that the selection of *SDT* value is important for the quantity of contacts. However, regardless of the selected *SDT* value, the number of contact points is a very good indicator of compactive effort in different asphalt mixtures.

5. Calculation of the orientation of aggregates

First step in computation of the orientation is the determination of the major principal axis (D_{max}) of an aggregate. The D_{max} was determined using the formula:

$$D_{max} = \max\left(\sqrt{(x_i - x_{-i})^2 + (y_i - y_{-i})^2 + (z_i - z_{-i})^2} \right) \qquad [1]$$

where x_i, y_i and z_i are x-, y- and z-coordinate of a surface voxel with respect to the centroid of the image, x_{-i}, y_{-i} and z_{-i} are x-, y- and z-coordinate of a surface voxel (with respect to the centroid of the image) at the opposite side of a line going through the centroid of the aggregate (Figure 5a).

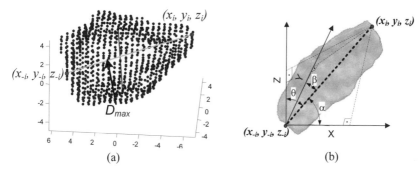

Figure 5. *(a) The illustration of the coordinates used in the D_{max} and orientation calculation and (b) the orientation angles α, β and θ*

Once the D_{max} was determined, the orientation angles were computed using the surface voxels in the opposite ends of the D_{max} line as follows:

$$\alpha = \cos^{-1}\left(\frac{x_i - x_{-i}}{D_{max}}\right) \quad \beta = \cos^{-1}\left(\frac{y_i - y_{-i}}{D_{max}}\right) \quad \theta = \cos^{-1}\left(\frac{z_i - z_{-i}}{D_{max}}\right) \quad [2]$$

where α, β and θ are the orientation angles of the principal axis with respect to the x-, y- and z- axes, respectively, as shown in Figure 5b. Physically, these angles represent how the major principal axis is oriented with respect to the x-, y- and z-axes. For example, if an aggregate has $\theta = 90$, it is lying on x-y plane. If both $\theta = 90$, and $\alpha = 90$, then the aggregate is parallel to the y-axis.

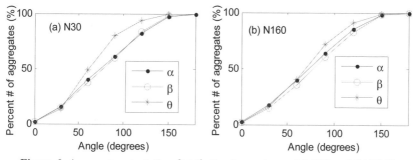

Figure 6. *Aggregate orientation distribution in specimens (a) N30 and (b) N160*

Figures 6a and 6b show the cumulative distribution of the α, β and θ orientation angles for specimens N30 and N160, respectively. For random orientation distribution of an angle, this graph should be linear as in the case for angles α and β. As shown in Figure 6a, in specimen N30, the distribution of the angle θ is not uniform. The maximum deviation of the curve (of angle θ) from the uniform distribution is at around 90 degrees. This indicates that the number of aggregates with θ ~ 90 is larger than the other angles. It should be noted that the z-axis is the direction of compaction. This indicates that the maximum principal axis of the majority of aggregates lie on x-y plane (i.e. perpendicular to the direction of compaction). On the other hand, Figure 6b shows that, at higher compaction level (N160), the curve for the angle θ gets closer to the uniform distribution. This may indicate that the aggregates become more randomly distributed as the compaction level increases.

6. Calculation of the segregation of aggregates

The segregation of aggregates in the specimens was quantified by dividing the specimen into different zones (groups) and calculating the number of aggregates in each group for each size. Since the specimens used in this study were cylindrical, radial and vertical segregation was investigated. For radial segregation, the 3D x-ray CT image was divided into three groups as shown in Figure 7a.

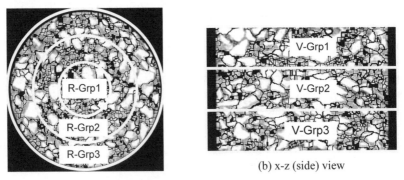

(a) x-y (top) view   (b) x-z (side) view

Figure 7. *Illustration of segregation groups for (a) radial and (b) vertical segregation*

In this figure, Group-1 is a cylinder with radius $R_{spec}/3$, Group-2 is a ring shape with inner radius $R_{spec}/3$ and outer radius $2*R_{spec}/3$, and Group-3 is a ring shape with inner radius $2R_{spec}/3$ and outer radius R_{spec}. The R_{spec} is the full radius of the specimen. Similarly, the vertical segregation was quantified by dividing the specimen into three zones in the vertical direction as shown in Figure 7b. In Figure 7b, each group is the 1/3 of the height of the specimen. It should be noted that image slices shown in Figure 7 are the processed images, showing the watershed lines separating the aggregates from each other.

Figure 8 shows the results of vertical and radial segregation analyses of specimens N30 and N160. Figure 8 can simply be interpreted as the aggregate gradation in each zone (i.e. group). As shown in Figures 8b and 8d, there seems to be no segregation in the vertical direction. On the other hand, as shown in Figures 8a and 8c, there are slight differences in gradation in radial groups (Figure 7a). In Figure 8a, the specimen N160 seems to have coarser gradation in Group 3 (outermost ring) as compared to the other groups. In specimen N30, as shown in Figure 8c, the Group 2 seems to have slightly finer gradation than the other groups.

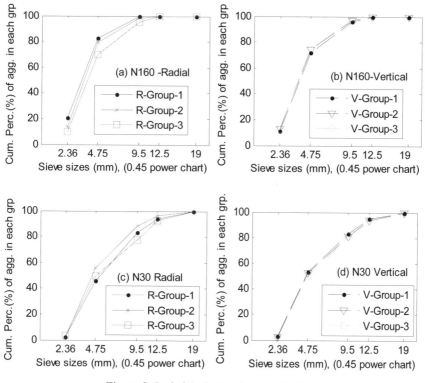

Figure 8. *Radial (a & c) and vertical (b & d)*
segregation of the aggregates

7. Conclusions

This paper presented several image processing and analysis procedures to extract aggregate packing characteristics in asphalt mixtures using x-ray CT images. First, the problem of touching-aggregates was addressed through an image processing procedure that utilizes Gaussian filtering, regional maxima filtering and watershed transformation. Once the aggregates are separated (i.e., each aggregate occupy distinctly separate regions), aggregate-to-aggregate contact points, 3D orientation angles and segregation (i.e. gradation of aggregates at certain zones of the specimen) was computed. Based on the foregoing, the following conclusions were drawn:

- The image processing technique presented in this paper can successfully be used to separate aggregates in compacted asphalt mixtures. The same technique can also be used in concrete images.

- It was observed that the number of contact points of aggregates in asphalt mixture vary greatly with the selected Surface Distance Threshold (*SDT*) value. However, regardless of the selected *SDT* value, the number of contact points is a very good indicator of compactive effort in different asphalt mixtures.

- Orientation analyses revealed that maximum principal axis of the majority of aggregates lie on x-y plane (i.e. perpendicular to the direction of compaction) in specimens analyzed in this study.

- Segregation analyses, comparing aggregate gradations at different zones of the specimen, indicated no segregation in vertical direction. Whereas, there was indications of segregation in radial direction in specimens (which were compacted using the gyratory compactor).

8. References

Kutay, M.E., Aydilek, A.H., Masad, E., and Harman, T. (2007a), "Computational and Experimental Evaluation of Hydraulic Conductivity Anisotropy in Hot Mix Asphalt", *International Journal of Pavement Engineering*, vol. 8 (1): 29-43.

Kutay, M.E., and Aydilek, A.H. (2007b), "Dynamic Effects on Moisture Transport in Asphalt Concrete", *ASCE Journal of Transportation Engineering*, vol. 133 (7): 406-414.

Masad, E., Jandhyala, V.K., Dasgupta, N., Somadevan, N. and Shashidhar, N. (2002), "Characterization of air void distribution in asphalt mixes using x-ray computed tomography", *Journal of Matls. in Civil Eng.*, vol. 4 (2): 122-129.

Tashman, L., Masad, E., D'angelo, J., Bukowski, J. and Harman, T. (2002), "X-ray tomography to characterize air void distribution in Superpave gyratory compacted specimens", *International Journal of Pavement Engineering*, vol. 3 (1), pp. 19-28.

Azari, H., McCuen, R., and Stuart, K. (2005), "Effect of Radial Inhomogeneity on Shear Properties of Asphalt Mixtures", *Journal of Matls. in Civil Eng.*, vol. 17 (1): 80-88.

Integration of 3D Imaging and Discrete Element Modeling for Concrete Fracture Problems

E. N. Landis* — J. E. Bolander**

*University of Maine
Department of Civil & Environmental Engineering
5711 Boardman Hall
Orono, Maine 04469, USA
landis@maine.edu

**University of California Davis
Department of Civil & Environmental Engineering
3121 Engineering Unit III
Davis, California 95616, USA
jebolander@ucdavis.edu

ABSTRACT. *This paper describes a collaboration where synchrotron-based x-ray CT images of micromechanical experiments on cement-based composites are used to develop discrete element computational models of the material. Preliminary results are presented showing how 3D images are used to create computational models, as well as the resulting simulations from those models. The results show a good qualitative agreement between simulations and experiments, opening a door that will allow us to establish previously difficult-to-characterize properties such as cement-aggregate interfaces.*

KEYWORDS: *synchrotron x-ray CT, 3D image analysis, lattice model, fracture*

1. Introduction

Predictive simulation of fracture in heterogeneous materials has traditionally been an ill-posed problem due to the complex microstructural features that do not lend themselves to simple geometric representations. Planer cracks, penny-shaped cracks, and spherical inclusions can provide useful insight into damage and fracture behavior, however, they can fall short as honest descriptors of real heterogeneous microstructure. The work described here is aimed at developing new ways to measure the physical microstructure of damage and fracture, and use those measurements to create realistic numerical models. Our approach is to identify key microstructural features in 3D tomographic images, and create lattice-type finite element models that match the measured microstructure (Landis and Bolander 2009). Below we provide an overview of the 3D imaging, the micromechanical experiments, and preliminary simulations results.

2. Instrumentation and experiments

The 3D imaging described herein was conduced at beamline 5BM of the Advanced Photon Source (APS). The high flux and monochromatic synchrotron x-ray source allows us to image subtle differences in x-ray absorption, and therefore subtle changes in microstructure. In all the tests here, an x-ray energy of 30 kV was used, tuned such that the specimen absorbed roughly 90% of the beam. Image acquisition and optics were set up such that acquired images had a spatial resolution of 6 microns per pixel.

The focus of our experiments was in situ fracture of small (4 mm diameter by 4 mm long) Portland cement-based specimens loaded in a split cylinder configuration. The materials were a conventional Portland cement paste (hardened cement and water), while the second was a mortar that had the same relative amount of cement and water as the cement paste but included fine aggregate particles not exceeding 80 microns. Specimens had a water-to-cement ratio of 0.42, corresponding to a conventional concrete with a moderate to high strength. For the specimens with aggregates, the aggregate volume fraction was 30%.

In situ imaging of fracture processes was made possible through the use of a load frame built to allow tomographic scans while the specimen is under load. An in-line transducer measures applied force, while a pair of capacitance transducers measures platen-to-platen displacement. Load was applied by both a screw actuator (coarse displacements) and a piezoelectric stack actuator (fine displacements). A schematic diagram of the load frame is shown in Figure 1.

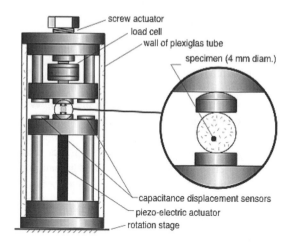

Figure 1. *Photograph of load frame in place at x-ray beam line*

We selected a split cylinder configuration, which produced predominately tensile stresses in an easily controlled compression loading. The experimental protocol was set up as follows. First a tomographic scan was made of an unloaded, undamaged specimen. Then we applied a load close to, but not exceeding, the failure load, and held that load while a second tomographic scan was made. Finally, the load was increased until the specimen ruptured, when a third tomographic scan was made. Tomographic scans consisted of 1500 views over a rotation of 180 degrees. Tomographic reconstructions were made using a filtered back projection algorithm for a parallel beam. The resulting 3D images were 1299 by 1299 by 800 voxels at 6 microns per voxels.

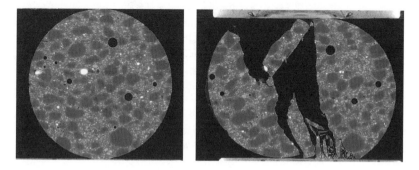

Figure 2. *Vertical slice images illustrating split cylinder fracture. Cylinder is 4 mm diameter. Note that due to non-uniform deformation, not all features in the damaged specimen image (right) appear in the same plane as the undamaged image (left)*

An example of pre- and post-fracture scans is shown in Figure 2, where a specimen with aggregate particles has been loaded to rupture. One of the significant aspects of this kind of test is that we are able to measure internal changes in structure that result from damage and fracture in a single specimen. One example of this is a measure of specific fracture energy, where we make the calculation based on the non-recoverable external work as measured in the load-deformation plot, and divide this amount by the new surface area created by fracture, as measured in the tomographic images (Landis *et al.*, 1999). Of particular note is the fact that, to the degree allowed by the spatial resolution of the scans, in this measurement we are able to account for the numerous crack branches and fragmentation that occurs as a result of heterogeneous fracture.

3. Lattice formulation and simulation

Lattice representations of heterogenous materials have been used for a number of years, emerging from work in statistical physics (Herrmann and Roux, 1990). The philosophy of the approach used here is that statistical variations and disorder can be explicitly represented in a way that resembles the actual material.

In the approach applied here, lattice topology is based on the Delaunay tessellation of nodal points within the specimen domain. The dual Voronoi tessellation defines the elastic and fracture properties of the lattice elements (Berton and Bolander, 2006). This discretization involves a three-phase representation of the material meso-structure as follows: hardened cement paste, aggregates, and cement-aggregate interface.

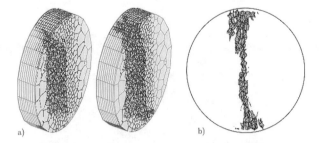

a)   b)

Figure 3. *Simulation of split cylinder test: a) dark regions show fracture development from central region of model toward the load platens; and b) frontal view of fracture surface*

Figure 3 shows the simulated fracture pattern for a split-cylinder without aggregate inclusions. Fracture initiates in the central region of the specimen, where the tensile stress is quasi-uniform, and propagates toward the load platens. Qualitatively, the general pattern of fracture is consistent with that observed in the in situ tomographic scans (Figure 4). Specifically, while the mesh was relatively coarse, we obtained qualitative agreement in both the physical features of the fractured specimen (appropriate crack branching and crack spatial distribution) and the shape of the load deformation curve.

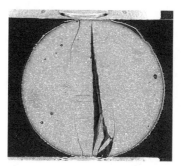

Figure 4. *Vertical tomographic section of fractured cement paste specimen. Image highlights multiple cracks that appear at the load points after the main vertical tensile crack forms*

4. Examination of specimens with spherical inclusions

Of particular interest in random heterogeneous materials such as concrete are the properties of the cement-aggregate interface. Presented here are the first steps of a process that will enable us to characterize interfaces *in situ*. A model specimen was developed in which the aggregate particles were spherical beads. This simplification was done to facilitate meshing of the lattice model, and it allows us to remove geometric variations in aggregates from the list of experimental variables. The selected aggregates were nominal 0.5 mm spheres made of soda lime glass. In order to vary the cement-aggregate interface bond strength, both smooth (as purchased) and etched beads were used. The specimens were etched using a very dilute HF acid. An example of a tomographic scan of such a specimen is shown in Figure 5, where the spherical aggregates appear as a relatively uniform graylevel in the image.

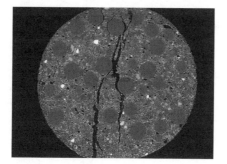

Figure 5. *Vertical tomographic section of fractured specimen with glass bead aggregates*

As the primary objective of this work was to match numerical to real specimens, we developed an image processing routine that allowed us to isolate the individual glass bead aggregates and determine the coordinates of the centroid so that a 3D discrete element model can be created from the images of actual specimens. Segmentation of the glass bead aggregates was accomplished through a simple 3D standard deviation analysis. Since the beads have a relatively uniform density, they can be distinguished from the more widely varying cement matrix. Once the aggregates are identified in the 3D images, we can apply routines that establish the centroid of each aggregate in the specimen. A table of aggregate locations can be then be used to create a lattice model of the specimen that has matching aggregate locations. Figure 6 shows a 3D rendering of the segmented tomographic image data along with the corresponding lattice model.

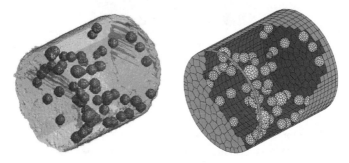

Figure 6. *Illustration of segmented tomographic image with isolated aggregates and corresponding lattice model discretization*

5. Summary

We have presented some preliminary work on the development of numerical models of random heterogenous composites that are meshed based on a direct correspondence between features of the numerical specimen and the real specimen. The work is presented as a first step in micromechanical characterizations of complex cement matrices and interfaces.

6. Acknowledgements

The authors gratefully acknowledge the support of a collaborative grant from the U.S. National Science Foundation (0625030 and 0625593).

7. References

Berton, S., and Bolander, J. E., "Crack band modeling of fracture in irregular lattices", *Computational Methods in Applied Mechanics and Engineering*, vol. 195, p. 7172-7181, 2006.

Herrmann, H. J., and Roux, S., *Statistical Models for the Fracture of Disordered Media*, North-Holland, Amsterdam, 1990.

Landis, E. N., Nagy, E. N., Keane, D. T., and Nagy, G., "A Technique to Measure Three-Dimensional Work-of-Fracture of Concrete in Compression", *Journal of Engineering Mechanics*, vol. 125 no. 6, 1999, p. 599-605.

Landis, E. N., and Bolander, J. E., "Explicit representation of physical processes in concrete fracture", *Journal of Physics D: Applied Physics*, (in press).

Application of Microfocus X-ray CT to Investigate the Frost-induced Damage Process in Cement-based Materials

M. A. B. Promentilla* — T. Sugiyama*

Graduate School of Engineering, Hokkaido University
Kita 13 Nishi 8, Kita-ku
Sapporo 060-8628, Japan
mabp@eng.hokudai.ac.jp
takaf@eng.hokudai.ac.jp

ABSTRACT. *This paper presents a methodology to investigate the damage process in cement-based materials that have been subjected to freezing and thawing action. The microfocus x-ray computed tomography (CT), a non-destructive and non-invasive imaging technique, allows us to examine the cracks in the damaged mortar at a resolution of the order of 10 micrometers. Image processing and analysis of the microtomographic images are also applied to gain three-dimensional (3D) information of the internal microstructure. This information can be used to quantify the damage parameters on the basis of the 3D characterization of cracks in the damaged mortar. Representative samples of the analyzed volumetric data are presented to demonstrate the feasibility of the proposed method.*

KEYWORDS: *freezing-thawing action, microfocus x-ray CT, image analysis, crack, mortar*

1. Introduction

X-ray computed tomography (CT) is a non-destructive technique to examine the internal structure of objects by creating three-dimensional (3D) images that map the variation of the x-ray attenuation coefficient. Although this technique is widely used as a medical diagnostic tool, it has also been shown to be very useful in material research. For example, medical and industrial CT scanners allow non-destructive imaging of many geomaterials like soils, concrete and rocks at spatial resolution that could range from an order of 100 micrometers to a few millimeters. With the recent development of microfocus x-ray CT (micro-CT) having spatial resolutions down to sub-micron to few μm, its potential to study the microstructure of geomaterials becomes more appealing to many. In this paper, we focus on the application of microfocus x-ray CT to investigate the internal structure of cement-based materials exposed to freezing-thawing action.

The freezing-thawing action is one of the most significant causes of deterioration in concrete structures particularly in cold regions such as that of Hokkaido, Japan. Concrete and mortars that are not resistant to cyclic freezing-thawing action manifest two principal types of damage namely surface scaling and internal cracking. Several methods have been developed to observe and quantify these types of damage caused by freeze-thaw (FT) cycles. For example, the surface scaling can be measured by weighing the scales formed from the specimen. As for internal cracks, the scanning electron microscopy (SEM) and optical microscopy (OM) have been the most widely used technique to characterize the cracks in cross-sections of mortar or concrete.

In contrast to the conventional 2D microscopy techniques, micro-CT can be used to examine the frost-induced cracks in three dimensions without the time-consuming and difficult sectioning of specimen. The purpose of this study is therefore to illustrate the feasibility of using the microfocus x-ray CT coupled with image analysis to visualize and quantify the air voids and cracks in a mortar specimen subjected to freezing-thawing action.

2. Experimental procedure

2.1. *Material tested for freezing-thawing experiment*

The material tested for this study was a non-air-entrained fly ash mortar with a fly ash replacement of 30%, a water-to-binder ratio of 0.50 and a sand-to-binder ratio of 2.50. The cement used was an ordinary Portland cement (OPC) designated as JIS R5210 whereas the fly ash is of JIS type II. The river sand used in the mixture is of sizes that range from 0.15 mm to 2 mm.

In accordance to JIS R5120, the casting in 4 cm x 4 cm x 16 cm steel mold was performed; and then after removing the specimen from the mold, the specimen was cured under water. After 14 days of curing, a 12-mm cored sample was obtained from the hardened mortar specimen. A cylindrical specimen with aspect ratio (L/D) of 2 was prepared for the freezing-thawing experiment.

In the freezing-thawing experiment, the cylindrical specimen was cyclically exposed to freezing in water at -20° C and then thawing in water at 18° C in an air-temperature controlled room. One cycle of freezing and thawing took about 12 hours. After a specified number of freeze-thaw cycles, the specimen was removed from the controlled room, and then the central portion of the specimen (about 12 mm and 8 mm in diameter and height, respectively) was scanned with x-ray CT (see Figure 1).

2.2. Image acquisition using microfocus x-ray CT

Acquisition of 3D images of the internal structure of the specimen subjected to cyclic freezing-thawing environment was done using a desktop microfocus x-ray CT system (TOSCANER-30000μhd, Toshiba IT & Control Systems Corporation, Japan). The micro-CT consists of a microfocus x-ray source, a specimen manipulator, an image intensifier (II) detector coupled to CCD camera, and an image processing unit.

In this experiment, a power setting of 130 kV and 124 μA was used for a full (360^0 rotation) cone-beam scan with 1,500 projection views. The specimen was set in a holder mounted on a precision rotation table, and then the table position was adjusted to fit the image within the field-of-view. Calibrations and setting of scan parameters and conditions prior to scanning were performed to obtain high resolution images, as well as to reduce the noise and artifacts during image acquisition.

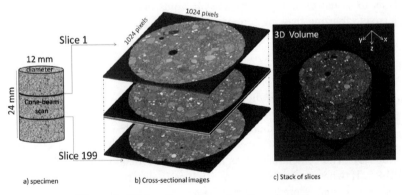

Figure 1. *CT data from scanned section of specimen*

3. Results and discussion

3.1. *Microtomographic images from cone-beam scanning*

Figure 1 depicts the reconstructed CT data obtained from cone-beam scanning of the central portion of the cylindrical specimen. From the reconstructed images, 199 contiguous slices were obtained such that each slice has a thickness of 40 μm. Stacking up these slices creates the 3D image of the scanned section of the specimen. Each slice or cross-sectional image consists of 1024 x 1024 pixels with a square pixel size of 12 μm. Accordingly, the voxel (volume element or volumetric pixel) of the CT image has anisotropic size of 12 x 12 x 40 μm. Each voxel is associated with the so-called CT number (CTN) which is proportional to the average x-ray attenuation coefficient of the material in the said voxel. Since the attenuation coefficient is a sensitive measure of atomic composition and density, the voxels associated with low-density phase such as air or water have much lower CT number as compared with that of the solid matrix of the mortar. By convention, the higher CT numbers were shown in the images as brighter voxels relative to that of the lower CT number.

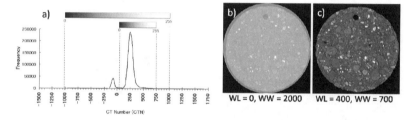

Figure 2. *CT histogram of a cross-section and the corresponding 8-bit images*

Figure 2 shows an example of an 8-bit grayscale CT image obtained from rescaling the 14-bit raw CT data. As shown in the histogram of the CT number, there are two distinct peaks which are associated with the air and the solid matrix, respectively. Depending on the setting of the window level (WL) and width (WW) or the minimum and maximum CTN cut-off, different brightness and contrast of the image can be obtained from the raw CT data. For this study, the CT data were transformed to 8-bit grayscale images (e.g. TIF images) by setting the window level and width to 400 and 700, respectively. In other words, we used 50 and 750 as the minimum and maximum cut-off CTN, respectively to give a better contrast between the air (first peak) and the mortar matrix (second peak). In this case, any voxel with CTN less than or equal to 50 was transformed to a grayscale value (GSV) of zero (black) whereas any voxel more than 750 was transformed to a GSV of 255 (white). For voxels with CTN between 50 and 750, the following scaling was used in transforming the CTN to 8-bit (0-255) GSV: *GSV = (255/700) (CTN – 50)*.

3.2. *Image processing and analysis of scanned section of specimen*

In this study, the freely-available programs such as ImageJ (Rasband, 2007) and SLICE (Nakano *et.al*, 2006) were used for image processing and analysis. For example, anisotropic diffusion filter was applied to the stack of CT images for denoising prior to subsequent analysis. Sample cross-sectional images of the specimen after 3, 9 and 35 freeze-thaw (FT) cycles are shown in Figure 3. In this figure, the entrapped air voids and cracks were shown as black whereas the cement pastes were imaged as dark gray. On the other hand, the aggregates (sand) were imaged as patches of different shades of gray in the microtomographic images. Notice that most of the cracks meander around the aggregates suggesting that the internal cracks attempt to follow the weaker interfacial transition zone (ITZ) between the sand and the cement paste in the frost-damaged mortar.

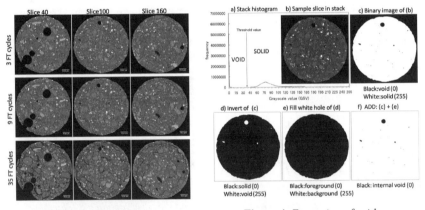

Figure 3. *Sample slices at different FT cycles*

Figure 4. *Extraction of void space from micro-CT image*

3.2.1. *Extraction of air void or cracks from the microtomographic images*

To quantify the air void or cracks, image segmentation using global thresholding was used to separate the void space (air void or crack) from the solid mortar matrix. The grayscale threshold value was selected as one half of the grayscale value (GSV) of the peak associated with the solid matrix as shown in Figure 4a. Thus, any voxel with a GSV below or equal to this threshold value is considered void space and colored as black (or white). Otherwise, any voxel with GSV above this threshold value is considered as solid and made as white (or black). Such thresholding routine results to binary images (e.g. Figure 4c and 4d) that can be used to further analyze the void space. However, to separate the surrounding background from the void inside the specimen, additional image operations were done. By applying a fill-hole binary operation, a solid foreground image (Figure 4e) was produced and then used to separate the interior void space from the exterior space. The image arithmetic operation of addition was applied to the said image (Figure 4e) and the binary image (Figure 4c), resulting to an image (Figure 4f) that includes only the internal void space. The void fraction of the scanned specimen can be easily quantified by dividing the total number of internal void voxels (e.g. black in Figure 4f) with the total number of foreground voxels (e.g. black in Figure 4e).

3.2.2. Visualization and quantification of void space

Figure 5 shows the slice-wise void fraction along the height of the scanned specimen at different freeze-thaw cycles (FTC). Note that at the early stage of FTC, no clear cracks were observed and the estimated void fraction was found to be 1.3%. This value seems to be reasonable for the initial air content of non-air-entrained mortars. As expected, a significant increase of void fraction (6.3%) was then observed at 35 FTC because of internal crack formation.

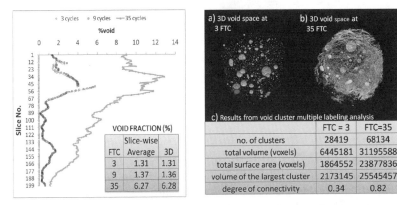

Figure 5. Void fraction at different FT cycles

Figure 6. Volume rendering and connectivity analysis of the 3D void space

Figure 6 shows the volume rendering of the 3D void space at 3 FTC and 35 FTC, respectively, as well as, some results from connectivity analysis. Such connectivity analysis (Nakashima and Kamiya, 2007; Promentilla et. al, 2008) through multiple cluster labeling algorithm allows us to identify, label, and then measure the individual void objects (or clusters). Prior to this analysis, the stack of 3D image was also rescaled along the z-axis using cubic-spline interpolation to create an isotropic voxel of 12 μm. Volume was determined by the number of void voxels in the cluster while surface area is determined by the number of interfacial voxel faces that are in contact with solid voxel. The degree of void connectivity was obtained from the volume ratio of the largest void cluster to the total void space. The results suggest that the generation of frost-induced cracks had caused the increase of the total volume and surface area of the void space. Moreover, the degree of void connectivity increased dramatically from 0.34 to 0.82 at 35 FTC, indicating that the individual cracks form a well-connected crack system in three dimensions.

Furthermore, studies are currently being conducted to characterize this connected crack network in terms of crack width distribution and tortuosity. Figure 7 shows some preliminary results of the internal crack width distribution and crack tortuosity in a selected cubic volume of interest (VOI = 6^3 mm^3). The crack width distribution was measured based on the local thickness calculation using direct distance transformation method developed by Hilderbrand and Ruegsegger (1997). As for crack tortuosity, the 3D medial axis method (3DMA) developed by Lindquist (1999) was used. The geometric tortuosity (τ) was calculated as the shortest paths along the void medial axis for x, y, and z-directions using $\tau = l/\Delta$ where l is the actual path length and Δ is the linear separation of the two parallel planes of the VOI.

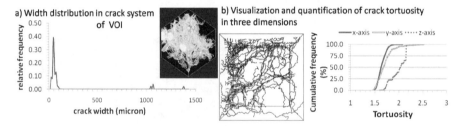

Figure 7. *Crack width distribution and crack tortuosity in a given volume of interest*

4. Conclusion

Although this study is limited in its extent, the findings demonstrate the promising potential of micro-CT as a noninvasive and nondestructive technique to investigate the damage caused by freezing-thawing action in cement-based materials. The micro-CT with the aid of image analysis allows us to visualize and quantify air void and crack in the mortar specimen at spatial resolution of 10 μm order. Measurement of void fraction including the volume, surface area, and degree of connectivity of void space in three dimensions could provide useful insight into the damage evolution caused by freeze-thaw cycles. Quantification of 3D parameters such as crack tortuosity (Promentilla and Sugiyama, 2009) based on random walk simulation may also be used in material models that are based on a realistic physical picture of the microstructure of the frost-damaged mortar.

5. Acknowledgments

Part of this research is funded by Research Grant from the Japan Society for the Promotion of Science (Kiban Kenkyu B, Sugiyama, T. Research No.19360193). The authors would like to thank Prof. Kaneko and Dr. Kawasaki for allowing them to use the micro-CT scanner and also to Mr. Hatakeda and Dr. Shimura for his assistance in the preparation of the specimen. The first author also acknowledges the postdoctoral fellowship support from the Japan Society for the Promotion of Science.

6. References

Hilderbrand, T., Ruegsegger, P., "A new method for the model-independent assessment of thickness in three dimensional images", Vol. 185, pp.67-75, *Journal of Microscopy*, 1997.

Lindquist, W.D., "3DMA General Users Manual", State of New York at Stony Brook: New York, 1999.

Nakano T., *et al.*, "SLICE –Software for basic 3-D image analysis." *Japan Synchrotron Radiation Research Institute*, 2006. Available from: http://www-bl20.spring8.or.jp/slice/.

Nakashima Y., Kamiya, S., "Mathematica programs for the analysis of three-dimensional pore connectivity and anisotropic tortuosity of porous rocks using X-ray microtomography", Vol. 44, No. 9, pp.1233-1247, *Journal of Nuclear Science and Technology*, 2007.

Promentilla, M.A.B., Sugiyama T., Hitomi, T., Takeda, N., "Characterizing the 3D pore structure of hardened cement paste with synchrotron microtomography", Vol. 6, No. 2, pp. 273-286, *Journal of Advanced Concrete Technology*, 2008.

Promentilla, M.A.B., Sugiyama T., "Computation of crack tortuosity from microtomographic images of cement-based materials", In press, *Proceedings of the Twelfth International Conference on the Enhancement and Promotion of Computational Methods in Engineering and Science*, China, 2009.

Rasband, W., "Image J: Image processing and analysis in Java", *National Health Institute*, 2007, available from: http://rsb.info.nih.gov/.

Evaluation of the Efficiency of Self-healing in Concrete by Means of μ-CT

K. Van Tittelboom* — D. Van Loo** — N. De Belie* — P. Jacobs***

*Magnel Laboratory for Concrete Research
Department of Structural Engineering, Ghent University
Technologiepark Zwijnaarde 904
B-9052 Ghent, Belgium
kim.vantittelboom@UGent.be
nele.debelie@UGent.be

**Centre for X-ray Tomography
Department of Subatomic and Radiation Physics, Ghent University
Proeftuinstraat 8
B-9000 Ghent, Belgium
denis.vanloo@UGent.be

***Center for X-ray Tomography
Department of Geology and Soil Science, Ghent University
Krijgslaan 281 S8
B-9000 Ghent, Belgium
patric.jacobs@UGent.be

ABSTRACT. *It has been estimated that, in Europe, 50% of the annual construction budget is spent on refurbishment and remediation of the existing structures. Therefore, self-healing of concrete structures, which are very sensitive to cracking, would be highly desirable. In this research, encapsulated healing agents were embedded in the concrete matrix in order to obtain self-healing properties. Upon crack appearance, the capsules break and the healing agent is released, causing crack repair. It was found that more than 80% of the original strength could be regained after self-healing. μ-CT was used as non-destructive test method, to visualize the interior of the beams after crack formation. It was shown that after breakage of the capsules, the glue was released from the tubes and sucked into the crack due to the capillary forces.*

KEYWORDS: *concrete, cracks, self-healing, computed tomography*

1. Introduction

Crack formation often appears in concrete because of its low tensile strength. As these cracks may lead to concrete deterioration and reinforcement corrosion, they need to be repaired. Usually cracks are treated from the outside by injection of epoxy or polyurethane resins into the crack. However a lot of labor is needed to restore these cracks; moreover crack repair is practically impossible for some structures, such as buried pipes, undersea structures, foundations, etc.

In this research, autonomous crack healing upon crack appearance is monitored with healing agents that are released from inside the structure. The advantage of this technique is that costs may be reduced and that undetected or unreachable cracks are also treated. Cementitious samples with hollow tubes, consisting of two compartments, that were embedded in the matrix were used for this experiment. Each tube compartment contained one component of a 2-component repair material. When a crack forms, the tubes break and the healing agent flows into the crack. When both components of the healing agent make contact, they polymerize and the crack faces are bound together.

To verify the healing efficiency, cracked samples were subjected to a bending test and the restoration of mechanical properties, due to self-healing, was evaluated. To visualize the interior of the concrete, cone beam micro-computed x-ray tomography (μ-CT) was used as a non-destructive test method. Computed tomography is often used in combination with cementitious materials, mostly to study concrete attack by aggressive liquids or to investigate the microstructure of hydrating cement paste. In this research, μ-CT was used to visualize crack formation and crack repair and to verify the amount of crack filling through the release of the embedded healing agent. Landis *et al.* (1999) also made use of this non-destructive test method to visualize cracks during a compression test of concrete specimens. In the present study it was not possible to scan the specimens during crack formation, so samples were scanned after cracks were obtained.

2. Materials and methods

2.1. *Preparation of the samples*

First, the tubes, which were used to carry the healing agent and which were embedded inside the samples, were prepared. The efficiency of glass tubes, each with a different diameter but with the same internal volume, was compared. In Table 1 the dimensions of the tubes are shown.

First the tubes were sealed at one end with glue. Then the compartments were filled with the healing agent which was injected by means of a syringe. In this experiment polyurethane (MEYCO MP 355 1K) was used as healing agent. MEYCO is a low viscous polyurethane that foams upon contact with water or air. As the expanding reaction products occupy a larger volume than the original product, a bigger crack space may be filled. The reaction time of this product may be shortened by addition of an accelerator. Here, one tube compartment was filled with polyurethane and the other tube compartment was filled with a mixture of accelerator and water. When all tubes are filled, the open end was sealed with glue. Then two types of cementitious specimens were made.

Name	Material	$\emptyset_i$	$\emptyset_o$	Length	Volume
	[-]	[mm]	[mm]	[mm]	[mm^3]
GLA-2	glass	2.00	2.20	82.6	259.2
GLA-3	glass	3.00	3.35	36.7	259.2

Table 1. *Dimensions of the tubes used for encapsulation of the healing agent*

For the first experiment, cement paste specimens, from which the composition shown in Table 2 was prepared. The samples had dimensions of 25x25x70 mm and were provided with two notches at the side surfaces. Glass tubes with two compartments with an inner diameter of approximately 2 mm were partly exposed to the surface of the specimens for direct observation during crack formation.

Material	Cement paste	Mortar
	Amount [g]	
Sand 0/4	-	1350
Cement CEM I 52.5 N	700	450
Water	210	225

Table 2. *Composition of the mixes*

Subsequently, mortar specimens were made, according to the composition displayed in Table 2. Moulds with dimensions of 60x60x220 mm, provided with a small notch at the lower side, were used. First, a 1 cm mortar layer was brought into the moulds. When this layer was compacted by means of vibration, two reinforcement bars (diameter 2 mm) and two 2-compartment tubes were placed on top of it. Afterwards the moulds were completely filled with mortar and vibrated. For both types of tubes, mentioned above, three specimens, in which the tubes were embedded, were prepared.

After casting, all moulds were placed in an air-conditioned room with a temperature of 20°C and a relative humidity of more than 90% for a period of 24 hours. After demoulding, the specimens were placed in the same air-conditioned room for at least 6 subsequent days.

2.2. Crack formation

In the cement paste specimens, cracks were created by performance of a splitting test. Two metal rods were positioned in the notches and a splitting force was introduced by means of a bench vice as shown in Figure 1.A. This test was done to prove the sensing and actuation mechanism and is inspired by the experiment performed by Li *et al.* (1998). As the tubes are partly exposed to the surface, breakage of the tubes and start of the foaming reaction may directly be observed.

The mortar prisms were cracked, and thus the healing mechanism was triggered, by means of a crack width controlled three point bending test. In Figure 1.B, the test setup is shown. Crack widths were measured by means of a linear variable differential transformer (LVDT) with a range of 1 mm and an accuracy of 0.0001 mm. Specimens were subjected to three loading cycles, each time with an interspace of one day. During the first cycle a crack was created and during the second and the third cycle the healing efficiency was evaluated. During each test, the crack width was increased with a velocity of 0.0005 mm/sec until a crack of 0.4 mm was reached. However, during the last cycle, loading was stopped when an additional crack width of 0.3 mm was reached.

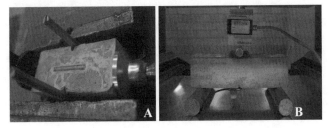

Figure 1. *Crack formation: (a) performance of a splitting test and (b) performance of a three point bending test*

2.3. Evaluation of healing by means of μ-CT

Samples were scanned using the in-house developed μ-CT system of the Center for x-ray Tomography of Ghent University (UGCT). The system is composed of a Feinfocus x-ray tube with tungsten transmission target, a Varian Paxscan 2520 A-Si flat panel detector and a Micos air-bearing rotation stage. For each of the samples a region of 1880x1496 of the detector was used in non-binned mode of 127 µm pixel size and 1,200 projections were taken at 0.3° interval over 360° of rotation. The mortar samples were scanned at 150 kV and 8 W with a spot size of approximately 8 µm and a 2 mm Cu filter. The voxel size of the obtained data is 50 µm. The raw data were reconstructed to CT-slices using Octopus (in-house developed software). The analysis and visualization was performed using Morpho+ and VGStudio.

3. Results and discussion

3.1. Sensing and actuation

Cement paste prisms were subjected to cracking in a splitting test. Upon crack formation, the glass tubes broke and both components of the glue flowed into the crack due to gravitational and capillary forces. Figure 2(a) shows polyurethane foam that formed when both components made contact. Also in the case the tubes were embedded inside the matrix, evidence was given that the glue leaked out of the tubes. In Figure 2(b) it is seen that, after performance of the bending test, the adhesive entered the crack plane. When all bending tests were fulfilled, cracked mortar prisms were completely broken and the cross-sections were examined. The cross-section shown in Figure 2(c) illustrates visual confirmation of glue flow into the crack plane.

Figure 2. (a) Foaming of polyurethane after breakage of the cement paste specimen, (b) mortar specimen after performance of the three point bending test showing that the glue flowed into the crack and (c) cross-section of a specimen after breakage showing the glue migration (migration front is indicated with a dashed line).

3.2. Regain in mechanical properties

In the graph in Figure 3(a) the crack width is plotted against the load for each of the three loading cycles, to which the mortar prisms were subjected.

The first loading cycle is indicated with the black curve. It is seen that the load increases first until a crack appears in the mortar beam. After reaching the peak load, some distinct drops in load are seen. These may be related to breakage of the tubes as these drops were accompanied with certain "pop" sounds. When the crack reaches a width of 100 μm, the load starts to increase again, this was caused by the reinforcement inside the specimens. When the crack width reaches 400 μm, the specimens are unloaded and the crack width decreases because of the elastic properties of mortar and steel. During the first reloading, indicated with the dark grey curve in Figure 3(a), a new peak is observed, indicating that strength was regained due to crack healing. It is also seen that before the peak load was reached, the curve is steeper compared with the unloading part during the first loading cycle. This indicated that the material regained stiffness.

During the second reloading cycle, indicated with the light gray curve, again a small peak is observed. However this time the peak load is much smaller than during the first reloading cycle, which indicates that during the second reloading a small regain in mechanical properties was obtained.

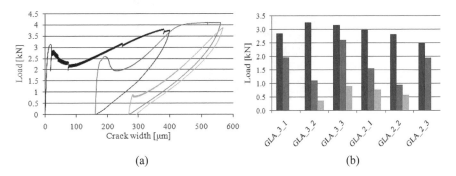

(a)   (b)

Figure 3. *(a) Load [kN] versus crack width [μm] during crack formation (black curve) and first (dark grey curve) and second (light grey curve) reloading cycle; (b) comparison of the initial strength (black bars) of the mortar beams and the strength after crack healing (first reloading: dark gray bars; second reloading: light gray bars).*

In Figure 3(b), the healing efficiency, indicated by the regain in strength, is shown for all test series. Prisms with embedded glass tubes are indicated by means of "GLA" followed by "_2" or "_3" depending on the diameter of the tubes. Suffixes 1, 2 and 3 are used to describe the different specimens of one test series. Test series were subjected to three loading cycles, except GLA_3_1 and GLA_2_3.

Figure 3(b) displays that during the first reloading cycle 30% to 80% of the original strength is regained due to self-healing. The diameter of the tubes seems to play an unimportant role in the extent that strength is regained. This finding conflicts with the expectations that larger diameter tubes would release the glue more easily.

3.3. *Amount of crack filling*

In order to find an explanation for the counter-intuitive results of the first reloading cycle, μ-CT scans of the samples were performed. Figure 4 illustrates the 3D organization of the mortar, glass tubes, crack and steel reinforcement. The polyurethane has very low x-ray absorption compared to the mortar and steel; it is difficult to distinguish in a quantitative way but it permits qualitative comparison.

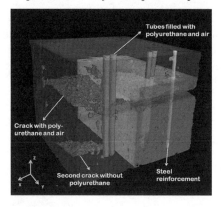

Figure 4. *3D CT image of the mortar sample, glass tubes, one of the two steel rods and crack (right). Crack and glass tubes with air and polyurethane displayed together (left)*

The scans resulted in more than a thousand slices in the Z-orientation, these were resliced in the Y direction (Figure 5) to get a cross-section in the same plane as the four glass tubes. This makes it possible to compare tube drainage efficiency during the formation of the crack and how the polyurethane has infiltrated. It becomes obvious that neither the 2 nor 3 mm tubes have completely been emptied during formation of the crack. The efficiency of the emptying polyurethane is dependent on several parameters, including capillary forces, but the diameter of the glass tubes, either 2 or 3 mm, does not seem to give significant differences in outcome. Polyurethane can be visualized in the tubes as well as in the cracks with μ-CT. Figure 5(c) illustrates a magnified sub-region of a cross-section of the sample with 2 mm tubes. Although the difference in contrast is small, it is clearly visible where polyurethane has filled the crack compared to regions where it has not.

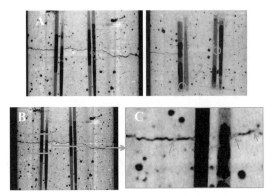

Figure 5. *Y-direction μ-CT cross-section through the glass tubes: (a) sample with 2mm diameter tubes, (b) sample with 3 mm diameter tubes and (c) magnified region of interest of the 2mm sample μ-CT data. Filling of the crack with polyurethane is indicated with arrows. Air voids are colored black as indicated by region (1), polyurethane is colored light gray as indicated by (2) and mortar is colored darker gray as indicated by region (3).*

4. Conclusions

The proposed technique of using encapsulated healing agents that are embedded inside the matrix may be useful in self-healing of concrete. Bending tests proved that after automated crack healing up to 80% of the initial strength is regained. Computed tomography has proved to be a very useful technique to visualize the interior of the concrete and to evaluate the crack filling by polyurethane by different tube diameters. However, regain in mechanical properties and μ-CT analysis showed no difference in healing efficiency for different tube sizes.

5. Acknowledgements

Financial support from the Research Foundation Flanders (FWO-Vlaanderen) for this study (Project No. G.0157.08) is gratefully acknowledged.

6. References

Landis, E.N., Nagy, E.N., Keane, D.T., and Nagy, G., "Technique to measure 3D work-of-fracture of concrete in compression", *Engineering Mechanics*, vol. 125, p. 599-605, 1999.

Li, V.C., Lim, Y.M., Chan, Y-W., "Feasibility study of a passive smart self-healing cementitious composite", *Composites Part B: Engineering*, vol. 29, p. 819-827, 1998.

Quantification of Material Constitution in Concrete by X-ray CT Method

T. Temmyo* — Y. Obara**

*Hazama Corporation
2-2-5, Toranomon, Minato-ku Tokyo
105-8479
Japan
temmyo@hazama.co.jp

**The University of Kumamoto
2-39-1, Kurokami, Kumamoto City
Kumamoto 860-8555,
Japan
obara@kumamoto-u.ac.jp

ABSTRACT: The x-ray CT (Computed Tomography) technology used for non-destructive testing techniques is able to visualize the internal structure of material objects without damaging them and also to analyze its images quantitatively by using the CT value, a numerical value for a scan image as taken by an x-ray CT scanner. By using these characteristics, this research paper will propose a method for setting a "new boundary CT value (threshold value)" to separate each of the constituent materials, such as aggregate, mortar, void from a CT value histogram of a sectional plane image of concrete. It will also apply this method to a concrete test specimen so as to clarify distribution characteristics of void ratio, aggregate ratio, and the relative density of mortar.

KEYWORDS: x-ray CT, scanner, quantitative evaluation, concrete constitution

1. Introduction

The x-ray CT method is an effective method for concrete research and diagnosis. Its effectiveness shows promise for enabling the 3D solid display of steel bar, void, or cracking inside concrete columns.

In contrast, concrete consists of water, cement, sand and gravel. Its mechanical characteristics and durability are influenced by the constituent materials or the ratio of compounds. X-ray computed tomography (CT) provides information about the density of each of a material compound. We have conducted studies on aggregates and voids on the vertical profile in the concrete and on method to evaluate concrete with different water:cement ratios by using average CT values of the mortar (Temmyo *et al.* 2003, 2006)

To quantitatively study concrete composition of each material, i.e. density or size, it is important to set an appropriate threshold values in the CT for each material during image processing. In fact, we set the threshold value so that objectives are "the most easily viewed" in CT images. Although this method is realistic for human evaluation, a more objective method of image processing is required for engineering. Here, we divide concrete into populations of large aggregates, mortar and small aggregates, and large voids, and propose a method to quantify material composition by using a histogram of the CT values obtained from planar images to quantify each population in terms of density and volume.

2. X-ray CT method

We used an industrial x-ray CT Scanner, TOSCANER20000RE made by Toshiba (Figure 1). In the x-ray CT scanner, the x-ray tube can be adjusted to 300, 200, and 150kV, It has 176 detectors fixed on the same horizontal plane. These can be moved in a vertical direction. Additionally, a rotating turntable is ued to rotate the sample as necessary. Five slice thicknesses can be selected; 0.3, 0.5, 1.0, 2.0 and 4.0 mm. The scan area can be selected from two areas, either φ150 mm or φ400 mm. The highest resolution of image display is 2048×2048 pixels. For this research, the scan is set to parameters displayed in Table 1.

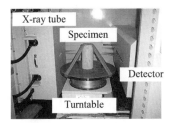

Scan type	Traverse/Rotation
X-ray tube voltage	300kV
X-ray tube current	2mA
Nos. of detectors	176
Thickness of slice beam	2.0mm
Scan area	150mm (Diameter)
Nos. of display pixels	1024×1024 pixels

Figure 1. *X-ray CT scanner*   **Table 1.** *Specification of x-ray CT scanner*

3. X-ray CT image and CT value

An example of a CT image for a φ125 mm concrete column specimen is shown in Figure 2. The resolution of the x-ray CT image is 0.146 mm in the horizontal. The CT value of each voxel (volume of $0.146 \times 0.146 \times 2.0$ mm^3) is determined as defined by the following formula:

$$CT \quad value = \frac{\mu_t - \mu_w}{\mu_w} K \qquad (1)$$

where μ_t is the x-ray absorption coefficient and μ_w is the water absorption coefficient. K is an arbitrary coefficient set to 1000 in this research. As the x-ray absorption coefficient is almost proportional to the density of an object, the CT value is considered to be a value proportional to the density with high CT values correlating to high density (white in images) low CT values corresponding to low density (black in images). Three materials are shown concrete specimen image (Figure 3); the white area is aggregate, grey area is mortar and a black area is void. Here we will section the concrete into these three materials. Figure 3 is a histogram of the CT values used to extract the three materials found in Figure 2.

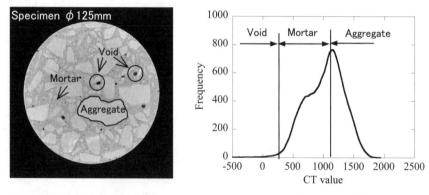

Figure 2. *X-ray CT image of concrete*   **Figure 3.** *Histogram of CT value*

4. Quantification method for material constitution in concrete

4.1. *CT value of the void-mortar boundary*

4.1.1. *Finding the CT value of the void-mortar boundary*

The amount of void in the concrete specimen is minute when one considers its ratio. It can also be considered to be included in mortar. However, it is possible, to enhance precision in the quantitative assessment of the average CT value.

While the theoretical CT value of air was set to −1000, however CT values similar to this value are rate in the histogram, and the ratio of pixels with a CT value of 0 is about 0.7%, or much lower than the 4% included in the concrete specimen during mixing. This may be because the air was not fully resolved by the voxel size used here. Hence, considering the theoretical amount of air as shown in the histogram in Figure 3, the CT value of void is around 0 to 500 due to voxel effects. When dips in the histogram are present, it is easier to decide the boundary than when the curve is smooth at an obvious boundary between the materials. So to decide the characteristic point from the smooth histogram in Figure 4(a) objectively, the threshold value was determined by differentiating histogram function. First, set a function that shows a histogram of the CT value as $f(CT)$; two apply a third order differential to it and evaluate the threshold value (Figure 4b). As a characteristic point rapidly increases in CT value, $d^3f/d(CT)^3=0$, we set the void-mortar boundary CT value, as 361 in the Figure 2 example.

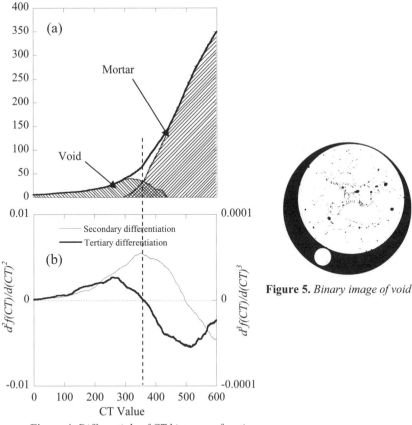

Figure 5. *Binary image of void*

Figure 4. *Differentials of CT histogram function*

4.1.2. *Evaluation of void ratio*

The image was converted to a binary image, where solids are 1 and voids are 0 to determine the void fraction (Figure 5) by using the threshold value between void and solids. Then all the void is assigned a value of 0 and the solid is assigned a value of 1. Here, if we define a value of the divided total sum of black area by the sectional plane area of specimen as void area rate, the rate will be 2.0% for this image. While this is for a plane image, it serves as an index for the composite material.

4.2. *CT value of the aggregate-mortar boundary*

4.2.1. *Phantom*

In the materials that constitute concrete, highest density material is aggregate followed by mortar. These CT values are plotted in the high-value area in the histogram of Figure 3. However, it does not show the characteristic bimodal shape that can be used to differentiate the material types in the area.

Hence it is impossible to use this method to decide the void-mortar boundary CT value here. So we adopted a method called the "Percentile (P-tile) Method". The percentile method decides the threshold value objectively and stably i.e. scanning an objective whose sectional plane area is already known, that is a phantom, together with the specimen, then decides threshold value by using the given value. In this percentile method, a planar section of the material shown in Figure 6 was scanned. At the same time, we scanned the specimen and the phantom to obtain values for the percentile method (Figure 7 and 8). Then a histogram was used to determine the CT value (Figure 9) from which a value of 1,092 was determined.

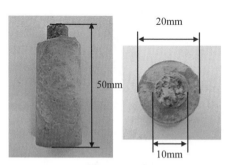

Figure 6. *Phantom*

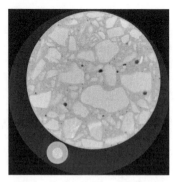

Figure 7. *Specimen and phantom*

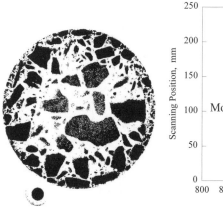

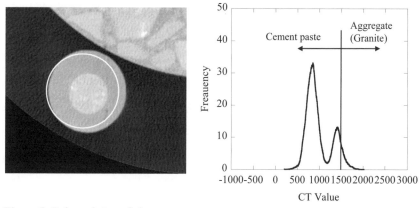

Figure 8. *Enlarged view of phantom*

Figure 9. *CT value histogram of phantom*

4.2.2. Evaluation of aggregate ratio

A binary image (white and black) to differentiate aggregate from mortar is shown in Figure 10, which is used to confirm the shape of aggregates in the concrete. In this image, if we define a value for the total area divided by the black parts in the concrete specimen, the percent will be 45.1%.

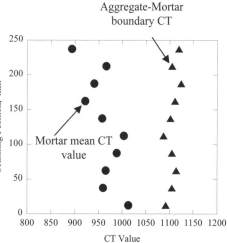

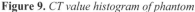

Figure 10. *Binary image of aggregate*

Figure 11. *Boundary CT value of aggregate and mortar by phantom and mean CT value of mortar specimen*

The aggregate area ratio is a two-dimensional aggregate ratio calculated from the sectional plane image. This differs from the general aggregate ratio, yet it is an index of the aggregate proportions.

5. Application to concrete specimen

5.1. *Boundary CT value in scan sectional plane of mortar specimen*

In case that x-ray CT, is applied to the concrete specimen to conduct its quantitative assessment at a range of points, sometimes sample density is not readily obtained, especially when a homogenous specimen is scanned in several sectional planes, CT values often differ slightly.

To demonstrate this effect, we scanned a mortar specimen for indoor testing of φ125 mm, height 250 mm and a phantom simultaneously per 25 mm. We assessed the average CT value and aggregate-mortar boundary CT value of the specimen for ten sectional plane images. The result is shown in Figure 11. The mean CT values vary at different heights. The main cause for this may be the inhomogenous nature of the specimen, but CT values for the aggregate-mortar boundary are somewhat uneven. As shown in the experiments described above, it is possible to enhance quantitative precision by scanning every sectional plane and comparing this to a phantom similarly scanned.

5.2. *Boundary CT value of concrete specimen*

We created a concrete specimen of φ125 mm, height 250 mm and scanned 10 sectional planes increasing the height by 25 mm each time along with a phantom. The concrete mix is a water-cement ratio 55%, fine aggregate ratio 40% with this broken into a volume ratio of fine aggregate with grain diameter 5 mm or large to be 43%, and that smaller than 5 mm to be 29%. We assessed each area ratio by applying the "boundary CT value assess method".

The scan station and void area ratio, aggregate area ratio that we assessed are shown in Figure 12. Though the void area ratio shows constant values not dependent on area, the aggregate/area ratios are uneven. It can be thought that this is due to the aggregates settling to the lower part of the specimen during material preparation, such that CT values are larger at the base of the specimen.

Lastly, we examined the mean CT value distribution of mortar. The range between the void-mortar boundary CT value and the aggregate-mortar boundary CT value can be considered as the mortar's CT value distribution, so we adopted the average CT value of this range as a mortar average CT value (Figure 13). As the CT value is proportional to density, Figure 14 can be regarded as relative mortar density distribution; density increases with depth.

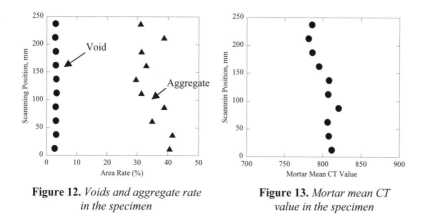

Figure 12. *Voids and aggregate rate in the specimen*

Figure 13. *Mortar mean CT value in the specimen*

From this finding, it is clear that even specimens created using formwork, have distributions in the specimen. Hence when material assessment are collected prior to core compression tests, it is desirable to scan several sectional planes, assess their CT value, and adopt these values as the material constituent characteristics.

6. Conclusions

In this paper, concrete material constituents were evaluated using x-ray CT and we proposed a method for quantifying material constituents using sectioning of the CT images into the three concrete components: aggregate, mortar and void.

A threshold value was chosen to asses the boundary values within the concrete by finding the characteristic point from histogram of CT value and then applying a differential processing to it as well as by scanning the sample in the presence of a phantom of known properties. As a result, it becomes possible to determine the boundary CT value precisely and objectively and that the void area ratio, the aggregate area ratio and the relative density distribution of mortar can be quantified.

7. References

T. Temmyo, T. Tsutsumi, Y. Murakami, Y. Obara. (2003), "Estimation of structural characteristics of RCD by an X-ray CT method", *Geo-X 2003, X-ray CT for Geomaterials, Soiles, Concrete, Rocks*, 199-205.

T. Temmyo, Y. Murakami, Y. Obara. (2006), "Evaluation of water cement ratio of hardened concrete by x-ray CT method", *Advances in X-ray Tomography for Geomaterials*, 443-449.

Sealing Behavior of Fracture in Cementitious Material with Micro-Focus X-ray CT

D. Fukuda* — Y. Nara* — D. Mori** — K. Kaneko*

*Hokkaido University
Kita 13 Nishi 8, Kita-ku, Sapporo 060-8628
Japan
daichang@geo-er.eng.hokudai.ac.jp
nara@geo-er.eng.hokudai.ac.jp
kaneko@geo-er.eng.hokudai.ac.jp

** Taiheiyo Consultant Corporation
2-4-2, Osaku, Sakura 285-8655
Japan
daisuke_mori@grp.taiheiyo-cement.co.jp

ABSTRACT. *High strength and ultra low permeability concrete (HSULPC) has been studied for an alternative technology for long term confinement of Carbon14 (C-14) in geological disposal in geological disposal of transuranic radio waste (TRU waste). Thus knowledge of time-dependent fracturing and permeability is important. For cementitious materials, precipitation of calcium compound can occur on the surface. This can affect the time-dependent crack growth and permeability. In this study, the sealing behavior of cracks and pores in HSULPC submerged in water was observed by using micro-focus x-ray CT. It was shown that cracks and pores were sealed by precipitation of the calcium carbonate. Furthermore, it was shown that the sealed volume increased with increasing elapsed time and sealing rate was calculated. It is concluded that x-ray CT is useful for quantitative estimation of sealing of cracks and pores in cementitious material.*

KEYWORDS: *fracture sealing, cementitious material, micro-focus x-ray CT, image subtraction*

1. Introduction

For geological disposal of radioactive wastes, the intensity of radioactivity of radionuclides can be reduced by engineered barriers such as bentonite buffer and by natural barriers such as a rock mass. If the repository of radioactive wastes is located in an area where the hydraulic gradient and the permeability are high, the retardation of migration of radionuclides by these barriers may not be enough. Therefore, the influences of nuclides with low adsorption, such as I-129 and C-14, will be large. In order to retard the migration, several alternative concepts of radioactive waste packages are being developed (Owada *et al.*, 2005). It is planned that high-strength and ultra low-permeability concrete (HSULPC) will be used for a radioactive waste package for the geological disposal of transuranic radionuclides (TRU waste) (Kawasaki *et al.*, 2005; Owada *et al.*, 2005). Figure 1 shows a schematic illustration of this alternative concept using HSULPC. The aim of this scheme is to confine radionuclides, especially C-14, for a long term. It is supposed that C-14 is confined for 60,000 years, which corresponds to 10 times of the half-life of C-14. If a crack nucleates and propagates in HSULPC, however, the confining ability of the package can decline after crack growth for the long term even though the crack propagates slowly. Therefore, time-dependent fracturing of HSULPC has been studied (Nara *et al.*, 2008).

In the case of cementitious materials, it is generally known that precipitation of calcium compounds on the surface occurs in water. Thus sealing of pores and cracks by precipitation is possible. This may affect the crack growth and permeability.

In this study, we investigated the sealing of cracks and pores. Specifically, the sealing of cracks and pores in HSULPC in water by calcium precipitation was observed using micro-focus x-ray CT. Then, the change of the sealed volume by precipitation with elapsed time was estimated.

2. Sample

In this study, HSULPC (made by Taiheiyo Consultant Corporation) was used as a sample. The composition of HSULPC is shown in Table 1. The bulk density was 2460 kg/m^3. The uniaxial compressive strength was 203 MPa. The porosity was 5%. The water permeability coefficient was 4×10^{-19} m/s (Kawasaki *et al.*, 2005).

In Figure 2, a specimen of HSULPC used in this study is shown. The cylindrical test pieces of HSULPC was split into two halves and set in the acrylic cylinder in the way that the width of two split fracture planes became 0.1 mm. The height of acrylic cylinder was 35 mm, and the diameter and height of split HSULPC specimen were approximately 13 mm and 15 mm, respectively.

High-strength and ultra-low
permeability concrete

Figure 2. *Split specimen
of HSULPC*

Canister   Steel box

Figure 1. *Illustration of an alternative
concept of radioactive waste package
using HSULPC*

Contained amount [kg/m^3]	
Low-heat Portland cement	744~1014
Silica fume	158~496
Fillers (fly ash, blast furnace slag, etc.)	225~541
Aggregates	631~947
Water-reducing admixture	24
Water	180

Table 1. *The composition of HSULPC*

3. Methodology

3.1. Observation method of precipitation

The micro-focus x-ray CT system made by Toshiba IT & Control Systems Corporation (TOSCANER 30900 μhd) was used to observe the sealing process in HSULPC specimens. The focal spot size of x-ray tube was 5 μm.

At first, the specimens were immersed in water and kept for a selected term at 20 degree Celsius. Specimens were stored in one of two types of water, i.e., simulated fresh reducing high pH type groundwater (FRHP) and simulated seawater. In Table 2, chemical composition of water is shown. Next, we took out a specimen from water at the desired term, dried it in a glove box filled with nitrogen gas and then observed it with x-ray CT. The observation was performed periodically over a 6 month term.

	Ca	Si	Al	SO$_3$	Na	K	Cl	Mg	HCO$_3$
Seawater	1.0×10^{-2}	-	-	2.9×10^{-2}	4.5×10^{-1}	1.9×10^{-2}	5.6×10^{-1}	5.5×10^{-2}	2.4×10^{-3}
FRHP	1.1×10^{-5}	3.4×10^{-4}	3.4×10^{-7}	8.1×10^{-5}	3.8×10^{-3}	6.2×10^{-5}	1.5×10^{-2}	5.0×10^{-5}	3.5×10^{-3}

Table 2. *Chemical composition of simulated seawater and simulated FRHP in mol/l*

In x-ray CT observation, at first, the cone beam scanning was used for the 3D reconstruction of whole specimen with a resolution of 1024×1024 pixels in each slice. In this scanning mode, the slice thickness was set to be 0.04 mm and 402 slice images were obtained along the direction of the sample height. Then multi-slice scanning was conducted to obtain more precise images than those for cone beam scanning. This scanning mode can achieve a high resolution of 2048×2048 pixels which corresponds to a resolution of 7.6 μm. For the quantitative estimation by image analysis, images from multi-scanning were used.

3.2. *Image analysis method*

As mentioned before, images obtained from multi-slice scanning were used for image analysis because they had higher resolution. Specifically, the images of the samples obtained before the immersion were compared to those obtained after the immersion using the image subtraction method. In this study, the CT image before the immersion was defined as initial image. In Figure 3, an initial image is shown. It is necessary to correct the unavoidable alignment gap between each time of scanning so that the same regions of the initial and immersed core can be compared. For this purpose, affine transform was used. To obtain the optimal parameters for affine transform (displacements in two orthogonal directions (u_x and u_y) and rotation (ω_{xy})), following two steps were used. No distortion between initial image and images after immersion is assumed.

First, by comparing the initial image and post-immersion images for several representative points, approximate values of u_x, u_y and ω_{xy} were obtained. Then, by changing these three parameters and calculating the cross correlation value, we found the maximum value of cross correlation and the optimal parameters of u_x, u_y and ω_{xy} for affine transform. The value of cross correlation, C_c, is calculated as:

$$C_c = \iint f_0(x,y) \cdot f_R(x - \omega_{xy}y + u_x x, y + \omega_{xy}x + u_y y)\,dxdy \qquad [1]$$

where f_0 and f_R are the CT values of individual pixels within the initial image and image after the immersion, respectively. Then the image subtraction analyses between the initial image and the images after the immersion were conducted.

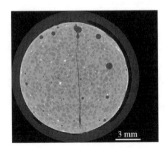

Figure 3. *An initial image of a specimen obtained by multi-slice scanning*

4. Results

4.1. *Results of observation*

Images of x-ray CT of a specimen immersed in simulated FRHP obtained by cone beam scanning are shown in Figure 4 (radioactive waste management funding and research center (RWMC), 2009). In these images, dark colors correspond to cracks and pores.

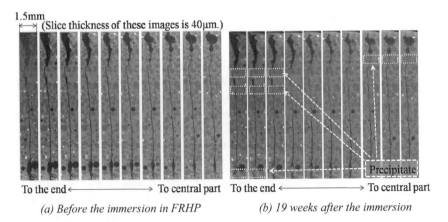

(a) Before the immersion in FRHP   (b) 19 weeks after the immersion

Figure 4. *X-ray CT images at the fracture obtained by cone beam scanning*

Figure 4(a) shows the initial images (before immersion) and Figure 4(b) shows images of the same positions of the specimen immersed in water for 19 weeks. It is clear that precipitation happened and the precipitate sealed the fracture plane. The same sealing behavior was found in the specimen immersed in simulated seawater. From observations obtained with x-ray CT, the sealed parts of the specimen were distinguishable in approximately 3 weeks of submergence in water. It was also found that sealing completed more quickly for the fracture with a smaller aperture. In Figure 4(b), sealing is apparent in the lower portion of the figure, yet it is not as complete as in the upper portion. Sealing was not observed in central part of the specimen. From these results, it is likely that the sealing rate decreased with the elapsed time after the immersion. The precipitate was mainly $CaCO_3$ (RWMC, 2009) as determined by an Electron Prove Micro Analyser (EPMA).

4.2. Results of image analysis

In Figure 5, a result of image subtraction is shown. The parts with a lighter color indicate the sealed parts.

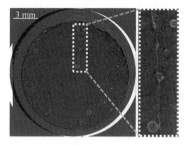

Figure 5. *A result of image subtraction for a specimen immersed in simulated seawater*

By using image subtraction analysis, the parts of the sample that were sealed by precipitation were revealed. As shown in Figure 5, the change of volume percentage of precipitate in a pore in case of simulated seawater was calculated by the result of image subtraction and threshold processing. Although the quantitative estimation for the pore resulted in success, there needs further study for the quantitative estimation of split part, because left hand side of Figure 5 has several pixels of non-zero grayscale values except in the area where precipitation of $CaCO_3$ occurred. This is due to the inevitable difference of the scanning position. Since this difference was not considered in this study, it will be the future subject to conduct image analysis considering the difference of the scanning position.

5. Discussion

From the observation with x-ray CT, the sealed parts of fractures in HSULPC increased with elapsed time. It is important to estimate the sealed volume quantitatively in order to clarify the progress of sealing behavior. By using the result of image subtraction, we estimated the temporal change of the sealed volume. In particular, after using binary conversion to the subtraction image, we calculated the number of the pixel of sealed parts. Then, we determined the percentage of the sealed parts by using the following equation:

$$P_s = N_s \div N_t \qquad\qquad\qquad [2]$$

where P_s is the percentage of the sea the sealed parts, N_s is the number of the pixel P_s means the percentage of the sealed volume by precipitation. In Figure 6, the temporal change of P_s for one pore in the specimen immersed in simulated seawater is shown. From this figure, it is shown that P_s increased with elapsed time.

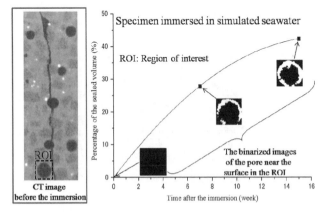

Figure 6. *Temporal change of the sealed volume of a pore in specimen immersed in simulated seawater*

Thus P_s means the percentage of the sealed volume by precipitation. In Figure 6, the temporal change of P_s for one pore in the specimen immersed in simulated seawater is shown. From this figure, P_s increased with elapsed time. Additionally, it is shown that the change of P_s decreased. For precipitation of $CaCO_3$, migration of water into and through HSULPC is necessary. Fracture and pore networks provide the principal pathways for fluid flow. If the pathway is partly sealed, water migration will be limited and precipitation may decrease. Therefore, it is considered that the decrease in the change of P_s (Figure 6) is caused by a temporal increase of the sealed volume by precipitation. Furthermore, the temporal increase of the sealed volume suggests that HSULPC possesses the property of self-sealing, which can delay fluid flow into and through these structures and ensure long-term integrity.

6. Conclusions

We have conducted the observation of sealing behavior in split HSULPC in water using x-ray CT and estimated the sealed volume with image analysis. The sealing behavior of cracks and pores by precipitation in HSULPC was visualized using a micro focus x-ray CT scanner. Sealing was observed mainly near the ends of the specimen. However, sealing was not observed in the central part of the specimen. The sealed volume increased with time while changes in sealed volume decreased with time due to a decrease of water migration in the specimen.

7. Acknowledgements

We would like to thank Radioactive Waste Management Funding and Research Center for the HSULPC samples used in the research project "Research and development of processing and disposal technique for TRU waste containing I-129 and C-14", the Ministry of Economy, Trade and Industry and the Research Fellowships of the Japan Society for the Promotion of Science for Young Scientists.

8. References

Kawasaki, T., Asano, H., Owada, H. Otsuki, A., Yoshida, T., Matsuo, T., Shibuya, K., Takei, A., "Development of waste package for TRU-disposal (4) – Evaluation of confinement performance of TRU waste package made of High-Strength and Ultra Low-Permeability concrete", *Proc. GLOBAL* 2005, No. 254, 2005.

Nara, Y., Mori, D., Owada, H. and Kaneko, K., "Study of subcritical crack growth and long-term strength for rock and cementitious material for radioactive waste disposal", *Proc. SHIRMS 2008*, Vol. 2, p. 135-147, 2008.

Owada, H., Otsuki, A., Asano, H., "Development of waste package for TRU-disposal (1) – Concepts and performances", *Proc. GLOBAL* 2005, No. 351, 2005.

Radioactive waste management funding and research center, *Research and Development of Processing and Disposal Technique for TRU Waste Containing I-129 & C-14*, Vol. 2, 2009.

Extraction of Effective Cement Paste Diffusivities from X-ray Microtomography Scans

K. Krabbenhoft — **M.R. Karim**

Centre for Geotechnical and Materials Modelling
University of Newcastle, NSW, Australia
kristian.krabbenhoft@gmail.com

ABSTRACT. *The problem of extracting effective diffusivities of cement pastes on the basis of X-ray microtomography images is considered. A general computational homogenization framework has previously been developed and is here applied to a variety of cement paste whose microstructure has been digitized to a resolution of 1 μm. With this resolution, important submicron features are not resolved. Consequently, we propose a methodology whereby the pore space is ascribed a diffusivity less than the free diffusivity. For this purpose a simple rule that incorporates microtomography data is proposed and shown to yield satisfactory results.*

KEYWORDS: *X-ray microtomography, cement paste, homogenization, numerical methods.*

1. Introduction

The transport properties of cement pastes are of crucial importance to the durability of much of our civil infrastructure. Due to its corrosive effects and abundance in many natural environments, considerable efforts have been made to quantify particularly the transport of chloride. To date, most research has been concerned with the diffusive transport through fully saturated cement pastes and a number of specialized testing methods for evaluating the effective diffusivity have been developed (Tang and Nilsson 1992; Friedman et al., 2004; Krabbenhoft and Krabbenhoft 2008).

As an alternative to traditional experimental determination of transport properties of porous materials, a number of studies have recently demonstrated the feasibility of 'virtual experiments'. The basic idea is here to obtain detailed three-dimensional images of the microstructure of the material after which the relevant tests are simulated and the quantities of interest, for example effective diffusion coefficients, are extracted in much the same way as in traditional physical laboratory tests. For the former task, X-ray microtomography (XMT) is typically used while the latter is carried out using either finite element/difference analyses or more specialized techniques such as lattice-Boltzman methods and Brownian motion simulations. Representative works combining both elements, i.e. imaging and simulation, include (Bentz et al., 2000; Knackstedt et al., 2006; Krabbenhoft et al., 2008; Promentilla et al., 2009).

Although the methodology described above is conceptually simple, it does involve a number of significant challenges. Firstly, one has to deal with very large data sets. Thus, digitized material elements consisting of more than one billion voxels ('volume pixels') is not uncommon. Regardless of which method of simulation is chosen, it is necessary to pay particular attention to this reality. In the present study where the finite element method is used, specialized iterative solvers for solution of systems of linear equations, which at times are severely ill-conditioned, have been developed. Furthermore, in order to limit the size of the problems solved in the actual simulations, a statistical homogenization procedure has been formulated. The resulting theoretical and computational framework is generally applicable to mass diffusion (as well as heat conduction and related physical phenomena) in porous and other heterogeneous materials.

Another, and in many ways more serious, complication with the virtual testing methodology, is that important microstructural characteristics may remain unresolved in the XMT scans. In the present study where XMT images of cement pastes at a resolution of approximately 1 μm are used, this is very much the case. As will be discussed in detail later on, this level of resolution does not account for submicron features that are of crucial importance to mass diffusion processes. Consequently, it is necessary to make a less sharp distinction between the solid and pore phases than would ideally be desirable. More concretely, we find it necessary to ascribe part of pore space a microscopic diffusion coefficient less than the free diffusion coefficient which would normally be expected to characterize the transport within the individual pores. A new rule that incorporates XMT data has been developed for this purpose.

The approach of not distinguishing sharply between the pore and solid phases has a number of similarities with other recent work, in particular that of Bentz (2000) who obtained microtomographic images of Fontainebleau sandstone and determined effective diffusivities numerically by means of a finite difference method. However, owing to the resolution of the images (19 μm, enhanced to 6.65 μm by optical magnification), the 'solid phase' was not considered impermeable but ascribed a not insignificant microscopic diffusivity. Furthermore, recent work of Promentilla et al. (2008, 2009) on cement pastes scanned at a resolution of 0.5 μm show that it generally is non-trivial to segment the raw grayscale XMT data into solid and pore phases. Moreover, it is demonstrated that the end result, in terms of effective diffusivity, depends strongly on the particular way that the segmentation is carried out. These findings are confirmed by the present study where the NIST Visisble Cement Data Set (Bentz et al., 2000) is used.

2. Numerical results

In the following, the results of diffusion through the pore space of the cement paste are presented. For the present study, the NIST Visible Cement Data Set (Bentz et al., 2000) was used. This data set contains images of cement pastes with difference water/cement ratios and at various stages of hydration. To illustrate the procedures developed, we will in the following use a particular data set which describes the microstructure of a cement paste with an initial water/cement ratio of $w/c = 0.45$ and a hydration time of 137 hours.

Applying the computational homogenization procedure described by Krabbenhoft et al. (2008) and Karim and Krabbenhoft (2009), the diffusivities shown in Figure 1 are obtained. Also shown are the standard deviations. For each computed diffusivity

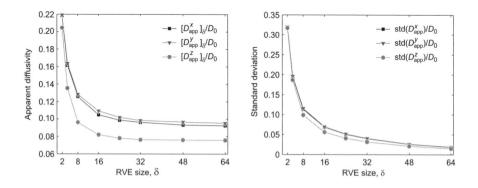

Figure 1. *Computed apparent diffusivities: stochastic averages (left) and standard deviations (right). The number of voxels (and hence finite elements) in each RVE is given by δ^3. The slight anisotropy between the x, y, and z directions is most likely a result of the preparation procedure (Karim and Krabbenhoft 2009).*

a total of 10,000 tests were performed. The diffusivity of the pore space was set equal to $D_0 = 1$ while the diffusivity of the solid phase was set equal 10^{-10}.

The cement samples scanned were in the form of cylinders with z denoting the longitudinal direction and x and y denoting to arbitrary directions in the cross section of the cylinder.

As seen, the computed mean values appear to attain near-asymptotic values at around $\delta = 30$ while the standard deviation continues to decrease, though at a decreasing rate, as δ increases. The asymptotic values of the effective diffusion coefficients are:

$$D_{\text{eff}}^x = [D_{\text{app}}^x]_\infty \approx [D_{\text{app}}^x]_{64} = 0.095D_0$$
$$D_{\text{eff}}^y = [D_{\text{app}}^y]_\infty \approx [D_{\text{app}}^y]_{64} = 0.095D_0 \qquad (1)$$
$$D_{\text{eff}}^z = [D_{\text{app}}^z]_\infty \approx [D_{\text{app}}^z]_{64} = 0.075D_0$$

Concerning the absolute magnitude of diffusivities, Garboczi and Bentz (1992) have proposed the following expression:

$$D_{\text{eff}}/D_0 = H(\phi_p - 0.18)1.8(\phi_p - 0.18)^2 + 0.07\phi_p^2 + 0.001 \qquad (2)$$

where H is the Heaviside function. This formula is based on microstructural considerations and has been calibrated against experiments. By direct comparison to the above formula, we see that the computed effective diffusivities significantly overestimate what can be expected for a cement paste of the type considered. Indeed, for a cement paste with a porosity of around 29%, we should expect a value of $D_{\text{eff}}/D_0 \approx 0.028$, which is a factor of three to four less than obtained in the present study. The possible reasons for this deviation will be discussed in the next section.

3. Correction for unresolved submicron features

The most immediate explanation to the deviation between computed and experimentally determined effective diffusivities is that the threshold level, $T_{ps} = 42$, is too high. This threshold level was chosen on the basis of achieving a porosity of approximately 29%. However, in principle, a small change in threshold level could reduce the effective diffusivity significantly without altering the porosity correspondingly. If, for example, the chosen threshold is close to the percolation threshold, a small change in T would imply a potentially very large change in the effective diffusivity without affecting the porosity significantly. However, from Figure 2 we see that this scenario is not responsible for the deviation between computed and experimental determined diffusivities. Thus, in order to achieve an effective diffusivity of $D_{\text{eff}}/D_0 \approx 0.028$, the threshold would have to be reduced to a level corresponding to less than 20% porosity, which for the given water/cement ratio is unacceptably low. We also see that the percolation threshold corresponds to a porosity of approximately 10% while (2) operates with a percolation threshold of 10%.

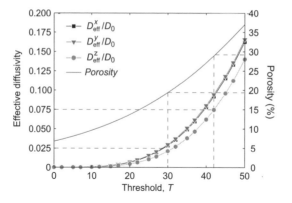

Figure 2. *Effect of thresholding on porosity and computed effective diffusivities.*

The second hypothesis is based on the observation that of the total pore space, constituting 28.8% of the volume, only 6.7% is recognized as a having a gray scale value of 0. The remaining 22.1% is attributed voxels with a grayscale value of up to 42. This obviously poses the question as to how representative the final digitized volume really is. In particular, it is somewhat worrying that a very considerable part of the pore space is not immediately recognized as such from the raw XMT data. The most reasonable explanation to this failure to clearly identify the pore space is that resolution of the XMT scans (approximately 1 μm) was insufficient. Figure 3, which shows the submicron structure of two different hydrated cement pastes, offers some support of this hypothesis. From these figures it appears that a resolution in the micrometer range is insufficient to capture the complex dendritic structures that form as a result of hydration. Moreover, this particular structure suggests that a voxel which has been identified as belonging to the pore space due to a low grayscale level in fact may contribute significantly to the effective resistance against diffusion. As

Figure 3. *Submicron structure of two different hydrated cement pastes. After Buenfeld et al., 2007 (left) and Tritthart and Haussler, 2003 (right).*

Dataset	w/c	T_{ps}	T_{hu}	ϕ	$D_{\text{eff}}^{\text{av}}/D_0$ $D_0^*/D_0 = 1$	$D_{\text{eff}}^{\text{av}}/D_0$ $D_0^*/D_0 = 0$	D_{eff}/D_0 Analyt. (2)
cez16_sld_2mmv1c300	0.40	151	48	0.157	0.0189	0.00641	0.00272
cez16_d_6dv1c300	0.40	112	39	0.222	0.0375	0.00953	0.00767
pt045_sld_2mmv1c300	0.45	142	53	0.245	0.0583	0.0145	0.0129
p35h40v1c300	0.35	117	49	0.283	0.0907	0.0378	0.0256
pt045_sld_7dv1c300	0.45	114	42	0.288	0.0832	0.0260	0.0278
pt045_sld_7dv1c300	0.45	113	43	0.297	0.0952	0.0316	0.0319
p35h12v1c300	0.35	109	57	0.362	0.175	0.0651	0.0700
p35h08v1c300	0.35	106	48	0.381	0.198	0.0884	0.0836
p35h25v1c300	0.35	114	78	0.393	0.202	0.0513	0.0933
pate035_8_5dv1c300	0.35	143	102	0.400	0.204	0.0597	0.0992
pate03_1_7dv1c300	0.30	103	74	0.424	0.246	0.121	0.121
pt045_b_5dv2c300	0.40	99	58	0.468	0.277	0.158	0.166
pt045_h_5dv2c300	0.45	103	62	0.489	0.290	0.162	0.189

Table 1. *Cement pastes analyzed. The data sets are from from http://visiblecement.nist.gov. The average effective diffusivities are calculated as* $D_{\text{eff}}^{av} = \frac{1}{2}\left[\frac{1}{2}(D_{\text{eff}}^x + D_{\text{eff}}^y) + D_{\text{eff}}^z\right]$.

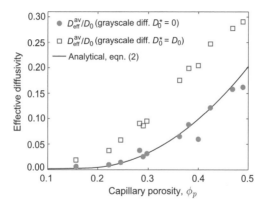

Figure 4. *Comparison between computed effective diffusivities and analytical estimate* (2).

such, assigning a constant diffusivity to the 'pore space' voxels may lead to significant overestimates of the effective diffusion coefficient.

In an attempt to remedy this predicament, local diffusivities less than the free diffusivity were assigned to the intermediate gray scale voxels. This was done via a simple linear relationship so that the local pore diffusivities were taken as

$$D_p/D_0 = 1 - (1 - D_0^*/D_0)\frac{G}{T_{ps}}, \quad 0 \le G \le T_{ps} \quad (3)$$

where D_p is the pore diffusivity, D_0 is the free diffusivity, D_0^* is a new constant that will be referred to as the 'linear grayscale diffusivity', G is the grayscale ($0 \le G \le 42$), and $T = 42$ is the pore/solid threshold value. In other words, pores with a grayscale count of 0 (6.7% in total) are assigned the free diffusion coefficient whereas the magnitude of the local diffusivities will decrease linearly to D_0^* at the threshold level.

In order to probe the validity of the proposed grayscale-pore diffusivity rule (3), we have analyzed a total of 13 different cement pastes as summarized in Table 1 and Figure 4. Remarkably, for the most part, the diffusivities resulting from setting the grayscale diffusivity equal to zero, i.e. by letting the pore space diffusivity vary linearly with the grayscale value from $G = 0$ to $G = T_{ps}$, are in excellent agreement with the analytical/empirical formula (2) of Garboczi and Bentz (1992). It this appears reasonable to conclude that the proposed empirical rule has general validity, at least for the cement pastes in the NIST data set. As for application to other data sets, the rule would probably in most cases furnish a reasonable first estimate.

The success of the proposed procedure for correcting for submicron features motivate the search for a more physical means of constructing such laws. From the wide range of pastes considered (see Table 1) it appears that the rule in the present case is relatively insensitive to porosity, water/cement ratio, hydration time, etc. We would, however, expect it to be a function of the particular type of cement paste (chemical composition of the cement, additives, etc) and of the XMT settings (resolution, beam energy, etc). Ultimately, however, it could be envisioned that the submicron features actually are resolved. With recent advanced in nano-CT this could well be a reality in coming years. One would then use the effective nanoscale diffusivity as input to microscale models such as the ones considered in the present paper.

4. Conclusions

The problem of extracting effective diffusivities of cement pastes on the basis of X-ray microtomography images has been considered. A previously developed computational homogenization framework has applied to selected data sets. With the resolution of the selected data sets (approximately 1μm), important submicron features are not resolved and it is necessary to perform a 'correction' to account for unresolved submicron features. This is done be ascribing the pore space a diffusivity less than the free diffusivity. In particular, it is found that a linear variation of the pore phase diffusivity with grayscale level furnishes reasonable results.

5. References

Andrade C., "*Calculation of chloride diffusion coefficients in concrete from ionic migration measurements*", vol. 23, p. 724–742, Cement and Concrete Research, 1993.

Barberon F., Korb J. P., Petit D., Morin V., Bermejo E., "*Probing the Surface Area of a Cement-Based Material by Nuclear Magnetic Relaxation Dispersion*", vol. 90, no. 116103, Physical Review Letters, 2003.

Bentz D. P. et al., "*Microstructure and transport properties of porous building materials. II: Three-dimensional X-ray tomographic studies*", vol. 33, p. 147–153, Materials and Structures, 2000.

Bentz D. P. et al., "*The Visible Cement Data Set*", vol. 107, p. 137–148, Journal of Research of the National Institute of Standards and Technology, 2002.

Buenfeld N., "*Concrete Durability Group. Secondary Electron Imaging*", 2007.

Friedmann H., Amiri O., Ait-Mokhtar A., Dumargue P., "*A direct method for determining chloride diffusion coefficient by using migration test*", vol. 34, p. 1967–1973, Cement and Concrete Research, 2004.

Garboczi E. J., Bentz D. P., "*Computer Simulation of the Diffusivity of Cement-Based Materials*", vol. 27, p. 2083–2092, Journal of Materials Science, 1992.

Karim M. R., Krabbenhoft K., "*Extraction of effective cement paste diffusivities from X-ray microtomography scans*", under review, 2009.

Knackstedt M. A. et al., "*Elastic and transport properties of cellular solids derived from three-dimensional tomographic images*", vol. 462, p. 2833–2862, Proceedings of the Royal Society A, 2006.

Krabbenhoft K., Hain M., Wriggers P., "*Computation of Effective Cement Paste diffusivities from Microtomographic Images*", p. 281–297, Kompis V., Ed., Composites with Micro- and Nano-Structure, Spinger, 2008.

Krabbenhoft K., Krabbenhoft J., "*Application of the Poisson-Nernst-Planck equations to the migration test*", vol. 38, p. 77–88, Cement and Concrete Research, 2008.

Promentilla M. A. B., Sugiyama T., Hitomi T., Takeda N., "*Characterization of the 3D pore structure of hardened cement paste with syncrotron microtomography*", vol. 6, p. 273–286, Journal of Advanced Concrete Technology, 2008.

Promentilla M. A. B., Sugiyama T., Hitomi T., Takeda N., "*Quantification of tortuosity in hardened cement pastes using synchrotron-based X-ray computed microtomography*", vol. 39, p. 548–557, Cement and Concrete Research, 2009.

Tang L., Nilsson L. O., "*Rapid determination of the chloride diffusivity in concrete by applying an electric field*", vol. 89, p. 49–53, Cement and Concrete Research, 1992.

Torquato S., Random Heterogeneous Materials, Springer, 2002.

Tritthart J., Häußler F., "*Pore solution analysis of cement pastes and nanostructural investigations of hydrated C3S*", vol. 33, p. 1063–1070, Cement and Concrete Research, 2003.

Contributions of X-ray CT to the Characterization of Natural Building Stones and their Disintegration

J. Dewanckele*,** — D. Van Loo** — J. Vlassenbroeck** — M. N. Boone** — V. Cnudde*,** — M. A. Boone* — T. De Kock* — L. Van Hoorebeke** — P. Jacobs*,**

*Department of Geology and Soil Science, Ghent University
Krijgslaan 281/S8, B-9000, Ghent, Belgium
veerle.cnudde@ugent.be
jan.dewanckele@ugent.be
patric.jacobs@ugent.be

** Department of Subatomic and Radiation Physics
Center for X-ray Tomography, Ghent University
Proeftuinstraat 86, B-9000 Ghent, Belgium
jelle.vlassenbroeck@ugent.be
yoni.dewitte@ugent.be
matthieu.boone@ugent.be
luc.vanhoorebeke@ugent.be

ABSTRACT. This paper highlights the use of the high resolution scanner at the Center for X-ray Tomography in Ghent, Belgium (UGCT), for the 3D quantitative evaluation of the disintegration of some French natural building stones. Rocks deteriorate when they are exposed to extreme weathering factors such as a combination of water and freeze-thaw cycles or high pressure. The results of those processes can be very diverse: from element migration to crust formation to the origination of micro-cracks. Thanks to its non-destructive character, high resolution computed tomography (CT) turned out to be an excellent monitoring tool as it contributes to the characterization of the internal structure of the natural building stone. X-ray CT also provides a better insight into the micro-structural durability properties of the building stone.

KEYWORDS: characterization, building materials, high resolution x-ray CT scanner, micro-cracks, rock failure

1. Introduction

Many rock types are formed under high temperature and/or high pressure conditions in the upper part of the earth's crust. Some of these are used as natural building stones for various purposes such as house fronts, floors, roof tiles and street pavements. In most cases they are therefore in a thermodynamically instable condition in the environment where they are used nowadays. Exposure of rocks to new exogenous parameters such as fluctuating temperature, pressure and the presence of water/salt may in some cases lead to their deterioration. Some rock types are more susceptible to weathering processes than others, in which fluctuating environmental factors as well as the endogenous or geological parameters of the stone itself (porosity, fossil content, mineral structure, etc.) play an important role.

In order to evaluate rocks used as building material, several characterization tests (petrographic analysis, determination of the porosity, pressure resistance, etc.) and durability tests (freeze-thaw resistance, thermal shock resistance, etc.) exist. The significance of the latter lies in the predictability of the stone's weathering behavior or resistance under known external conditions. After the tests, the stone's new physical properties are evaluated with regard to their initial condition. There is thus a registration of both the initial "fresh" rock's situation before the test and the final "failed" situation at the end of the test series. The main shortcomings in these traditional durability tests are that these evaluations are mainly based on visual inspection and that there is no quantification of internal micro-structural reorganization. However, these internal micro-structural weathering processes are of great importance to understand the deterioration of the stone in its entirety. Internal quantification of dynamic processes still remains difficult with destructive analysis tools. For that purpose, in this study non-destructive high resolution x-ray CT is combined with image analysis to visualize, characterize and quantify freeze-thaw cycles of the Noyant Fine limestone and pressure test on the Euville, Noyant Fine and Savonnières limestone.

2. Methods and instrumentation

2.1. *Outline of the UGCT scanner and scanning conditions*

The experiments were carried out at the Center for X-ray Tomography at Ghent University (UGCT, www.ugct.ugent.be). The x-ray tube of this high-resolution scanner is an open-type device with a dual head (Feinfocus®, FXE-160.51), consisting of a transmission head for small samples down to sub-micron scale and a high power directional head for relatively large samples. An open-type scanner offers the important advantage that different add-ons can be applied and different detectors can be used to optimize the scanning conditions. In this study the Varian PaxScan 2520 with a pixel size of 127 μm and a dimension of 1880 × 1496 pixels was used. The freeze-thaw cycle tests on the Noyant Fine limestone are performed with the transmission head and a voltage of 80 kV with a power of 8 W and 100 μA tube current. A thin (550 μm) Al-filter was used to block the low-energetic x-rays to prevent beam hardening. For each scan, 1000 projections were registered over an angle of 360° with a source-detector distance of 890 mm and a source-object distance of 55 mm, resulting in a voxel pitch of around 7.8 μm. For each projection 8 frames were taken with an exposure time of 300 ms. The total period of time for one scan was around 1 hour and 30 minutes.

The scans for the pressure tests on the Noyant Fine, Euville and Savonnières limestone are taken with the same parameters, but with a tube voltage of 110 kV instead of 80 kV and an average of 4 frames for one projection.

2.2. Software

The obtained raw CT data is afterwards processed with the in-house developed reconstruction software Octopus. The projection data is filtered by removing bright and dark spots, normalized and regrouped into sinograms. The latter are used to calculate virtual cross-sections through the object. For 3D visualization, VGStudio Max and VGStudio 2.0 (Volume Graphics) have been used. For details about the 3D quantification UGCT software tool Morpho+, the reader is referred to Vlassenbroeck *et al.,* 2007, Brabant 2009 and Cnudde *et al.,* 2009. Some of the main features of the program concern selecting a volume of interest, dual thresholding of porosity/grains, labeling, distance transform and finally calculation of the equivalent diameter and maximum opening. The latter corresponds to the diameter of the maximum inscribed sphere, while the equivalent diameter refers to the diameter of a sphere with the same volume of the selected pore/grain. In this way, quantitative data is obtained together with the 3D visualization of the changing geological parameters.

2.3. Characterization and durability tests on natural building stones

Generally, characterization tests are performed to determine the intrinsic material properties of natural building stones, independent of its future use and environment. The most common characterization test for natural building stones is the use of the petrographic microscope for mineralogical description. As Figure 1 shows, micro-x-ray CT is an excellent complementary technique of petrographic microscopy to visualize, quantify and render the rock's volume in order to get a deeper insight into its structure. Micro-CT possesses the possibility to calculate porosity and pore interconnectivity in a non-destructive way, parameters that normally are much more difficult to obtain by traditional microscopy. Also pressure tests are applied to characterize natural building stones. In this study an external load is applied on top of the Euville, Noyant Fine and Savonnières limestone samples with a pressure stage until visual deterioration occurred. All samples were cylindrical with a diameter of 8 mm. Afterwards, the deteriorated cores were scanned at the UGCT facility.

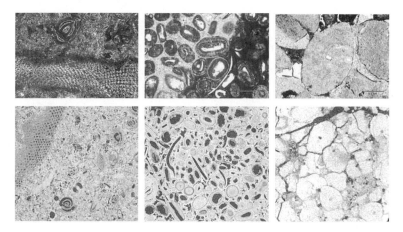

Figure 1. *Petrographic recording under the petrographic microscope (top row) compared with tomographic slices (bottom row). From left to right Noyant Fine, Savonnières and Euville limestone*

Durability tests analyses the behavior of the rock's intrinsic parameters in function of time. Evaluation of rock stability is based on those tests that are therefore considered as one of the most important to assess natural building stone behavior for the construction industry. Eleven freeze-thaw cycles were carried out on a cylindrical core (diameter 9 mm) of the Noyant Fine limestone. Each cycle consists of an immersion of the sample in water for two hours, followed by storage for 6-7 hours in a cold room at −15°C. After the sample was brought back to room temperature, a high resolution CT scan was made, after which a new freeze-thaw cycle starts. In total 11,000 projections were made, resulting in more than 60 Gbyte of raw projection data.

3. Materials

The Noyant Fine is a Lutetian (Eocene) limestone originating from the north eastern part of the border of the Paris Basin (France). It is composed of numerous Foraminifera and Bryozoa (tiny colonial animal) fragments. The hollow chambers of the Miliolids (a group of foraminiferans) largely contribute to the total rock porosity. The Savonnières and Euville limestone on the other hand have a Late Jurassic age. The former is an oolitic rock with shell fragments. Remarkably most of the ooids (small, spheroidal layered sedimentary grains) have a hollow core. Most of the shell fragments are dissolved leaving behind a secondary or moldic porosity. The Euville limestone on the other hand is composed of large crinoid fragments with a syntaxial overgrowth of calcite.

4. Results and discussion

4.1. *Characterization test*

The green structures in the three different stone types in Figure 2 correspond with the micro-cracks induced by the pressure test. The pore structure not affected by the micro-cracks (i.e. the "isolated porosity") is colored red. If a micro-crack affects an interconnected pore network the whole network is considered as being induced by the micro-crack and thus rendered green. Therefore, the term "micro-crack porosity" would be more appropriate. Visualization of both types of pore structures provides an excellent overview of the stones' internal behavior by increasing external pressure.

(a)   (b)   (c)

Figure 2. *Rendered volume of the induced micro-crack porosity (green) and the isolated porosity (red) of the Noyant Fine (a), Savonnières (b) and the Euville (c) limestone*

The distinction between the micro-crack porosity and the surrounding but not affected pore structure is obtained on the basis of the ratio of the equivalent diameter to the maximum opening, calculated in Morpho+. The micro-crack porosity will have a large equivalent diameter but a small maximum opening (sphericity $\ll$ 1), while the pores will be more spherical (sphericity $\leq$ 1). Besides visualization, the micro-crack porosity and the amount of isolated porosity have also been determined with Morpho+. The micro-crack structure of the Noyant Fine limestone is irregular and amounts to 55% of the total porosity. The crack-structure porosity of the

Savonnières stone amounts to 42%. The cracks in the Euville stone amounts to 97% of the total porosity. Figure 3 gives an overview of the ratio between the micro-crack porosity and the isolated porosity for every 10 slices for each stone. For each sample the crack density decreases in function of the height. x-ray CT not only makes it possible to visualize the internal reorganization but also offers the possibility to calculate the ratio of the crack density and the isolated but not affected pore structure.

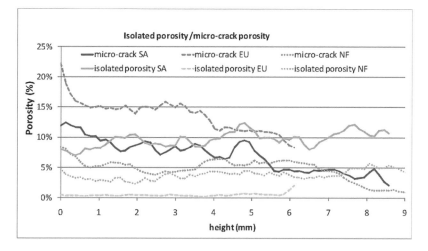

Figure 3. *Trends of induced micro-crack porosity and isolated porosity of the three rocks*

4.2. Durability test on the Noyant Fine limestone

Figure 4 shows the deterioration of the Noyant Fine building stone from the 6th till the 11th cycle. After cycle number 7 a micro-crack structure (diameter of approximately 78 μm) appears in the Bryozoa fragment, as clearly can be seen in the figure. After cycle 9 the crack already reached a diameter of around 187 μm and one cycle later it already showed an opening of more than 280 μm. At the end of cycle 11, a part of the stone flakes off. Freeze-thaw tests of the Noyant Fine limestone monitored with high resolution CT have already demonstrated that the crack initiation is located in the micro-porous Bryozoa fragments (Boone 2009). Most of the cycles are fully characterized in 3D with the aid of Morpho+. The total amount of material (fossil fragments, matrix, etc.) with exception of the pores has been calculated before and after the tests by defining the same scan, reconstruction and threshold parameters. It turned out that the material loss after 11 freeze-thaw cycles amounted to around 6.2% of its initial volume, largely attributed to the fragmentation at the end of cycle 11.

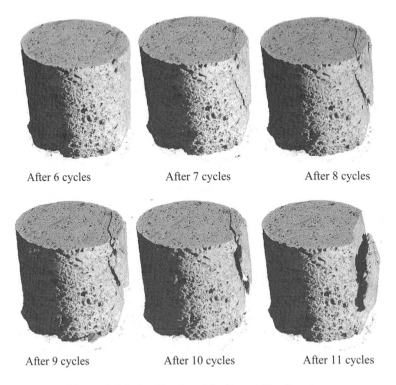

After 6 cycles   After 7 cycles   After 8 cycles

After 9 cycles   After 10 cycles   After 11 cycles

Figure 4. *3D visualization of the Noyant Fine limestone
after 11 freeze-thaw cycles*

5. Conclusion

Some French limestones were used to explore the potential of the technique of
high resolution CT for the quantification of the deterioration of building stones and
to test and evaluate their durability. With the aid of the in-house built software
package Morpho+ it was possible to calculate the percentages of the induced micro-
crack porosity structure and of the pore structure not affected by the pressure test.
After 11 freeze-thaw cycles, the aperture and the evolution of the micro-cracks
could been visualized in three dimensions and calculated. The material loss after the
last cycle could be evaluated by taking into account constant scan, reconstruction
and threshold parameters and also the same volumes of interest. Clearly high-
resolution CT combined with image analysis largely contributes to better insight into
the characterization and durability assessment of natural building stones. The
complex processes of deterioration are recorded in a non-destructive way and finally
4D quantitative data has been obtained.

6. Acknowledgements

The Institute for the Promotion of Innovation by Science and Technology in Flanders, Belgium (IWT) is acknowledged for the PhD grant of J. Dewanckele. The Fund for Scientific Research-Flanders (FWO) is acknowledged for the post-doctoral research grant to V. Cnudde.

7. References

Boone, M.A. 4D-monitoring van fysische verweringsprocessen in carbonaatgesteenten met X-stralen computergestuurde microtomografie (micro-CT), MSc thesis, Ghent University, Belgium, 2009.

Brabant, L. Geavanceerde algoritmes voor 3D-analyse van micro-CT data, MSc thesis, Ghent University, Belgium, 2009.

Cnudde, V., Vlassenbroeck, J., De Witte, J., Brabant, L., Boone, M.N., Dewanckele, J., Van Hoorebeke, L., Jacobs, P. "Latest developments in 3D analysis of geomaterials by Morpho+." *GeoX2010 conference* (submitted).

Vlassenbroeck, J., Dierick, M., Masschaele, B., Cnudde, V., Van Hoorebeke, L., Jacobs, P. "Software tools for quantification of X-ray microtomography at the UGCT", *Nuclear Instruments and Methods in Physics Research Section A: Accelerators, Spectrometers, Detectors and Associated Equipment*, vol. 580, no. 1, p. 442-445, 2007.

Characterization of Porous Media in Agent Transport Simulation

Examining building materials using x-ray computed tomograhy

L.-B. Hu* — C. Savidge* — D. Rizzo* — N. Hayden* — M. Dewoolkar* — L. Meador** — J. W. Hagadorn***

* *School of Engineering, University of Vermont*
Burlington, VT 05405, USA
liangbo.hu@uvm.edu,
csavidge@uvm.edu
drizzo@cems.uvm.edu
nhayden@cems.uvm.edu
mandar@cems.uvm.edu

***Department of Anthropology, University of Massachusetts*
Amherst, MA 01003, USA
lmeador@anthro.umass.edu

****Department of Geology, Amherst College*
Amherst, MA 01002, USA
jwhagadorn@amherst.edu

ABSTRACT. *Microscopic geomorphic structure is critical to the process of transport in porous building materials. X-ray scans were obtained on a variety of building materials to both qualitatively and quantitatively evaluate their pore structures. Scanned images were subsequently processed using a random walk analysis to estimate the macroscopic transport properties that are useful for numerical simulation of transport phenomena. 3D image reconstruction was also performed to provide better visualization of the pore structures and a basis for 3D simulation.*

KEYWORDS: *porous media, characterization, x-ray CT, transport, random walk*

1. Introduction

An understanding of geomorphic pore structure of building materials (natural and man-made) in terms of pore size, shape and connectivity is important in contaminant transport studies involving porous building materials. Examples include the transport, fate and remediation of chemical and biological contaminating agents and contamination from fires and acid rain in porous building materials. It is also important to understand the relationship between the geomorphic structure of porous building materials and the properties governing fluid transport within them. Transport in porous media has been investigated theoretically in the context of multiphase flow theory, and is commonly modeled using extended Darcy's law, or alternatively, modeled directly as a diffusion process. An extensive discussion of the significance of these approaches is beyond the scope of this paper. However, it is evident that pore structures, often characterized quantitatively, are critical for understanding and modeling the transport process in the porous media. Although traditional macroscopic experiments provide an averaged estimate for properties such as porosity, many details of the pore structures cannot be obtained easily with these methods, and consequently, are often neglected or assumed for consideration in numerical simulation. Microfocus x-ray computed tomography allows exploration of building materials and quantification of their geomorphic properties (e.g. Cnudde and Jacobs, 2004; Roels and Carmeliet, 2006).

This paper presents a study on quantifying geomorphic pore structures of various building materials using x-ray tomography. A random walk-based analysis to quantitatively evaluate the transport properties of these materials using their x-ray images is also presented. In addition, 3D image reconstruction of the x-ray images is briefly discussed.

2. X-ray CT scan of porous building materials

2.1. *Materials and experimental set-up*

A variety of natural and man-made building materials were investigated ,including four different types of concrete, three types of sandstone, a limestone, and two types of brick. All x-ray scanning was performed with a micro-focus x-ray CT scanner (Skyscan) at the Geology Department, Amherst College, MA. Specimens of different sizes, ranging from 4 mm to 25 mm in diameter, were scanned at 74 keV (133 μA), with 540 equally spaced images collected across 180°. Three frames were averaged for each image to improve the signal-to-noise ratio, ring artifacts were corrected using a random movement algorithm, and x-ray attenuation slices were generated using a modified filtered backprojection of shadow images. Resulting stacks of cross-sectional images had voxel sizes ranging from 1.5 to 7 μm.

2.2. *Scanned images of specimens and image processing*

A typical cross-sectional image of a concrete specimen is shown in figure 1a. The grayscale values vary from white to dark gray and the pixel values range from 0 (black) to 255 (white). These colors represent the range of x-ray attenuation through the sample, with black pixels representing the pore space (air) and the white or gray pixels representing the solid space.

To assist in better distinguishing (visually) pore spaces from solid spaces, some image processing measures were employed to convert the original grayscale image to black and white (B/W). We first plotted the pixel distribution (histogram), and then identified the histogram peaks to establish a threshold value that separated the voids from the solids in the original grayscale image. Finally, all the pixels with values less than the threshold were changed to "0" (black); and the pixels with values above the threshold are converted to "1" (white), resulting in a B/W image with pores represented by black and solids represented by white. Lastly, we switched these colors to better match the conventional visualization of x-ray images, using black for solids and white for pores.

This procedure was summarized in Nakashima and Watanabe [2002] and applied to CT scans of an assembly of beads, where clear distinction between beads and pores is readily visible prior to any processing. However, for scanned images of building materials, it is not as easy to distinguish the histogram peak values representing the solids and pores, because it is not an assembled porous system of beads in which the two peaks are well separated.

The pore structure of concrete is a good example of this phenomenon. An area of interest is cropped from the original image (red square of the CT image in Figure 1a). Before the image processing procedure described above is applied, an alternative way to tentatively explore the geomorphic structure is to produce a grayscale picture where the contrast of the image is enhanced, while the histogram of the output image is preserved. Such a picture is shown in Figure 1b, which clearly reveals the coarse aggregates (the lightest gray part) and some small pores (the darkest part) between these aggregates. Most of the space between the coarse aggregates could be finer sand- to clay-size aggregates or cement.

Figure 1c is the histogram of the cropped area of the original cross-sectional x-ray image of 5,000 psi concrete (Figure 1a) where most of the pixels are values less than 100. A closer look at the histogram distribution reveals there are two peaks. The first peak, pixel value 0, represents the pores. The second peak is around the value of 30, which has the largest population, about 1.1×10^5. Consequently, the midpoint, 15 is chosen as the threshold for separating the solid and pore spaces. The resulting BW picture is presented as Figure 1d after the black and white pixels have been switched. Clearly the pore space represented by white is rather small and primarily reflects the large pores captured by the grayscale image (Figure 1b).

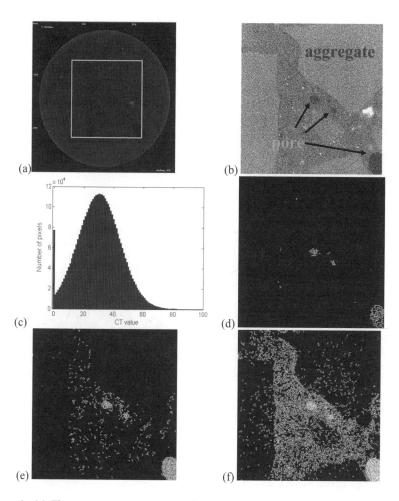

Figure 1. *(a) The raw x-ray cross-sectional image of 5,000 psi concrete. (b) A grayscale image for the cropped area of x-ray image in (a). The contrast is enhanced with the original histogram maintained. (c) The histogram of the cropped area (red in (a)). (d) The final BW image of processed x-ray image with a threshold value of 15. (e) The final BW image of processed x-ray image of 5,000 psi concrete with a threshold value of 20. (f) The final BW image of processed x-ray image of 5,000 psi concrete with a threshold value of 25.*

Clearly the amount of pores, in the final B/W picture, depends on the selection of the pixel threshold value. As a result, the threshold value dictates how much of the space between aggregates gets classified as solid or pore space. Figure 1e and 1f are obtained with two different threshold values, 20 and 25, respectively. In comparison with Figure 1d (threshold value of 15), the pores are more abundant when a greater threshold value is used. The boundary of the large coarse aggregates is also evident.

In Figure 1f, more of pixels between the coarse aggregates are classified as pores as opposed to fine aggregates or cements in Figure 1d.

The image is controlled by the physical nature of the porous material, although the processing procedure, quality and limitations of experimental device and set-up are all potentially significant. See, for example, the Ohio sandstone sample shown in Figure 2a. The final cropped BW image is presented in Figure 2b. The boundaries of sandstone grains are evidently distinguishable from the pores.

One approach is to measure the porosity and adjust the contrast threshold until the porosity (estimated using the image alone) reasonably matches the measured porosity. In this study, pixel histograms were generated; however, the threshold was selected also by considering the measured porosity.

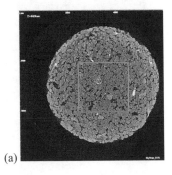 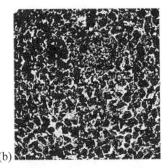

(a)   (b)

Figure 2. *(a) A sample x-ray image of Ohio Sandstone, cropped area is outlined by the red square (contrast enhanced). (b) The final B/W image of Ohio sandstone after the completion of the imaging processing*

3. 3D image reconstruction

The stacks of cross-sectional x-ray attenuation images were volume rendered in order to reconstruct and analyze the 3D geometry of the pore structures of scanned specimens. This was done using Avizo™ 6.0 software, which converts post-processed cross-sectional image data into 3D volumes or meshes. Through automatic or interactive image slices alignment, registration and segmentation capabilities, Avizo can reconstruct and visualize the 3D volume or the 3D surface of the specified material. Some reconstructed 3D images are presented (Figure 3) and provide good visual evidence of the pore structures in terms of the size, amount and connectivity.

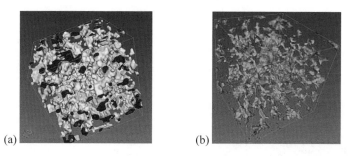

(a)   (b)

Figure 3. *(a) A 3D reconstructed image of the pore structures of a portion of an Ohio sandstone specimen, 0.3 mm × 0.3 mm × 0.3mm. (b) A 3D reconstructed image of the pore structures of a portion of a 5,000 psi concrete specimen, 0.3 mm × 0.3 mm × 0.3mm. The solid colors indicate pores.*

4. Quantitative estimate of material properties based on x-ray scans

The simplest property obtained from the CT scan images is the porosity. This may be calculated by counting the number of 1's (white pixels) in the matrix representing the BW 2D image. The methods for estimating other transport properties such as permeability and tortuosity are more complicated but nevertheless well documented (e.g. Coker *et al.* 1996, Maier *et al.* 1998). Here, the random walk algorithm discussed by Nakashima and Watanabe [2002] was adopted to estimate these properties and compare them with macroscopic laboratory test results. The specimens used for CT scans were cored out of the larger specimens used for macroscopic testing. Because the properties of the outer surfaces of concrete and brick used in building materials differ substantially from the properties of their interiors, we sampled the same exterior surfaces for both CT-scan core samples and for macroscopic testing samples.

The relationship of random walk simulation to the Wiener process, a similar process to Brownian motion as the physical phenomenon of a minute particle diffusing in a fluid, has been established in probability and statistics theory (e.g. Fuller, 1968). A detailed discussion of its application can be found in Nakashima and Watanabe (2002). For simplicity, the presented results are also based on the 2D images rather than the 3D reconstructed images since the theoretical background of the random walk algorithms is independent of the image dimensions. However, a 2D analysis would include even those pores that are not connected, whereas the physical measurement would indicate effective porosity (of connected pores).

In the B/W image of the CT scan (e.g. Figure 2b), a walker is initially randomly placed in a pore (white) pixel. It then migrates randomly on discrete pixels with the progress of the dimensionless integer time τ. Usually a large number of these walkers must be simulated and the average of the square distances of *all* these random walkers are computed as mean-square displacement $\langle r(\tau)^2 \rangle$, as a function of τ. If this random walk is completely unrestricted, i.e. in a perfectly full pore space without any solid particles present, $\langle r(\tau)^2 \rangle$ is proportional to τ (solid line in Figure 4b), and the proportionality constant will reflect the diffusion coefficient of the walker in the free space without solids (water diffusivity in bulk water).

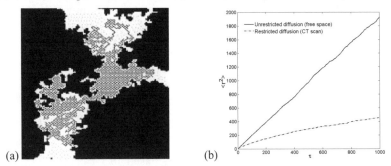

(a)   (b)

Figure 4. *(a) A 2D random walk trajectory in an image of Ohio sandstone;*
(b) mean-square displacement of random walk in (a)

However, because real porous media contain solids (black pixels in Figure 4a), $\langle r(\tau)^2 \rangle$ is reduced (solids act as obstacles and whenever a walker encounters a black pixel, it will stay at that same place at that moment). The change in the function of $\langle r(\tau)^2 \rangle$ provides a measure for the diffusivity in the real porous materials, and thus, tortuosity may be obtained by comparing the gradient to that of a free random walk $\langle r(\tau)^2 \rangle_{\text{free}}$. The random walk path must be limited within the pore space, and consequently the gradient of the restricted diffusion is substantially smaller in Figure 4b.

	Sandstone	5000psi Concrete
Porosity	0.15	0.13
	(0.12)	(0.14)
Specific surface area [1/m]	7×10^5	1.8×10^5
Tortuosity	4.4	16.0
Permeability [m^2]	7×10^{-14}	2.6×10^{-13}
	(1.5×10^{-13})	(7.9×10^{-14})

Table 1. *Estimate of transport properties*
(values in brackets are physical measurements)

Subsequently, specific surface area and permeability may also be estimated based on the Kozeny-Carman equation. Two sample results are summarized in Table 1. The numbers in brackets are properties measured in the laboratory. The image-based estimates are close to the measured values, and may be considered a success at this stage of the work, although simulations based on 3D images may be more appropriate for our comparison.

5. Conclusions

A number of common building materials were subjected to x-ray CT scanning and used for subsequent 2D and 3D image analyses. 2D image analysis had varied success. For concretes tested in this work, the sensitivity of the final output to the specific processing technique was quite high. Overall the quality of scanned images and the subsequent imaging processing is sufficient for 3D image reconstruction, which provides a satisfying 3D visualization of the pore structures.

Estimates of relevant transport properties are successful to some degree. These values were further used for numerical simulations that are beyond the scope of this paper. Current and future work on CT scans of samples with permeating fluids is expected to reveal more details of fluid wicking process inside the porous building materials.

6. Acknowledgements

This work is supported by Defense Threat Reduction Agency (#HDTRA1-08-C-0021).

7. References

Cnudde, V., Jacobs, P. J. S., "Monitoring of weathering and conservation of building materials through non-destructive X-ray computed microtomography", *Environmental Geology*, Vol. 46, 2004, p. 477–485.

Coker, D. A., Torquato, S., Dunsmuir, J. H., "Morphology and physical properties of Fontainebleau sandstone via a tomographic analysis", *Journal of Geophysics Research*, Vol. 101, No. 17, 1996, p. 497–506.

Feller, W., *An Introduction to Probability Theory and its Applications*, Vol. 1. New York, John Wiley & Sons, 1968.

Maier, R. S., Kroll, D. M., Kutsovsky, Y. E., Davis, H. T., Bernard, R. S., "Simulation of flow through bead packs using the lattice Boltzmann method", *Physics of Fluids*, Vol. 10, 1998, p. 60–74.

Nakashima, Y., Watanabe, Y., "Estimate of transport properties of porous media by microfocus X-ray computed tomography and random walk simulation", *Water Resources Research*, Vol. 38, No. 12, 2002, p. 1272, doi:10.1029/2001WR000937.

Roels, S., Carmeliet, J., "Analysis of moisture flow in porous materials using microfocus X-ray radiography", *International Journal of Heat and Mass transfer*, Vol. 49, 2006, 4762-4772.

Two Less-Used Applications of Petrophysical CT-Scanning

R. P. Kehl — S. Siddiqui

KehlCo Inc.
6435 N.Haywood
Houston, TX 77061, USA
bob.kehl@kehlco.com

Petroleum Engineering Department
Texas Tech University
Box 43111, 8th and Canton Avenue
Lubbock, TX 79409, USA
shameem.siddiqui@ttu.edu

ABSTRACT. Computerized Tomography (CT) has been used by oil and service companies for over 25 years for looking inside rocks. Although the first applications were for viewing multiphase fluid flow, qualitative and primarily visual characterization, such as examination for mud invasion, inclusion and heterogeneity, soon took precedence. This paper discusses two CT applications which yield quantitative 3D maps of physical attributes pertinent to the core analyst. A bulk density map is derived from dual-energy scans and saturation and/or porosity maps are derived from multiple scans of multi-phase flow. While these applications are more complex than the qualitative ones, careful scanning and appropriate software can get rewarding results. Step-by-step instructions are shown in this paper for applying these two features with appropriate examples and also common pitfalls with precautions and some remedies are discussed.

KEYWORDS: CT, tomography, dual energy, core characterization, saturation mapping

1. Introduction

The petroleum industry has used Computerized Tomography (CT) scanners to evaluate petrophysical and fluid flow properties of reservoir rocks in a non-destructive and cost-effective manner. CT helps identify lithology, measure density and porosity, evaluate formation fluid damage; and it allows viewing of multiphase flow in rocks. Detailed reviews of the applications of CT in rock and fluid research can be found in the literature (Vinegar and Wellington, 1987; Wellington and Vinegar, 1987; Kantzas, 1990; Akin and Kovscek, 2001). Overall, the petroleum industry applications of CT can be divided into two major areas: core characterization and fluid flow visualization. CT serves as a powerful tool for fluid flow visualization in cores and it is widely used in coreflooding tests to quantify multiphase flow behavior and to observe the effect of treatment fluids. In this paper we focus mainly on two aspects of CT-scanning: 1) dual-energy CT for core characterization and 2) saturation calculation during multiphase flow.

2. Dual energy CT for core characterization

In dual-energy CT-scanning, the object is scanned twice at the same location, using a high- and a low-energy setting. By selecting the high- and low-energy settings, one can take advantage of the two predominant x-ray interactions with matter namely, the photoelectric absorption (predominant at low-energies) and the Compton scattering (predominant at high-energies). These, in turn, have different dependence on effective atomic number (Z_{eff}) and electron density. It is possible to calculate the Z_{eff} and electron density of an object by scanning it at two x-ray energies with sufficient energy separation. Wellington and Vinegar (1987) suggested using 100 kV as the threshold value for the two effects. Since most medical scanners are limited to a maximum of 140 kV, the latter can be used as the high-energy setting for scanning rocks. Although many medical scanners have settings of 90, 80, 70 or even 60 kV as the lowest energy, it is very difficult to pass enough x-rays through a typical reservoir rock such as a carbonate at energies below 70 kV. The same is true for most microfocus CT-scanners. The following equation, modified from Wellington and Vinegar (1987), is most widely used in calculating dual-energy based true density (ρ) and atomic number (Z).

$$\mu = \rho \left[a + \frac{bZ^n}{E^{3.2}} \right] \qquad [1]$$

where μ is the linear attenuation coefficient (the parameter measured by a CT-scanner), E is the X-ray energy, a is the Klein-Nishina coefficient, n is an exponent for the Z (with different values used by researchers for n such as 3, 3.1, 3.6 and 3.8), and b is a constant. Equation [1] is highly non-linear and calculation of the energy-dependent coefficients a and b becomes uncertain due to the polychromatic nature of the x-ray beams and due to unknown effective energies at the high and low voltage settings. Vinegar and Kehl (1988), Coenen and Maas (1994), Siddiqui and Khamees (2004) suggest using techniques involving scanning "standards" of known ρ and Z_{eff} (applicable to compounds and mixtures) at the same conditions and their procedures yield true density of the unknown material and the Z_{eff} that can be used for identifying major minerals in core samples or rock fragments (cuttings). The dual-energy CT-scanning involves calculating, for a fixed (usually circular) region of interest (ROI) of each slice, the average μ (attenuation) or CTN (normalized attenuation, in Hounsfield Units, output of a medical CT-scanner), at two energies, which are then converted to true ρ and Z_{eff}. Software toolkits are now available that can make it convenient to interact with variables in the dual-energy equations for the two- or three-dimensional objects. PV-Wave and Matlab are examples of languages which can do this and in addition, they have the means of making a sophisticated graphical user interface (GUI) and they have extensive math libraries for computation. The code can readily be used on Windows, Linux, and UNIX operating systems. With this approach, CT-scan data analysis programs are developed that are portable, convenient to use and rapid to develop, with good performance. One such program is *VoxelCalc*, is used here to demonstrate the ease of processing dual-energy and saturation data.

2.1. *Scanning considerations-spatial repeatability*

Since dual-energy CT-scanning involves scanning the core at the same exact locations twice, each time with a different energy, and since most scanners are not capable of taking these two scans simultaneously, it is very important to have a scanning table (or a rotation stage, for micro CT) with good repeatable positioning accuracy. Also, in order to prevent displacement due to vibration during table (or rotor) movement, the sample itself or its holder (core holder) must be harnessed firmly for the scans. Additionally the table (or rotating stage) should have sufficient capacity to support heavy loads with minimum deflection.

2.2. *Scanner calibration for rock – beam hardening*

Calibration of the scanner is performed to reduce the artifact of "beam hardening", a cupping effect seen on objects due to preferential absorption of the lower energy X-rays from a polychromatic source. Figures 1a illustrates beam hardening, caused by inappropriate calibration, before (left) and after (center) post-scan adjustment. The corresponding profiles are also shown in Figure 1a (right). The surface renderings in Figure 1b show the same cupping effect before and after the adjustment. The software used applies a radial curve fit on the slices which generally works well if the objects within the slices are not too heterogeneous.

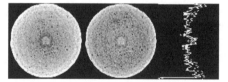 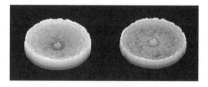

Figure 1a. *Example of beam hardening, before adjustment (left), after adjustment (center), and the corresponding profile images (right)*

Figure 1a. *Surface renderings of image with beam hardening artifact before (left) and after (right) adjustment*

Phantoms with densities and chemistry approximating rocks to be scanned are used to measure the amount of correction, or flattening applied to each slice. Typically, the industry uses fused quartz phantoms for sandstones and Macor phantoms for carbonates, rather than the medical water phantom. Calibration should be performed for each of the major mineral types, for each energy level, and then applied at scan time. Placing metal filters in front of the X-ray source, packing the sample in some special media, and using a core holder can reduce these artefacts; however, one must take care to assure good penetration through the sample. It should be noted that even with all the precautions beam hardening cannot be totally avoided. Post-scan adjustments by software can flatten the slices without a phantom image, but this pre-scan calibration is far better.

2.3. *Core preparation*

For dual-energy analysis, which is very helpful to understand the mineral contents and distribution in rock samples, cylindrical shape is preferred (to avoid the X artefacts, seen in Figure 2a).

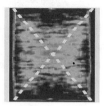 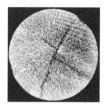

Figure 2a. *The X-artifact (after Akin and Kovscek, 2001)*

Figure 2b. *A spotty image due to insufficient X-ray energy*

If it is desired to measure the properties at reservoir condition (at higher temperature and pressure than ambient), then the core must be put inside a core holder, which is typically made of aluminum or aluminum wrapped with carbon fiber tapes. In general the core holder pre-hardens the beams but they can potentially reduce the penetration capability in the case of large or dense samples. For larger samples and for samples kept inside core holders, the ability of X-rays to take sharp images, especially at the low-energy setting must be tested. Figure 2b shows a CT slice for a 4" diameter sandstone core for which the X-ray energy was not enough, resulting in a spotty image. For these cases either choosing a smaller sample size if possible, or switching to a higher energy setting, while still staying within the domain of photoelectric dominance, is recommended.

2.4. Standard materials

For success of the dual-energy procedure, a good set of dual-energy "standards" must be used. These standards are homogeneous, solid materials (non-porous), for which the bulk density and Z_{eff} must be accurately known. The standards should preferably have the same diameter as the core sample. The exact bulk density can be easily calculated using the weight and bulk volume of each standard. For pure materials with known compositions Z_{eff} can be calculated (Siddiqui and Khamees, 2004). *VoxelCalc* lists several dual-energy standards that have been used successfully for analyzing reservoir cores. They are – calcite, dolomite, quartz (both crystalline and fused), NaCl, Teflon, Kel-F, aluminum 6061, water, doped water for which typical values of Z_{eff} and ρ are provided. There is an option to add up to three new standards and to edit the provided Z_{eff} and ρ data to incorporate data for the actual standards used. If the standard is not homogeneous and if it contains impurities, serious errors may result from it. Composition information can sometimes be verified by using X-ray diffraction and scanning electron microscopy techniques.

   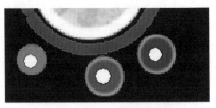

Figure 3a. *Scanning of dual energy standards*   **Figure 3b.** *Reading of CT numbers of dual energy standards*

Current industry practice is to scan these standards, which are typically in the form of ½" (1 cm) diameter cylinders (the liquid standards such as water and doped water are usually placed inside small cylinders) by placing them around the outside of the core holder (Figure 3a). This configuration can yield good results if extreme care is taken in applying small centered regions as indicated by the white areas in Figure 3b when extracting the mean CT values. The regions must be congruent for the high and low energy CT values obtained. An alternative way is to use short calibration standards (about ½" to 1" long, preferably all having the same length), taped together and scanned at the same conditions as the core sample (at the same energy and 'environmental' conditions, i.e., the same calibrations and during the same scanner power-up session).

2.5. Obtaining coefficients and solving for density and effective atomic number

The first step in conducting dual-energy data analysis is to generate the dual energy coefficients. The procedures used by Vinegar and Kehl (1988) and Siddiqui and Khamees (2004) both use six coefficients (three for Z_{eff}, three for ρ), which are calculated based on the attenuation data for the standards at the high- and low-energy settings. Although several standards are scanned, data from a minimum of three standards are needed to generate the six coefficients. There is no standard procedure for selecting three standards out of the six or seven scanned during the process. Ideally, any three standards that have properties similar to the minerals present in the core should give the best result. In reality, finding three such standards is quite difficult and sometimes an odd combination may end up giving better results. We suggest using a trial-and-error procedure in which the coefficients are used to predict the Z_{eff} and ρ for a homogeneous core plug (Berea and Fontainebleau sandstone, Indiana limestone, etc.) for which these data are known, before using the coefficients for finding these properties for the unknown samples. *VoxelCalc* offers the option to calculate the coefficients quite easily. CT values for standards are obtained by using the ROI tool which obtains the mean with a circle on a selected slice. This is done for both high and low energies, and the data are then entered into the calibration table for the appropriate material (Figure 4). At least three standards should be used; however, *VoxelCalc* allows data from more standards in which a regression-based method is used to determine the coefficients.

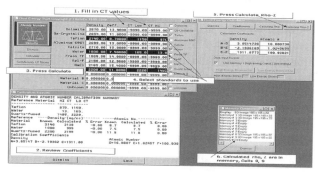

Figure 4. *Calculating and applying coefficients for determining Z_{eff} and ρ data*

3. Saturation calculation during multiphase flow

Visualizing saturation changes in core samples during a laboratory coreflooding test was almost unknown before the use of CT. Fluid flow visualization with CT almost always involves the use of one or more radiopaque tracers (also called dopants) containing high atomic number ions (e.g. I⁻, Br⁻, etc.). The dopants provide sufficient contrast between the various fluid phases which allows us to view and quantify saturation distribution and its changes, gravity and viscous effects, trapping and bypassing, effects of heterogeneity on flow, etc. The concentration of the dopant depends on the porosity of the core, typically a higher concentration is needed for cores with a lower porosity. For two phase (oil-water) flow typically the water phase is doped with NaI, KI, KBr, NaBr, etc. but for three phase flow, two of the phases are doped. Vinegar and Wellington (1987) provide details of the use of dopants.

3.1. Scanning considerations

All saturation monitoring experiments require the use of a core holder. Like the dual-energy case, having good positioning accuracy for the table and keeping the core holder fixed with the table are of utmost importance for accurate calculation of saturations. For slight movements, some post-scan alignment capability is always useful. Although the presence of core holder pre-hardens the x-rays, beam-hardening can still potentially affect the data, especially when the pore space is saturated with a high concentration of dopant. During scanning after each stage of coreflooding, the core plug or plugs should be always scanned in the same direction (inlet to outlet), with full coverage. The scanning speed should be fast enough to capture all the fluid related changes, instructing the scanner to avoid reconstructions generally speeds up image acquisition. Before or after the coreflooding, all calibration standards (typically plastic bottles containing oil and water with the exact concentration of dopants used for flowing through the core) must be scanned using the same operating parameters and under same environmental conditions. A typical two-phase coreflooding experiment can have several stages – pulling vacuum, saturating with doped water, flowing doped water through core to ensure 100% saturation (images from this stage are typically used as reference for the image subtraction calculations later), displacing water by oil (drainage) in order to bring the core to its irreducible water saturation (S_{wir}), displacing oil by water (imbibition) to bring it to its residual oil saturation (S_{or}), etc. The core is scanned at the end of each displacement stage and also at several intermediate stages. The images from scans at various stages of water saturation are used to calculate in-situ saturations.

3.2. *Image processing*

Although average attenuation data for slices from different scan sequences are sufficient to calculate in-situ saturations, the ability of some image processing programs such as *VoxelCalc* to conduct a pixel-by-pixel subtraction opens many possibilities. Matrix-subtracted images allow the observation of minor flow anomalies, which helps our understanding of multiphase flow, especially in heterogeneous porous media.

Figure 5a. *Wet scan slice*   **Figure 5b.** *Dry scan slice*   **Figure 5c.** *Difference slice*

Typical image processing workflow includes subtracting from the slices of the core when it is 100% saturated with doped brine (e.g. the wet scan slice in Figure 5a), the slices with the under vacuum (e.g. the dry scan slice in Figure 5b), which results in the matrix-subtracted slices (e.g. the dfference slice in Figure 5c). The latter can give a clear picture of fluid distribution in cores, separating the more-porous areas from less-porous areas, which are otherwise very hard to detect from simple slice images at varied stages. In addition to providing various options for array calculations, *VoxelCalc* also offers circular cropping, generation of horizontal and vertical slabs (Figures 6a and 6b) and saturation profiles (Figure 6c) and saving of the matrix-subtracted images as image stacks, which can then be brought into *VoxelCalc* or other advanced image processing programs to make animations of multiphase flow in porous media.

  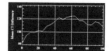

Figure 6a. *Horizontal slab*   **Figure 6b.** *Horizontal slab*   **Figure 6c.** *Saturation profile*

4. Conclusions

In this paper we have demonstrated the use of two applications – dual-energy CT-scanning and saturation calculations using CT with the help of appropriate image processing software. These data are very useful in petrophysical studies of reservoir rocks and small cuttings. Although examples shown are for medical scanners, the procedures can also be applied to micro CT-scanning. The ability to estimate the mineral content and to observe flow and distribution of fluids during coreflooding operations can enable better decisions about production and environmental remediation.

5. References

Akin, S. and Kovscek, A.R., "Use of Computerized Tomography in Petroleum Engineering Research," Annual Report of SUPRI TR 127, Stanford University, Stanford, CA, August 2001, pp. 63-83.

Coenen, J.G.C., Maas, J.G., "Material Classification by Dual-Energy Computerized X-ray Tomography," *Proceedings of the International Symposium on Computerized Tomography for Industrial Applications*, Berlin, Germany, 1994, pp. 120–127.

Kantzas, A, "Investigation of Physical Properties of Porous Rocks and Fluid Flow Phenomena in Porous Media Using Computer Assisted Tomography," *In Situ*, Vol. 14, No. 1, 1990, p. 77.

Siddiqui, S. and Khamees, A.A., "Dual-Energy CT-Scanning Applications in Rock Characterization," *SPE paper No. 90520*, presented at the 2004 SPE Annual Technical Conference and Exhibition held in Houston, Texas, 26-29 September 2004.

Vinegar, H.J. and Kehl Jr., R.P., *User Guide for Computer Tomography Color Graphics System – CATPIX*, Shell Bellaire Research Center, Houston, TX, 1988, 85 pp.

Vinegar, H.J. and Wellington, S.L., "Tomographic Imaging of Three-Phase Flow Experiments," *Rev. Sci. Instrum.*, January, 1987, p. 96.

Wellington, S.L. and Vinegar, H.J., "X-Ray Computerized Tomography," *Journal of Petroleum Technology*, August 1987, p. 885.

Trends in CT-Scanning of Reservoir Rocks

Medical CT to Micro CT

S. Siddiqui — M. R. H. Sarker

Petroleum Engineering Department
Texas Tech University
Box 43111, 8th and Canton Avenue
Lubbock, TX 79409
USA
shameem.siddiqui@ttu.edu
md-rakibul.sarker@ttu.edu

ABSTRACT. Converted medical x-ray CT-scanners have been used since 1984 for obtaining important rock and rock-fluid properties in a non-destructive manner for core characterization and fluid flow visualization. Although medical CT-scanners have seen constant improvement in speed, the spatial resolution has not improved from the 350-450 μm range. Typical pore sizes of reservoir rocks vary from less than a micron to a few microns and therefore, the oil industry has been looking for alternatives such as micro and nano CT-scanners. Micro CT-scanners have been used successfully to extract information from cuttings-sized rock fragments and micro CT-based research can potentially eliminate the need for cutting cores. This paper presents a state-of-the-art review of micro CT-based petrophysical research and describes some the techniques used for building realistic pore network models and obtaining various petrophysical properties.

KEYWORDS: CT-scanning, micro CT, cuttings, core characterization, flow visualization, LBM

1. Introduction

Computerized Tomography (CT), introduced in 1972, is a non-destructive medical imaging technique that uses x-ray technology and mathematical reconstruction algorithms to view a cross-sectional slice of an object. With the improvements in speed of data acquisition and resolution, CT-scanners found use in disciplines other than medical radiology and these disciplines include material inspection, material development and evaluation, groundwater hydrology, petroleum engineering, civil engineering and mechanical engineering. In particular, converted medical x-ray CT-scanners have provided the petroleum industry with many important rock and rock-fluid interaction properties in a non-destructive manner since the early 1980s.

The overall petrophysical applications of a CT-scanner can be divided into core characterization and fluid flow visualization. CT-scanners are excellent density/porosity tools. In addition to their qualitative use in identifying formation tops and lithology changes and assessing the condition of preserved whole cores and plugs, CT data are used for depth matching (by comparing with density logs), estimating mineralogy using dual-energy scanning (Wellington and Vinegar, 1987), evaluating heterogeneity, and evaluating fracture volumes – all of which are core characterization applications. As a flow visualization tool, CT has been used successfully to calculate saturations and saturation distributions during multiphase flow processes (Wellington and Vinegar, 1987, Withjack, 1988), to observe flow-related heterogeneities such as viscous fingering, gravity segregation, etc. (Siddiqui et al., 1996), to evaluate pore volume compressibility, to observe wormhole development during acid flow (Bazin and Abdulahad, 1999) or blocking of acid flow by foam. Reviews of the various applications of CT-scanning can be found in the literature (Wellington and Vinegar, 1987, Kantzas, 1990; Akin and Kovscek, 2001 and Withjack et al., 2003).

While medical CT-scanners are still being used heavily in petroleum engineering research, some operational difficulties in obtaining cores, especially from deviated and horizontal wells, are forcing researchers to use micro CT scanners for understanding the underlying conditions that dictate flow in porous media by analyzing rock fragments (or drilled cuttings) – which are automatically generated during the drilling process. Siddiqui et al. (2005) used x-ray CT, x-ray diffraction (XRD), Environmental Scanning Electron Microscopy (ESEM), micro CT, Nuclear Magnetic Resonance (NMR), APEX mercury porosimetry and thin section petrography on cuttings of various sizes generated from a single core plug for which conventional core analysis data such as density, porosity and permeability data were available as a reference. They concluded that below the critical size of 2.5 mm (U.S. mesh size 8), medical CT-scanners fail to provide reliable density and porosity data for cuttings.

Micro CT-scanners, which belong to the industrial CT-scanner group, have been in existence since the early 1980s (Elliott and Dover, 1982). They are used mostly in materials evaluation and inspection and have typical resolutions in the 1-50 micron range. Microtomography, like tomography, uses x-rays to create cross-sections of a 3D-object that later can be used to recreate virtual models non-destructively. Micro CT-scanners are usually smaller in design compared to the medical version and are ideally suited for scanning smaller objects.

2. Petroleum engineering applications of micro CT-scanning

Jasti *et al.* (1993) were among the first to use a micro CT for petroleum engineering applications using cone-shaped beams. They scanned bead pack, Berea sandstone and chalk samples at a very high resolution and recommended the use of microfocus CT to resolve 3D microstructures present in reservoir rocks and also to characterize flow of multiphase fluids in porous media. Medical CT-scanners, in general, do not have the actual ability to directly observe the pores present within reservoir rocks. Their large focal spots cause "unsharpness" (due to penumbra effects) and therefore, the actual dimensions of features below the scanner's resolution cannot be ascertained from the medical CT images. For instance, a fine fracture would show as a broad fuzzy line and a small vug can show as a large cavity. The bulk density is usually calculated for a large region-of-interest (ROI) within a core slice by comparing its average CT number with that for a known "standard", and the density data are then converted to porosity data using appropriate grain and fluid density information. Image subtraction technique can also be used for the direct calculation of porosity from medical CT slices of saturated and dry core plugs (Alvarado *et al.*, 2003).

Micro CT-scanners, which typically have a resolution close to 1 μm, have the ability to "see" the pores in most reservoir rock samples and to distinguish between pore throats, pore bodies and rock grains. The information can then be used for detailed evaluation, modeling and other applications related to multiphase flow dynamics in porous media. In the last few years more and more effort has been put into micro CT-based research using moderate (< 50 μm) to high-resolution (≈ 1 μm) scanners.

Although the resolutions of most commercial micro CT-scanners are sufficient for samples from most sandstone reservoirs, there are very few scanners with resolutions sufficient for sub-micron size pores that are abundant in carbonates. Okabe (2004) proposed to reconstruct image features that are in the sub-micron range by using geostatistical techniques such as multipoint statistics. For these capturing features in the sub-micron range, it may also be worthwhile to use the newly emerging nano-tomography. There is always a tradeoff between resolution and sample size. A higher resolution requires the sample size to be very small. For obtaining one-micron resolution typically the sample has to be about 1 mm in size. Obtaining good images at the one-micron scale can be a challenge. Latham *et al.* (2008) developed a registration technique to combine low-resolution micro CT data with high resolution SEM image data that can greatly improve the segmentation process used for pore network generation. Recently Grader *et al.* (2009) conducted micro-CT based core panoscopy, or multi-scale imaging experiments and remarked that in heterogeneous rocks such as carbonates, proper digital rock physics may require even five or six registered resolutions to define mechanical and transport properties of core samples.

3. Typical workflow for determining petrophysical properties using micro CT

The first step in the determination of petrophysical properties from rock fragments is to acquire micro CT images. The raw images from micro CT-scanners may contain different image artifacts. They include the aliasing artifact (or streaks), the partial volume artifact, the ring artifact and the beam-hardening artifact. Several pre-processing and post-processing routines are available to minimize or eliminate these artifacts. After the artifacts are removed, the images need to go through some additional image processing before pore networks can be extracted. These generally include 3D cropping, and segmentation using either simple thresholding or indicator kriging (Oh and Lindquist, 1999). Some "cleaning" algorithms are available for removing isolated phase blobs. Segmentation is required to separate the rock and pore phases from the CT-scan images. Pore networks are generally extracted using either the medial axis (Lindquist *et al.*, 1996) or the maximal ball algorithm (Silin *et al.*, 2004). Figure 1 is a typical workflow diagram for using micro CT data for calculating important reservoir properties such as porosity, absolute and relative permeability, formation factor, capillary pressure, etc. through pore network modeling.

Porosity is calculated by dividing the total number of voxels representing pores by the total number of voxels representing the sample. Porosity calculation and pore network extraction is very subjective and depends on the threshold values used. Absolute permeability is generally determined from the image by solving Stokes' Law in discretized domain obtained readily from the digitized image. The Lattice-Boltzmann Method (LBM) is commonly applied as a solution technique (Auzerais *et al.*, 1996). Arns *et al.* (2001) calculated formation factor directly from microtomographic images of Fontainebleau sandstone. They selected Fontainebleau sandstone as it is homogenous, made up of a single mineral (quartz), does not contain clay and only displays irregular porosity. They assigned a conductivity value of 0 to the matrix phase of the sandstone and a normalized conductivity value of 1 to the pore phase. The conductivity calculation is based on a solution of the Laplace equation with charge conservation boundary conditions. To compute relative permeability from CT images, Auzerais *et al.* (1996) used a simulation technique based on a 3D immiscible lattice-gas model and they considered a two-fluid model where one fluid perfectly wets the solid boundaries and the viscosities and densities in each fluid are equal. They obtained a very good match for the end-point relative permeability values but did not get good results at other saturations. Arns *et al.* (2003) calculated relative permeability using the theory as given by Øren *et al.* (1998). Both Auzerais *et al.* and Arns *et al.* (2003) concluded that producing meaningful multiphase transport parameters would require considerably larger networks. To calculate capillary pressure curves from micro-CT images, Jin *et al.* (2007) assumed that the rock is water-wet and the wetting phase (water) occupies the corners of the large pores and small pores while the non-wetting phase (oil) occupies the central parts of the invaded pores. To assign spatial fluid distribution in the pore space, they implemented a simple percolation algorithm. They calculated fluid saturations from the calculation of the volumes occupied by each fluid phase at a given capillary pressure and then plotted the capillary pressure curve.

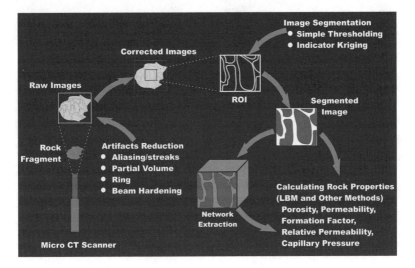

Figure 1. *Typical workflow used in micro CT-based pore network modeling*

4. Results for carbonate reservoir data

High-resolution micro CT data from a small rock fragment from a carbonate reservoir was used to generate a segmented image shown in Figure 2. The image shown represents a volume of about 0.6 mm × 0.6 mm × 0.6 mm (length of each voxel is about 0.2 μm in a 256^3 matrix) and the porosity is about 17%. The 3DMA Rock software was used to generate cleaned segmented image (Figure 3), medial axis (Figure 4) and pore throat distribution data (Figure 5). Lattice-Boltzmann simulation was then conducted to generate permeability data in three orthogonal directions and the values were 358, 642 and 358 mD, respectively. Although data were available for 512^3 and 1024^3 on the same rock fragment, the Lattice-Boltzmann simulation could not be run with those because of hardware and software limitations. Faster processors and parallel processing codes may improve the situation, especially for determining the multiphase flow parameters such as relative permeability and capillary pressure, which require more computational power.

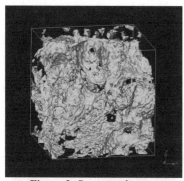

Figure 2. *Segmented image*

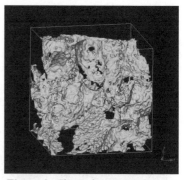

Figure 3. *Cleaned segmented image*

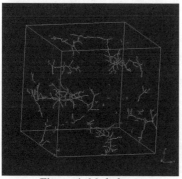

Figure 4. *Medial axis*

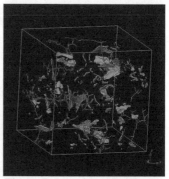

Figure 5. *Pore throat distribution*

5. Discussion

In recent years, mainly due to the interest of the oil industry in eliminating coring and obtaining most of the static and dynamic petrophysical parameters from drilled cuttings, lot of research efforts have been directed towards micro CT-based pore network modeling. Significant progress has been achieved in network extraction and petrophysical property prediction. There has been limited success so far in validating the pore network-based parameters, especially those related to multiphase flow and there is a need for conducting more research in this area. Having better models for predicting various rock related parameters from small rock fragments and having the ability to solve simulation problems for larger networks are two more important areas of research. Ultimately upscaling of all the pore network model generated data will become another focus area for research.

6. Acknowledgments

The authors acknowledge the High Performance Computing Center (HPCC), Texas Tech University for providing its Grendel cluster resources for this study. The authors also acknowledge Dr. W.B. Lindquist (SUNYSB), Dr. R. Vadapalli (TTU) and Mr. S. Addepalli (TTU) for their help with the computational part of this work.

7. References

Akin, S. and Kovscek, A.R., "Use of Computerized Tomography in Petroleum Engineering Research", *Annual Report of SUPRI TR 127*, Stanford University, Stanford, CA, August 2001, pp. 63-83.

Alvarado, F., Grader, A.S., Karacan, O. and Halleck, P.M., "Visualization of Three Phases in Porous Media Using Micro Computed Tomography", *SCA 2003-21*, presented at the International Symposium of the Society of Core Analysts held in Pau, France, September 21-24, 2003.

Arns C.H., Knackstedt, M.A., Pinczewski, W.V., Lindquist, W.B., "Accurate Estimation of Transport Properties from Microtomographic Images", *Geophysical Research Letters*, Vol. 28, 2001, pp. 3361-3364.

Arns, J-Y., Arns, C.H., Sheppard, A.P., Sok, R.M., Knackstedt, M.A. and Pinczewski, W.V., "Relative Permeability from Tomographic Images; Effect of Correlated Heterogeneity", *Journal of Petroleum Science and Engineering*, Vol. 39, 2003, pp. 247-259.

Auzerais, F.M., Dunsmuir, J., Ferreol, B.B., Martys, N., Olsen, J., Ramakrishnan, T.S., Rothman, D.H., Schwartz, L.M., "Transport in Sandstone, A Study Based on Three Dimensional Microtomography", *Geophysical Research Letters*, Vol 23, No 7, April 1996, pp. 705-708.

Bazin, B. and Abdulahad, G., "Experimental investigation of some properties of emulsified acid systems for stimulation of carbonate formations", *SPE 53237* presented at the SPE Middle East Oil Show held in Bahrain, February20–23, 1999.

Elliott, J.C. and Dover, S.D., "X-ray microtomography", *Journal of Microscopy*, Vol. 126, 1982, pp. 211-213.

Grader, A.S., Clark, A.B.S., Al-Dayyani, T. and Nur, A., "Computations of Porosity and Permeability of Sparic Carbonate Using Multi-Scale CT Images", *SCA 2009-31*, presented at the *International Symposium of the Society of Core Analysts* held in Noordwijk, The Netherlands, September 27-30, 2009.

Jin, G., Torres-Verdın, C., Radaelli, F., Rossi E., "Experimental Validation of Pore-Level Calculations of Static and Dynamic Petrophysical Properties of Clastic Rocks", *SPE paper No. 109547*, presented at the *ATCE 2007* Technical program held in Anaheim, CA, 11 – 14 November, 2007.

Jasti, J.K., Jesion, G. and Feldkamp, L., "Microscopic Imaging of Porous Media with X-ray Computer Tomography", *SPE Formation Evaluation*, September, 1993, pp.

Kantzas, A, "Investigation of Physical Properties of Porous Rocks and Fluid Flow Phenomena in Porous Media Using Computer Assisted Tomography", *In Situ*, Vol. 14, No. 1, 1990, p. 77.

Latham, S., Varslot, T. and Sheppard, A., "Image Registration, Enhancing and Calibrating X-Ray Micro-CT Imaging", *SCA 2008-35*, presented at the International Symposium of the Society of Core Analysts held in Abu Dhabi, UAE, October 29-November 2, 2008.

Lindquist, W.B., Lee, S.-M., Coker, D.A., Jones, J.W. and Spanne, P., "Medial Axis Analysis of Void Structure in Three-Dimensional Tomographic Images of Porous Media", *Journal of Geophysical Research*, 101B, (1996), pp. 8297-8310.

Oh, W. and Lindquist, W.B., "Image Thresholding by Indicator Kriging", *IEEE Transactions on Pattern Analysis and Machine Intelligence*, Vol. 21, (1999), pp. 590--602.

Okabe, H., Pore Scale Modelling of Carbonates, PhD dissertation, Imperial College, London, UK, June 2004.

Øren, P.E., Bakke, S. and Arntzen, O.J., "Extending predictive capabilities to network models", *Society of Petroleum Engineers Journal*, Volume 3, Number 4, December 1998, 324-336 pp.

Siddiqui, S., Hicks, P.J. and Grader, A.S., "Verification of Buckley-Leverett Three-Phase Theory Using Computerized Tomography", *Journal of Petroleum Science and Engineering*, Vol. 15, 1996, pp. 1-21.

Siddiqui, S., Grader, A.S., Touati, M., Loermans, A.M. and Funk, J.J., "Techniques for Extracting Reliable Density and Porosity Data from Cuttings", *SPE paper No. 96918*, presented at the 2005 *SPE Annual Technical Conference and Exhibition* held in Dallas, Texas, 9-12 October 2005.

Silin, D.B., Jin, G., Patzek, T.W., "Robust determination of the pore-space morphology in sedimentary rocks", *Journal of Petroleum Technology*, Vol. 56, No. 5, 2004, pp. 69-70.

Wellington, S.L. and Vinegar, H.J., "X-Ray Computerized Tomography," *Journal of Petroleum Technology*, August, 1987, p. 885.

Withjack, E.M., "Computed Tomography for Rock-Property Determination and Fluid-Flow Visualization", *SPE Formation Evaluation*, December, 1988, p.696.

Withjack, E.M., Devier, C. and Michael, G., "The Role of X-Ray Computed Tomography in Core Analysis", *SPE 83467* presented at the *SPE Western Regional/AAPG Pacific Section Joint Meeting* held in Long Beach, California, May 19-24, 2003.

3D Microanalysis of Geological Samples with High-Resolution Computed Tomography

G. Zacher* – J. Santillan ** – O. Brunke* – T. Mayer*

* GE Sensing & Inspection Technologies GmbH, phoenix|x-ray
Niels-Bohr-Str. 7
D - 31515 Wunstorf
Germany
Gerhard.Zacher@ge.com
Oliver.Brunke@ge.com
Thomas.Mayer2@ge.com

**phoenix|x-ray Systems, Inc., part of GE Sensing and Inspection Technologies
7007 Gateway Blvd.
Newark, CA 94560
USA
Javier.Santillan@ge.com

ABSTRACT: *During the last decade, Computed Tomography (CT) has progressed to higher resolution and faster reconstruction of the 3D-volume. More recently it has even allowed a three-dimensional look inside geological samples with submicron resolution. In recent years major steps in key hardware components like open microfocus or even nano-focus x-ray tube technology on the one side and the development of highly efficient and large flat panel detectors on the other, enabled the development of very versatile commercially available, high resolution laboratory CT systems. Electromagnetic focusing of the electron beam allows the generation of x-ray beams with an emission spot diameter smaller than 1 μm. The paper will showcase several geological applications performed with the nanotom, the first 180 kV nanofocus CT system with exceptional resolution less than 500 nm. The machine is capable of both high resolution nanofocus scans and high power mode scans (up to 15 W at the target), which can examine high-absorbing samples.*

KEYWORDS: *high resolution x-ray computed tomography, 3D micro-analysis, quantitative evaluation*

1. Introduction

High resolution CT is nowadays a well established method for numerous industrial applications [Roth *et al.* 2003; Moller-Gunderson 2007; Nier *et al.* 2003] as well as for a wide range of research areas [Bonse 2004 and 2006]. During the last decade, Computed Tomography (CT) has progressed to higher resolutions and faster reconstructions of the 3D-volume. More recently it has even allowed a three-dimensional look into the inside of geological samples with submicron resolution.

CT for geological purposes can lead to a new dimension of understanding of the distribution of rock properties. In particular, spatial distribution of pores and pore-connections as well as cementation properties are of utmost importance in the evaluation of reservoir properties. The possibility to visualize a whole plug volume in a non-destructive way and to use the same plug for further analysis is undoubtedly currently the most valuable feature of this new type of rock analysis and will be a new area for routine application of x-ray CT in the near future.

The paper will outline the hard- and software requirements for high resolution CT. It will showcase several geological applications which were performed with the nanotom, the first 180 kV nanofocus CT system tailored specifically for highest-resolution scans of samples up to 120 mm in diameter and weighing up to 1 kg with voxel-resolutions down to <500 nm (<0.5 microns).

2. High resolution CT

For many years, the only way to determine the interior structure of a sample with resolution in the sub-micron range was to section the part. This technique was not only time-consuming, but in this destructive process a valuable sample was lost. With advances in x-ray technology, however, this is no longer necessary. In many fields like biology, geology or engineering, CT with nanofocus x-ray sources allows the researcher to explore a sample's structure at the sub-micron level. In recent years major steps in important hardware components like open microfocus or even nanofocus x-ray tube technology (the later was commercially introduced the first time by phoenix|x-ray in 2001) on the one side and the development of highly efficient and large flat panel detectors (by e.g. GE, Perkin-Elmer, Varian or Hama-matsu) using CCD or CMOS technology on the other, have allowed the development of very versatile and high resolution laboratory CT systems like the nanotom (see next section) which are commercially available. Electromagnetic focusing of the electron beam allows the generation of x-ray beams with an emission spot diameter down to well below 1 µm which is essential for CT examination with voxels-sizes in the sub-micron range. These characteristics with respect to spatial resolution principally allow CT measurements which valuably complement many absorption contrast setups at synchrotron radiation facilities [Withers 2007; Brunke *et al.* 2008].

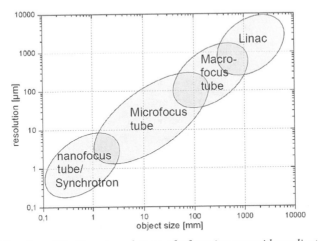

Figure 1. *Microfocus CT offers at resolutions of a few microns a wide application field for common 3D analyses. By use of nanofocus tube technology, CT systems are pushing forward into application fields that have been exclusive to expensive synchrotron techniques so far*

3. Possible resolution and detail detectability

The requested resolution from the worldwide market increases year by year. Applications are in the fields such as materials science, micro mechanics, electronics and geology looking for the inner structures in the micron or even sub-micron range.

In order to cover the widest possible range of samples the CT system has to be equipped with an x-ray tube, manipulation stage and detector which allows in the sum a detail detectability in the sub-micrometer range. Therefore the nanotom is equipped with a 180 kV/15 W x-ray tube with adjustable spot size < 0.9 μm. Due to the penumbra effect the spot size predominates the image sharpness for extreme magnifications (for details see e.g. Brockdorf *et al.*2008).

In Figure 2(a) the resolution capability of this high power nanofocus (HPNF) source is demonstrated. It shows that the 0.6 μm structure (line width) of the so called JIMA test pattern (Japan Inspection Instruments Manufacturers' Association) can clearly be resolved. Figure 2(b) shows that for isolated structures of high absorbing material on a low absorbing substrate it is even possible to detect details of 0.5 μm size and below. The limit for this so called detail detectability for the HPNF tube lies at about 200 – 300 nm.

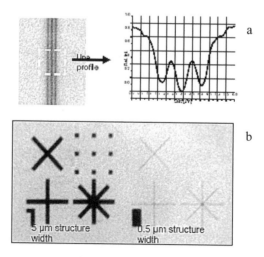

Figure 2. *X-ray images of test patterns showing the capabilities of resolution and detail detectability of phoenix|x-ray's high power nanofocus tube. In (a) the 0.6 μm line pair structure of the JIMA test pattern is clearly resolved. (b) shows the capability of the tube to resolve structures which are as small as 0.5 μm and below.*

On the other hand the x-ray tube generates up to 15 W at the target and enables the penetration of samples like copper, steel or tin alloys thus allowing the analysis of electronic devices or high absorbing geological samples etc.

The manipulation stage is build from granite stone. Together with a high precision rotation unit samples can be moved and rotated with highest precision. On the detection side a 5-megapixel flat panel CMOS detector with a GOS scintillator deposited on a fiber optic plate is used. The pixel size of 50 μm and a 3-position virtual detector (i.e. 360 mm detector width) give rise to a wide variety of experimental possibilities.

4. Results

4.1. *Oolithic carbonate: pore analysis*

The first example shows an oolithic carbonate (sample diameter Ø 5 mm) scanned with 6 μm voxel size to extract information about the pore network (size of pores, connectivity, size of restrictions etc.). In Figure 3b is shown the quantitative pore analysis. Light colors are small pores whereas in black actually many interconnected pores are interpreted as one large pore. Analyzing the volume in quantitative manner yields extremely valuable information for the petrologists.

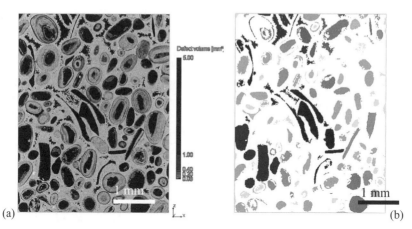

Figure 3 a+b. *Tomographic slice (a) of an oolithic carbonate (Ø 5 mm). Pore analysis (b) give quantitative distribution of pore sizes (different gray levels)*

4.2. Pyroclastic rock: pore network and surface extraction

As a next example a very porous pyroclastic rock (Ø 15 mm) from Etna (Sicily) has been examined at a resolution of 15 μm (see Figure 4) showing the possibility to study the spatial structure of the pore network. The resulting volume data can be used in general to produce extracted surface data for any CAD application and furthermore for FEM modeling for hydrogeological purposes.

Figure 4a+b. *3D images of the reconstructed volume of a pyroclastic rock (view width: 13.5 mm). White areas indicate the pore network relevant for permeability (a), black are closed pores. In the right image (b) is shown the extracted surface as polygon model (black and gray means front and back surfaces).*

4.3. *Shell limestone: micro fossils*

The last examples (Figures 5 and 6) are limestone samples called "tuffeau" largely used to build most of castles, cathedrals, churches and houses along the Loire valley (France). Due to weathering these buildings are often altered and eventually destroyed. Therefore it is important to understand the weathering mechanisms of building stones, i.e. to relate the microscopic mechanisms occurring at the pore scale (dissolution of minerals, transport, precipitation, etc.) to their consequences at the macro-scale (desquamation, powdering, etc.). Such a goal can be achieved with x-ray CT image analysis and numerical simulation of water flows within a realistic 3D image.

Samples were scanned with 1.2 μm voxel size and Figures 5 and 6 show impressively that such an investigation at pore scale is possible with no doubt and that evaluation of tomographic slices enables modeling at micron range.

Figure 5. *Virtual cut through the 3D reconstructed volume of a shell limestone sample. The pore network is faded out and color coding is used to visualize different rock densities*

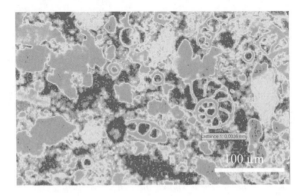

Figure 6. *Zoom into a tomographic slice of a shell limestone sample to measure the wall thickness (3.6 μm) of a small ammonite*

5. Summary

Since density transitions usually indicate boundaries between materials or phases, CT data is usually intuitive for geoscience professionals to evaluate. Due to the digital form, 3D data can be used for quantitative analysis as well as for a variety of measurement and visualization tasks.

Powerful software enables rapid reconstruction and visualization of the volume data allowing the user to extract and view internal features and arbitrary sectional views. The nanotom is the first nanoCT system featuring voxel resolutions of less than 500 nanometers (< 0.5 μm). The nanotom's ability to deliver ultra high-resolution images of any internal object detail at virtually any angle caters to even the most complex geological and petrological applications.

The nanotom was designed from the ground up with the primary goal of meeting the unique needs of high-resolution computed tomography and comes standard with a 180 kV high-performance nanofocus tube, 5-Megapixel digital detector, and 24 inch flat panel display. The 180 kV high power nanofocus tube enables the inspection of even the highest absorption materials such as metal, while the 5-Megapixel digital detector and a 3-position virtual detector enlargement enable the highest possible resolutions.

Today's high-resolution x-ray CT with its powerful tubes and great detail detectability lends itself naturally to geological and petrological applications. Those include the interior examination and textural analysis of rocks and their permeability and porosity, the study of oil occurrences in reservoir lithologies, and the analysis of morphology and density distribution in sediments – to name only a few.

6. Acknowledgments

We thank E. Rosenberg, IFP, France and O.Rozenbaum, ISTO Orleans France.

7. References

Bonse, U. (Editor.), "Developments in X-Ray Tomography IV", *SPIE*, Wellingham, (2004).

Bonse U., (Editor.), "Developments in X-Ray Tomography V", *SPIE*, Wellingham, (2006).

Brockdorf, K. *et al.*, "Sub-micron CT: visualization of internal structures" in *Developments in X-ray Tomography VI*, edited by Stuart Stock, *Proceedings of SPIE,* Vol. 7078, (2008).

Oliver Brunke, Kathleen Brockdorf, Susanne Drews, Bert Müller, Tilman Donath, Julia Herzen and Felix Beckmann: "Comparison between x-ray tube-based and synchrotron radiation-based μCT" in *Developments in X-Ray Tomography VI*, edited by Stuart R. Stock, *Proceedings of SPIE,* Vol. 7078, (2008).

Nier, E., Roth, H., "Analysis of Crimp Interconnections by Microfocus Computed Tomography", *QZ*, 9, 916-918, (2003).

Moller – Gunderson, D., "When 2D X-ray isn't enough", *SMT*, 8, (2007).

Roth, H. Mazuik B., "What you can't see can hurt you", *Quality Test and Inspection*, 5,(2003).

Withers, P., "X-ray nanotomography", *Materials Today*, 10(12), 26-34 (2007)

Combination of Laboratory Micro-CT and Micro-XRF on Geological Objects

M. N. Boone* — **J. Dewanckele**** — **V. Cnudde**** —
G. Silversmit*** — **L. Van Hoorebeke*** — **L. Vincze***** —
P. Jacobs**

*UGCT – Ghent University, Dept. Subatomic and Radiation Physics
Proeftuinstraat 86
9000 Gent, Belgium
Matthieu.Boone@UGent.be
Luc.VanHoorebeke@UGent.be

**UGCT – Ghent University, Dept. Geology and Soil Science
Krijgslaan 281/S8
9000 Gent, Belgium
Jan.Dewanckele@UGent.be
Veerle.Cnudde@UGent.be
Patric.Jacobs@UGent.be

***XMI – Ghent University, Dept. Analytical Chemistry
Krijgslaan 281/S12
9000 Gent, Belgium
Geert.Silversmit@UGent.be
Laszlo.Vincze@UGent.be

ABSTRACT: *Laboratory micro-CT scanning is a very useful tool in the characterization of geological samples. Due to its ability to visualize different phases and structures in 3D, many characteristics can be derived from the data. However, very limited chemical information on the different phases can be derived by CT. Micro-XRF (μXRF) images this compositional information for a wide range of samples. In μXRF, a 2D grid of fluorescence spectra is collected from the surface of the sample, generating 2D maps of the elemental composition of this surface. By extrapolating the gray values of the μCT data, this information is eventually known in the whole 3D structure.*

KEYWORDS: *μCT scanning, x-ray fluorescence, element mapping, granite*

1. Introduction

Although x-ray micro-CT is a very powerful tool for structural analysis of geological samples, it gives very little information on the chemical composition of the different structures visible in the sample. Since the attenuation coefficient strongly depends on the atomic number of the elements composing the material, an educated guess can be obtained for simple samples. For more complicated samples such as granite or very low concentrations, a more detailed analysis is necessary.

X-ray fluorescence (XRF) captures this information from the surface of a sample. In micro-XRF, the surface is scanned through a focused x-ray beam, and a fluorescence spectrum is measured for each point on a 2D grid. The intensity of the different fluorescence peaks is related to the abundance of the corresponding chemical elements. These peak intensities can then be visualized as elemental maps, showing the relative abundance of each element on the surface.

Although this is similar to what can be done in SEM-EDX, μXRF has some advantages over this technique such as simple sample preparation (Nicolosi *et al.,* 1998), lower detection limits and a less stringent limitation on the size of the sample (Newbury and Davis, 2009). The higher penetration depth of x-ray compared to electrons can be seen both as an advantage and a drawback.

When this information is combined with μCT data, it becomes possible to fully identify the different phases seen in the 3D structure.

2. Experimental setup

To obtain this chemical information, different approaches can be used. For this paper we have focused on XRF-based measurements, as this technique is the most versatile towards sample size, sample type and field of view.

Two different experimental setups were used, each with its advantages and limitations. In the first, the sample under investigation was scanned at the UGCT (Ghent University Center for X-ray Tomography, http://www.ugct.ugent.be) micro-CT scanners, and afterwards elemental surface maps were recorded with an EDAX Eagle III laboratory μXRF instrument at the XMI (X-ray Microspectroscopy and Imaging, http://www.xmi.ugent.be) research group. This combination offers high accuracy and reliable results, but it requires two separate measurements, implying long measurement times and manual alignment of the resulting data. In a second, more experimental setup, a Canberra X-PIPS™ spectroscopy detector is added to the UGCT μCT scanner, recording XRF spectra from the sample while illuminated for the CT projections. This parallel measurement is less time-consuming, but since the complete sample is illuminated, it has a very low spatial resolution and can only be used to obtain global information on the sample.

Three different samples were used for this research. A Precambrian granite from China (Yellow Rock) was used for identification of the different phases, seen both on the CT data and by visual inspection. A limestone from Maastricht (Netherlands) treated with a silicon-based consolidant at one side was analyzed by µXRF to determine the penetration depth of the consolidant, based on the silicon concentration in the rock. A third sample, a volcanic rock, was analyzed only at the CT scanner, where global information on the composition was obtained with the X-PIPS™.

3. Results

3.1. *Yellow rock granite*

A granite rock consists mainly of quartz, Na-Ca feldspar and K-Na feldspar, with a minor contribution of ferromagnesian phases and some other trace phases. This variety of mineral phases results in a typical heterogeneous look. Visually these phases are hard to distinguish exactly, especially in 3D, because of the optical transparency of some phases. In the 3D rendering of the CT data, the phases show sharp boundaries. The comparison between both can be seen in Figure 1. Except for the high-density features in the middle of the object, there is little resemblance between both.

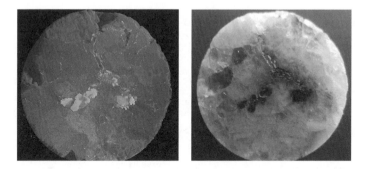

Figure 1. *Comparison between 3D rendering (left) and visual image (right) of the analyzed surface. Total diameter = 8 mm*

µXRF measurements (Figure 2) resulted in high silicon concentrations in most of the sample, but different concentrations can be easily distinguished on this map. Note in this respect the very low silicon count rate in the centre of the sample. The elemental map for aluminum is almost binary, showing regions with and without aluminum. A similar result is found for potassium. For the high density region, different concentrations of mainly iron, manganese and titanium can be found.

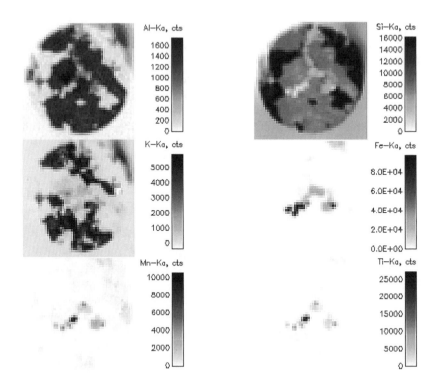

Figure 2. *Element maps for Al, Si, K, Fe, Mn and Ti for the granite surface shown in Figure 1 (map dimensions: 51×200µm by 51×200µm)*

With this information, one can identify the different minerals. The main mineral phases of this granite are biotite ($M(Mg,Fe)_3(Al,Si)_4O_{10}(OH)_2$), magnetite ($Fe_3O_4$), quartz ($SiO_2$), orthoclase ($KAlSi_3O_8$) and albite ($NaAlSi_3O_8$). Linear attenuation coefficients of these minerals show very similar values for albite and quartz (Figure 3), which explains the identical gray values in the CT data.

Due to the high penetration depth of x-rays, XRF radiation is created in a large part of the sample, especially compared to SEM-EDX. However, the thickness of the measured surface layer is limited by the absorption of the exiting radiation by the material. This can be self-absorption for a homogeneous surface layer, or absorption by a different material. In Figure 2, this latter can be expected around the very intense hotspots of the Fe mapping, where a large spot of lower intensity is seen. This is assumed to be magnetite covered by a thin layer of albite. In Figure 1, this covering can also be seen visually. Based on data extracted from the NIST XCOM photon cross sections database, it can be calculated that the measured Fe-Kα peak intensity (at 6.4 keV) of magnetite covered with 37 µm of albite is about 50% of the measured peak intensity for magnetite at the surface. Although this is a greatly simplified calculation, which ignores all effects such as the focusing of the x-ray beam, it is a good estimate of the thickness of the measured surface layer.

Granite

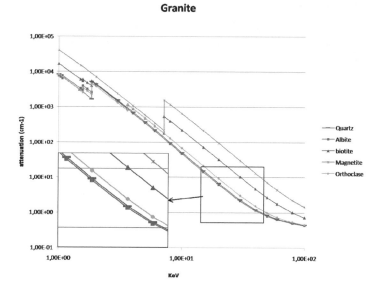

Figure 3. *Linear attenuation coefficients of different phases in granite. Region of interest for the CT scans is 30-80 keV*

3.2. *Maastricht limestone*

The Maastricht limestone is a very porous limestone with high concentration of calcite. A block of this material ($50 \times 50 \times 50$ mm^3) was treated with a silicon-based consolidant at one side. The penetration depth of this consolidant material was investigated, as this determines also the quality of the process (Cnudde *et al.*, 2004). To obtain higher CT resolution on this sample, a smaller subsample ($8 \times 8 \times 50$ mm^3) was taken for the analysis. Two CT scans were taken for achieving the highest resolution, one at the treated side of the sample and one at the un-treated side. A difference in porosity is expected between both sides, as pores are filled up by consolidant. A µXRF surface mapping (151×21 pixels, pixel size 200×200µm^2, resulting in an area of approx. 30×4 mm^2) was measured.

As expected from the $CaCO_3$ bulk composition, the surface map showed very high Ca abundances over the full sample. A much higher Si intensity can be seen at the treated side than inside the sample (Figure 4), showing a consolidant penetration depth of about 15mm.

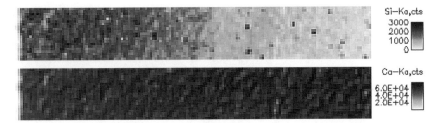

Figure 4. *Si and Ca elemental maps on the treated Maastricht limestone.*
The treated surface is at the left-hand side. Total area approx. $30 \times 4\ mm^2$

Analysis of the CT data using Morpho+ (Vlassenbroeck *et al.*, 2007) showed indeed a decrease in porosity between the two ends. In the scan of the treated surface, an average porosity of 42.5% was measured, to be compared with 46.2% on the other side. Although these porosities are no absolute figures due to the thresholding, they can be compared since scanning conditions and method of analysis were exactly the same. The consolidant can also be observed on the CT slices as regions of lower density between the grains of the limestone.

The same method was applied on a Bray sandstone. However, since this stone is almost pure quartz, the relatively small consolidant signal could not be distinguished from the intense SiO_2 matrix contribution. The porosity measurements reveal a similar effect, with 18.2% porosity at the bulk and 15.1% at the treated surface, indicating the presence of consolidant.

3.3. *Volcanic rock sample*

A series of scoria and pumice, coming from the area west of the Lac Pavin (lake in Auvergne, France) were scanned and analyzed at UGCT (Vandeputte, 2009). Due to their irregular shape, obtaining a 2D surface map was impossible. An experimental method was applied to these samples. During the CT scan, several XRF spectra were measured using the Canberra X-PIPS[TM] detector. Since the sample is rotating in a uniform way, angular information on the composition was obtained.

Two great advantages of this technique are the non-destructive nature, and the timesaving caused by the simultaneous measurement. However, the accuracy of the technique is very low. This is mainly due to the total illumination of the sample, which obstructs spatial resolution. This spatial resolution can be achieved by collimation, but this would reduce the measured XRF signal. Another problem is the irregular shape, causing the geometry to change for each angle. It would therefore be necessary to perform a normalization to correct for this. Two straight-forward methods are normalization based on the total background (scatter) counts measured, or based on the tungsten L-lines since no tungsten is expected in the sample, or a combination of both. These methods still need to be tested and verified on phantom samples.

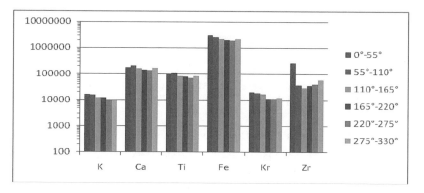

Figure 5. *Overview of XRF intensities for some elements found in the analysis, for 6 different angular intervals. Normalization has been performed based on W-Lα intensity*

Figure 5 shows the XRF intensities of some elements found in the spectral analysis. Normalization based on tungsten peak intensity has been performed. From this graph, it can be seen that Zr is particularly heterogeneous in the sample, while matrix material such as Ca has a relatively homogeneous distribution. Other detected elements have been left out for clarity. Note that this is only a preliminary result, and the normalization method has to be verified on phantom samples.

4. Conclusion

Combining CT data with μXRF measurements can provide useful information on the 3D structure of the different phases inside complex geological samples such as granite.

μXRF measurements can also be used for the detection of foreign substances in samples, such as consolidant in limestone.

For samples with an irregular shape, a good μXRF surface map can not be obtained. However, XRF measurement during the CT scan can already give information on the sample, although spatial resolution is very poor. This can be improved by a pinhole, but this comes with a reduced countrate. Based on phantom samples, this trade-off will be evaluated. Peak normalization will be tested and refined based on known samples.

5. References

Cnudde, V., Cnudde, J.P., Dupuis, C., Jacobs, P. "X-ray micro-CT used for the localization of water repellents and consolidants inside natural building stones", *Materials Characterization*, vol. 53 no. 2-4, 2004, p. 259 – 271.

Newbury, D.E. and Davis, J.M. "Solving the micro-to-macro spatial scale problem with milliprobe x-ray fluorescence/x-ray spectrum imaging", *Proceedings of the SPIE – The International Society for Optical Engineering, Scanning Microscopy 2009*, Monterey, CA, USA, 4-7 May 2009, p. 73780P

Nicolosi, J.A, Scruggs, B. and Haschke, M. "Analysis of sub-mm structures in large bulky samples using micro-X-ray-fluorescent spectrometry", *Advances in X-ray Analysis,* 1998, 41:227-233

Vandeputte, K. Master's thesis, Ghent University, Belgium, 2009.

Vlassenbroeck, J., Dierick, M., Masschaele, B., Cnudde, V., Van Hoorebeke, L., Jacobs, P. "Software tools for quantification of X-ray microtomography at the UGCT", *Nuclear Instruments and Methods in Physics Research Section A: Accelerators, Spectrometers, Detectors and Associated Equipment*, 2007, 580(1):442-445.

Quantification of Physical Properties of the Transitional Phenomena in Rock from X-ray CT Image Data

A. Sato* — K. Tanaka** — T. Shiote** — K. Sasa**

Dept. of Civil and Environmental Engineering
Graduate School of Science and Technology
Kumamoto University, Kurokami 2-39-1, Kumamoto 860-8555, Japan
asato@kumamoto-u.ac.jp

** *Graduate School of Science and Technology*
Kumamoto University, Kurokami 2-39-1, Kumamoto 860-8555, Japan

ABSTRACT. *X-ray CT method is generally used for the visualization of the geometry, the identification of defects and cracks, and the observation of transitional phenomena, such as water permeation and material migration process, in geo-materials. In this paper, examples of quantification of physical properties by CT values were introduced. The first is the analysis of the advection and diffusion process in the fractured porous rock sample. The tracer migration tests have been applied to the porous sandstone and the density distribution of the in the crack and the pores were evaluated. The second is the evaluation of the storage ability of CO_2 in the porous rock related to the geological storage of CO_2. The replacement ratio in the pores by CO_2 was estimated.*

KEYWORDS: *quantification, CT values, advection and diffusion, geological storage of CO_2*

1. Introduction

X-ray CT method is well known as one of the most useful visualization techniques not only the in medical use but also in the field of Geo-materials engineering (Otani *et al., 2003*, Desrues *et al.* 2007). Generally, x-ray CT method is used for the visualization of the geometry, the identification of defects and cracks, and the observation of transitional phenomena, such as water permeation and material migration process (Sato *et al.* 2009), in rock sample. The information obtained from the x-ray CT images is based on the density information, and it is convertible into to the various kinds of physical quantities. In this paper, we would like to introduce two examples of quantification of physical properties.

The first is the analysis of the advection and diffusion process in the fractured porous rock sample (Yonemura *et al.* 2008). The tracer migration tests have been applied to the porous sandstone and the tracer migration and diffusion process was visualized by x-ray CT scanner. Then the density distribution of the tracer in the cracks and pores are quantified from x-ray CT data.

The second is the evaluation of the storage ability of CO_2 in the porous rock related to the geological storage of CO_2 (Chadwick *et al.* 1993, Xue Z *et al.* 2004). When the CO_2 is injected into a water saturated rock sample, the nominal density changes since the water in the pore is replaced by low density CO_2 (Arimizu *et al.* 2008). This phenomenon is measured by x-ray CT, and the amount of the CO_2 stored in the sample is quantified.

Here the quantification process of those phenomena from x-ray CT data is introduced, and the applicability of the x-ray CT method not only as the visualization method but also as the measurement and quantification techniques is discussed.

2. Employed x-ray CT scanner

An x-ray CT scanner (TOSCANER-20000RE), manufactured by the Toshiba Corporation, has been operated by Kumamoto University. An x-ray bulb operating at 300 kV/2 mA provides the radiation source from which an x-ray beam is emitted. 176 detectors are aligned with the x-ray source in the horizontal plane to record attenuation data. The object to be scanned is positioned on a traversing turntable and the x-ray beam is projected through one plane of the object as it rotates and traverses. The beam thickness can be determined with collimators and the thickness is set to be 2 mm in this study. The pixel is a square of 0.072 mm x 0.072 mm, and the volume of a voxel is 0.0104 mm^3. Please refer to the references for a more detailed description of the specifications and principles (Sato *et al.*, 2009).

3. Analysis of tracer advection and diffusion process in porous rock

3.1. *Rock sample and tracer migration test*

A 100 mm x 100 mm x 100 mm cubic Kimachi sandstone sample shown in Figure 1 is used for the tracer test. The mean porosity of the sample is 26%. At first, the artificial crack was induced by shear test at the center of the sample.

Initially the rock sample was fully saturated by the water. Therefore, both the inside of crack and pores are completely filled with water. The initial x-ray CT image is shown in Figure 2(a). Then the tracer is injected into the rock sample from the bottom surface of the sample under the constant flow rate condition, and the injected tracer migrates upward in the crack. In this study, the mixture of Potassium Iodide (KI) solution and a medical contrast agent is applied, and its density is 1,254 kg/m^3.

The x-ray CT images during tracer migration test are shown in Figure 2 (b). As Figure 2 shows, x-ray CT images contain the influence of the heterogeneity of the rock sample and a various kinds of noise components unavoidable in CT images. The image subtraction between CT images is effective for elimination of those noise components (Sato *et al.* 2004). The image subtraction between initial CT image and the images after commencement of tracer tests are applied, and obtained subtracted images are shown in Figure 3.

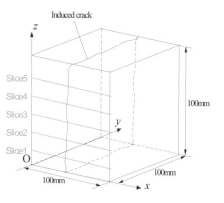

Figure 1. *Kimachi sandstone rock sample which has shear crack at the center*

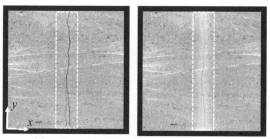

(a) t=0 minutes   (b) t=1361 minutes
Figure 2. *X-ray CT images after commencement of tracer test*

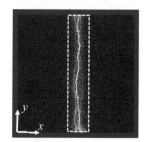

Figure 3. *X-ray CT images after image subtraction*

3.2. Coefficient of tracer density increment α

In order to evaluate the tracer density distribution from the CT images after image subtraction, the coefficient of tracer density increment is defined by Sato *et al.* 2009. In this study, this idea is expanded to the evaluation of tracer density both in the crack and pores. The CT value increment during tracer test from the initial condition is denoted by $\triangle C_l$, and the CT value increment between water and air is denoted by $\triangle C_{water}$. Firstly, let us focus on the CT value projection in the crack part. When the crack aperture is denoted by w, the projection p is given by

$$p = \Delta C_l \cdot w = \alpha \cdot \Delta C_{water} \cdot w \qquad [1]$$

where α is a constant of proportion between CT value increments obtained by different conditions, and this is defined as the coefficient of tracer density increment. If we focus on the CT value projection in the rock matrix part including pores, the projection p is also given by

$$p = \Delta C_l \cdot \phi = \alpha \cdot \Delta C_{water} \cdot \phi \qquad [2]$$

From the projection in the both crack part and rock matrix part, the coefficient of tracer density increment in both cases are given in the same form as

$$\alpha = \frac{\Delta C_1}{\Delta C_w} \qquad [3]$$

3.3. Analysis of density distribution of tracer

The distribution of coefficient α along the crack is shown in Figure 4. This represents the density distribution of the tracer which flows in the crack. For the comparison, original x-ray CT image and the crack aperture distribution are also shown in Figure 4(a) and (b), respectively. As this figure shows, the value of α is increasing with time as the water in the crack is replaced by the high density tracer.

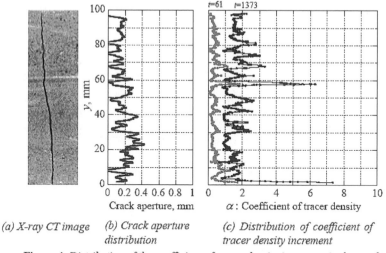

(a) X-ray CT image   (b) Crack aperture
distribution

(c) Distribution of coefficient of
tracer density increment

Figure 4. *Distribution of the coefficient of tracer density increment in the crack*

However, this distribution is not uniform. There is a tendency that the value of α become larger at where the crack aperture is small. This is because that water in the small space is easily replaced by the tracer. The distribution of coefficient α in the rock matrix part is shown in Figure 5. This distribution represents mean value of α each tomography region and the amount of tracer diffused into the rock. In this case also, the value of α is increasing with time. This is because the high density tracer in the crack diffused into the pores in the matrix part.

As these figures show, tracer density distribution both in the crack and the in the rock matrix part and the diffusion phenomena can be quantified by introduced x-ray CT image analysis.

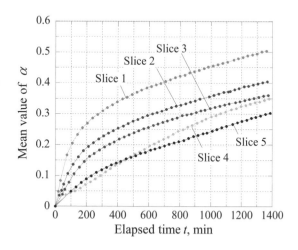

Figure 5. *Relation between the elapsed time and the mean value of α in the rock matrix part*

4. Analysis of CO₂ replacement ratio in porous rock

4.1. *Rock sample and CO₂ replacement test*

A ⌀50mm x 50mm cylindrical Berea sandstone sample shown in Figure 6 is used for the CO_2 replacement test. The side of the sample is completely sealed by acrylic resin. The drain pipe is installed on one end of the sample. Another end is free and the CO_2 is injected from this free surface into rock sample. The mean porosity of the sample is 24%. Initially the rock sample was fully saturated by the water. This rock sample is installed in the pressure vessels made by carbon fiber. The initial x-ray CT image is shown in Figure 7(a). X-ray CT images are taken every after 1ml liquid CO_2 (8MPa at 25 °C) is injected into the rock sample, and replacement process in the water saturated pores by CO_2 is visualized by x-ray CT scanner.

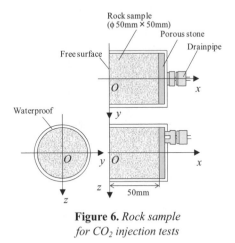

Figure 6. *Rock sample for CO₂ injection tests*

4.2. *Results*

The x-ray CT images during CO_2 replacement test are shown in Figure 7(b) and (c). However, it is hardly possible to identify the replacement process of CO_2 since the density change due to the replacement of water by liquid CO_2 is relatively small. Here also the image subtraction technique is applied between the initial image and the images during CO_2 injection test. Obtained images are shown in Figure 8 and the increment CT values are slightly confirmed. In order to clarify the CT value change, distribution of mean CT value increment along x-axis are shown as Figure 9. As this figure shows, CT value change due to the replacement by CO_2 is clearly confirmed.

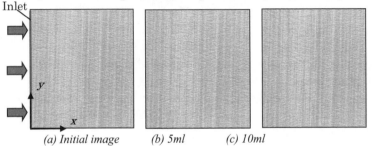

(a) Initial image   (b) 5ml   (c) 10ml

Figure 7. *X-ray CT images of Berea sandstone after CO₂ injection*

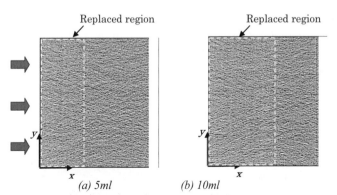

Figure 8. *X-ray CT images after the image subtraction by initial image*

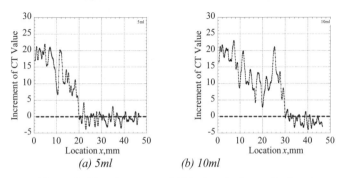

Figure 9. *Mean increment of CT values in the x-axis direction*

4.3. Replacement ratio R_v

CT values in this study means the value which density difference between the water and the air is divided by 1,000. Therefore, nominal density change due to the CO_2 replacement $\Delta\rho$ is given by

$$\Delta\rho = \frac{\Delta C_t}{1000} \qquad [4]$$

where ΔC_t is the increment of CT values obtained by image subtraction. Here, let us consider the specific volume ΔV of replaced CO_2 towards the total volume. When the density of the water and the CO_2 are denoted by ρ_w and ρ_{CO2} respectively, the specific volume ΔV is given by

$$\Delta V = \frac{\Delta\rho}{\rho_w - \rho_{CO2}} = \frac{\Delta C_t}{1000 \cdot (\rho_w - \rho_{CO2})} \qquad [5]$$

On the other hand, the maximum replaced volume by CO_2 is equal to the porosity ϕ. Here replacement ratio R_V is defined as the ratio of the replaced volume by CO_2 towards the total porosity, and given by

$$R_V = \frac{\Delta V}{\phi} = \frac{\Delta C_t}{1000 \cdot \phi \cdot (\rho_w - \rho_{CO2})} \qquad [6]$$

4.4. Analysis of replacement process of CO_2

The distribution of replacement ratio R_V is shown in Figure 10. As this figure show, the replaced region progresses gradually to the x-axis direction. In the case of Berea sandstone, the maximum replacement ratio is approximately 40%, that is, the 40% of pores are replaced by CO_2 and water still exist at 60% of pores. It is also confirmed that the replacement process is not uniform. In this study, CO_2 is injected perpendicular to the sedimentary layer. This is confirmed form Figure 6, and the value of R_V depends on each layer. The mean value of R_V of the rock sample is estimated as 30% approximately at the final stage of injection. This value is almost same as the estimation of RITE (RITE 2005). After that, the replacement was not occurred and the CO_2 simply flows through the rock sample.

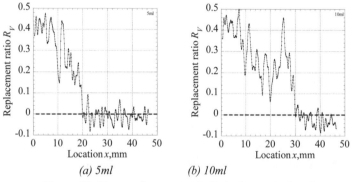

(a) 5ml   *(b) 10ml*

Figure 10. *Mean replacement ratio R_V in the x-axis direction*

5. Conclusions

X-ray CT images is based on the density information, and it is convertible into to the various kinds of physical quantities. In this study, two examples of quantification of physical properties was introduced. The first is the analysis of the advection and diffusion process in the fractured porous rock sample. The tracer migration tests have been applied to the porous sandstone, it was shown that the density distribution of the in the crack and the pores can be quantified from the x-ray CT image data. The second is the evaluation of the storage ability of CO_2 in the porous rock. Even though the density change is very small, it was shown that the quantification of the replacement ratio of CO_2 was possible from x-ray CT image data.

6. References

Arimizu T., Sato A., Tanaka K., Yonemura H., "Evaluation of replacement ratio and residual saturation in the water-air replacement process by means of X-ray CT", *Proc.12th Japan symposium on Rock Mechanics:JSRM2008*, 2008, CD-ROM.

Chadwick R.A., Eiken O. and Lindeberg E., "4D geophysical monitoring of the CO_2 plume at Sleipner, North Sea: Aspects of uncertainty", *Proc. Of the 7th SEGJ Int. Symp.*, Sendai, 1993, p.24-26.

Desrues J., Viggiani G. and Besuelle, *Advances in X-ray Tomography for Geomaterials*, 2007.

Otani J., Obara Y., *X-ray CT for Geomaterials, -Soilds, Concrete, Rocks*, 2003.

Sato A., Arimizu T., Yonemura H. and Sawada A, "Visualization and analysis of the tracer migration process in the crack by means of X-ray CT", *Journal of MMIJ*, Vol.125, 2009, p.146-155.

Sato A., Fukahori D., Sawada A. and Sugawara K., "Evaluation of Crack opening in the heterogeneous materials by X-ray CT ", *Shigen-toSozai*, Vol.120, 2004, p.365-371.

The Research Institute of Innovative Technology for the Earth (RITE), *Report of research and development of underground storage technology for carbon dioxide (2005)*, 2005, p.1104-1525.

Yonemura T., Sato A., "Visualization of advection-diffusion phenomena in the crack and the matrix inside by means of X-ray CT", *Proc.12th Japan symposium on Rock Mechancs:JSRM2008*, 2008, CD-ROM.

Xue Z., Ohsumi T, "Seismic wave monitoring of CO_2 migration in water-saturated porous sandstone", *Exploration Geophysics*, Vol.35, 2004, p.25-32.

Deformation in Fractured Argillaceous Rock under Seepage Flow Using X-ray CT and Digital Image Correlation

D. Takano* — P. Bésuelle* — J. Desrues* — S. A. Hall*

*Laboratoire 3S-R, CNRS and University Grenoble
BP 53, 38041, Grenoble, France
Daiki.Takano@grenoble-inp.fr
Pierre.Besuelle@grenoble-inp.fr
Jacques.Desrues@grenoble-inp.fr
Stephen.Hall@grenoble-inp.fr

ABSTRACT. *Argillaceous rock is a candidate for radioactive waste storage because of its extremely low permeability. However, during excavation of storage galleries, drilling can induce fractures that may modify (generally increase) the rock permeability. On the other hand, it is thought that stress redistribution after drilling and fluid seepage through time can lead to, at least partial, closure of these fractures. A number of studies reported previously have evaluated the changes in sealing capacity of these fractured argillaceous rocks, but such studies are mostly limited to 2D experiments or numerical modeling, and there are relatively few studies that consider the 3D behavior. The purpose of this paper is to evaluate, using x-ray computed tomography and 3D-volumetric Digital Image Correlation (DIC), the 3D deformation distribution in specimens of fractured argillaceous rock experiencing seepage flow. To achieve this, x-ray tomography images have been acquired of fractured rock specimens at several different stages of seepage flow experiments and 3D-DIC has been applied to evaluate the 3D displacement and strain fields. With these results, the deformation processes due to seepage flow is investigated quantitatively, including for example swelling of the clay matrix around the fracture. This new insight will lead to better understanding of the likely in-situ changes in sealing capacity of the fractured argillaceous rocks in waste storage sites.*

KEYWORDS: *fractured clay rocks, digital image correlation, swelling, seepage flow, x-ray*

1. Introduction

Argillaceous rock is a candidate for radioactive waste storage because of its extremely low permeability. However, during excavation of storage galleries, drilling can induce fractures that called Excavation Damaged Zone, in which the originally low permeability of the host rock may be modified locally by the development of strain localization. On the other hand, it is thought that stress redistribution after drilling and fluid seepage through time can lead to, at least partial, closure of these fractures (Davy *et al.*, 2007). After backfilling and closure of the storage, stress redistributions are shown to re-confine surface cracks and permeability decreases. To understand the deformation of the host rock due to excavation and seepage flow, it is necessary to investigate strain localization in argillaceous rocks.

The purpose of this study is to evaluate, using x-ray computed tomography and 3D-volumetric digital image correlation (DIC), the 3D deformation distribution in specimens of fractured argillaceous rock experiencing seepage flow. To achieve this, X-ray tomography images have been acquired of fractured rock specimens at several different stages of seepage flow experiments and 3D-DIC has been applied to evaluate the 3D displacement and strain quantitatively, including for example swelling of the clay matrix around the fracture.

2. Materials and methods

2.1. *Tested material*

The argillaceous rock tested in this study was provided by ANDRA (*Agence Nationale pour la Gestion des Déchets Radioactifs*) from their underground research laboratory site located at Bure (Meuse/Haute Marne, Eastern France) at approximately 550 m below the ground surface. The material has a compressive strength ranging from 20 to 30 MPa. Its water content is about 6%, clay content is in range from 40 to 45% and the other materials are calcite and quartz.

Cylindrical specimens with 10 mm in diameter and 20 mm height were prepared by cutting from the core by diamond wire saw in order to minimize the material disturbance during preparation (Lenoir *et al.*, 2007).

A fracture parallel to the axis of the samples is created by modified Brazilian test which loads the specimen by compression between two parallel. The fracture is kept open by plastic spacers between the halves of the specimen as shown in Figure 1.

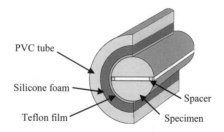

Figure 1. *Schematic view of the specimen in a PVC tube*

X-ray generator	
Type	Microfocus enclosed tube
Tube voltage adjustable range	40kV to 150 kV
Anode current adjustable range	0μA to 500μA
Maximum power	75W
Focal spot size	
Small spot mode	7μm (10W) or 5μm (4W)
Medium spot mode	20μm (30W)
Large spot mode	50μm (75W)
Beam open angle	43°
Position of the focal point	17mm behind the output window
Output window material	Beryllium (200μm thick)
Detector	
Type	X-ray flat panel detector
Matrix size	1920 x 1536 pixels²
Field of view	195.07 x 243.84 mm²
frame rate	0.2 to 30 frame per second
Data type	14bit gray levels

Table 1. Specifications of micro focus X-ray CT in Laboratoire 3S-R

2.2. Experimental procedure

The experiments described here were carried out using microscopic x-ray CT scanner at Laboratoire 3S-R in Grenoble. In this system, a microfocus x-ray source is used to achieve high spatial resolution. Table 1 gives some detailed information about the scanner. The x-ray source was operated at 130 μA of current and at energy level of 75 kV. A set of acquisition data was corrected by recording 1,200 radiographs at different angles equally spaced on 360°. The approximate time for each scan was 6.0 seconds. The actual size of reconstructed domain was 13 x 13 x 13 μm³.

The schematic diagram of the experiment is shown in Figure 2. The apparatus includes a small PVC tube and an electric pump and a reservoir tank. The specimen was wrapped with Teflon film and both ends of the specimen were covered with grease to avoid the contact of the seepage flow with any part of the specimen except the fracture. It was then inserted into a tube of silicone foam. Deionized (pure) water was used for seepage flow and the electric motor allows making constant flow rate in the value of 5 ml min^{-1}. The experiment was conducted in the following procedure. The first tomographic scan was made for initial conditions. Fluid seepage was then run for 6 h, followed by flow stoppage for a second scan (about 3 hours of acquisition). Fluid seepage was run again for 3 h, and last scan was conducted.

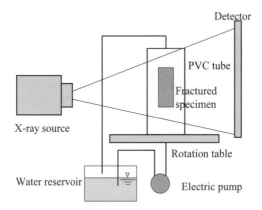

Figure 2. *Schematic view of the experiment setup*

2.3. *Image analysis*

Digital Image Correlation (DIC, hereafter), in this context, is a mathematical method that finds the displacement of a set of points between two consecutive digital images taken during a deformation process. The volumetric DIC analysis presented in this study was carried out using the code *TomoWarp* developed at Laboratoire 3S-R (see Hall *et al.*, 2009 for details). This DIC analysis involves the following steps: (1) definition of nodes and the correlation window centered on each node over the reference image; (2) calculation of a correlation coefficient over the correlation window for a series of displaced positions in the second image; (3) definition of the discrete displacement (integer number of voxels), given by the displacement with the best correlation; (4) sub-pixel refinement (because the displacements are rarely integer numbers of voxels); (5) calculation of the strains based on the derived displacements and a continuum assumption.

3. Results and discussions

3.1. *Direct observation from CT images*

Figure 3 shows the reconstruction of a CT slice perpendicular to the specimen's axis for each scanning step. The spacers placed in the main fracture are difficult to see because of their low X-ray absorption coefficient. As shown in this figure, secondary cracks have opened after 6 h of seepage flow. These cracks most likely already existed before the seepage and were opened by swelling effect of the material. In contrast, a few secondary cracks were closed during the seepage (Figure 3(b) and (c)). And also it is realized that the main fracture is partially closed in its centre position. Figure 4 shows the vertical slices of the sample. The re-closing the cracks can be observed from these figures.

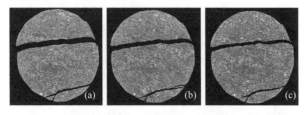

Figure 3. *Horizontal CT slices: (a) initial condition; (b) after 6h of seepage flow; (c) after 9h of seepage flow*

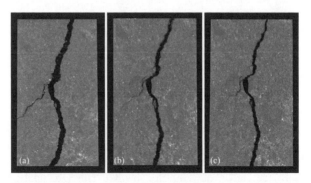

Figure 4. *Vertical CT slices: (a) initial condition; (b) after 6h of seepage flow; (c) after 9h of seepage flow*

	Initial	After 6 hours	After 9 hours
Area of cross section (mm²)	77.0	81.5	82.9
Percentage of variation (%)		5.8	7.6
Crack opening (mm)	1.1	0.8	0.6
Percentage of variation (%)		-27.9	-47.9
Height of specimen (mm)	20.0	20.2	20.5
Percentage of variation (%)		1.0	2.5
Total volume (mm3)	1519.2	1610.9	1681.9
Percentage of variation (%)		6.0	10.7

Table 2. *Results of geometric measurements*

Geometric measurements were carried out based on reconstructed slices as shown in Table 2. The measurements were made with thresholded images. The surface of the solid part in the slice orthogonal to the axis of the specimen (Figure 3) increased by 5.8% after 6 hours of flow and 7.6% after 9 hours. The total volume of specimen increases of 6% after 6 hours and 10% after 9 hours. The crack opening at the middle of the specimen decreases by 27% after 6 hours and 48% after 9 hours. It is considered that these results are due to swelling of the material and the opening of secondary cracks.

A simple way to measure the spatial distribution of the swelling would be to evaluate the grey level in CT slices. However this density change is too low to be significant. On the other hand, 3D-DIC provides a full displacement and deformation field for the pair of images as shown in the next section.

3.2. 3D-DIC results

Figure 5 (a) and (b) show the incremental volumetric strain field in a vertical slice through the specimen. The maximum strain plotted in these figures equals 0.3. In the increment from initial to after 6 hours of seepage flow (Figure 5 (a)), the development of dilatant volumetric strain (swelling in the bulk) can be observed in top-left half of the specimen (yellow). Also swelling can be observed around the secondary open cracks. The red fringes (compaction) around fractures are most probably DIC artefacts due to crack edge displacements.

The horizontal cut (Figure 5(c)) shows that the swelling starts from the edges of the main fractures (yellow spot above the crack). In the same cut, the contracting volumetric strain recorded at the ends of the fracture appears due to deformation of the spacer piles rather than of rock itself. In the increment from 6 hours to 9 hours seepage flow, Figure 5(b) and (d) show that swelling in the specimen starts developing extensively not only in the edge of the fracture but also inside the specimen.

These incremental volumetric strain fields suggest that the argillaceous rocks start swelling after only a few hours of water injection and then it develops extensively to inside the material. This means that micro-crack self healing has extremely rapid kinetics, which is an important factor to understand sealing capacity of the fractured argillaceous rocks.

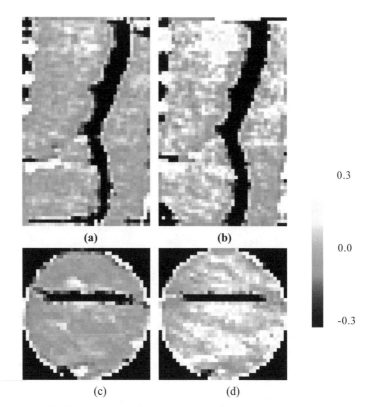

Figure 5. *Incremental volumetric strain (a) in vertical cut from initial to 6 h; (b) in vertical cut from 6 h to 9 h; (c) in horizontal cut from initial to 6 h; (c) in horizontal cut from 6 h to 9 h*

4. Conclusions

The deformation in fractured argillaceous rock under seepage flow was investigated using x-ray CT and digital image correlation. The swelling and secondary crack opening due to seepage flow can be visualized using x-ray CT scanner. It shows that self healing of a micro-crack has extremely rapid kinetics, which is an important factor to understand sealing capacity of the fractured argillaceous rocks in waste storage sites.

Future studies should attempt to conduct the tests using different kinds of solutions, for example site water from underground laboratory. Also, the experimental device will be improved to apply normal stress to the crack.

5. References

Davy C. A., Skoczylas F., Barnichon J. D., Lebon P., "Permeability of macro-cracked argillite under confinement: Gas and water testing", *Physics and Chemistry of the Earth*, vol.32, 2007, pp.667-680.

Lenoir N., Bornert M., Desrues J., Bésuelle P. and Viggiani G., "Volumetric digital image correlation applied to x-ray microtomography images from triaxial compression tests on argillaceous rock", *STRAIN*, vol.43, 2007, pp.193-205.

Hall S.A., Lenoir N., Viggiani G., Desrues J., Bésuelle P., "Strain localisation in sand under triaxial loading: characterisation by x-ray micro tomography and 3D digital image correlation", *Proceedings of Int. Symp. on Computational Geomechanics COMGeo09*, Cote d'Azur, France, April 29- May 1st, 2009.

Hall S.A., Bornert M., Desrues J., Pannier Y., Lenoir N., Viggiani, G., Bésuelle, P., "Discrete and continuum analysis of localised deformation in sand using X-ray micro CT and Volumetric Digital Image Correlation", *Géotechnique*, 2010, accepted.

Experimental Investigation of Rate Effects on Two-Phase Flow through Fractured Rocks Using X-ray Computed Tomography

C. H. Lee — Z. T. Karpyn

The Pennsylvania State University
Department of Energy and Mineral Engineering
110 Hosler Building
University Park, PA 16802-5000
USA
Cul169@psu.edu
ZKarpyn@psu.edu

ABSTRACT. *Capillarity, gravity and viscous forces control the migration of fluids in geologic formations. However, experimental work addressing the impact of injection flow rate in fractured core samples is limited. Understanding how injection flow rate affects fracture-matrix transfer mechanisms and invasion front evolution in fractured geomaterials are of crucial importance to modeling and prediction of multiphase ground flow. In this study, we monitor and analyze transfer mechanisms in a rock sample with a single tensile horizontal fracture using medical X-ray computed tomography. The impact of different injection rates on the resulting fluid recovery and saturation maps is evaluated through visual and quantitative analyses. Results from this investigation provide a comprehensive set of data for the validation of numerical models and strengthen fundamental understanding of multiphase flow in fractured rocks.*

KEYWORDS: *fractured rocks, multiphase flow, capillarity, x-ray computed tomography*

1. Introduction

Structural characteristics of fractures such as the aperture and orientation affect fluid migration in fractured systems. For instance, Karpyn, *et al.* (2009) found that bedding planes adjacent to fracture zones with higher aperture tend to have higher porosity, and higher permeability, thus affecting the overall hydraulic conductivity of the system. Firoozabadi and Markeset (1992) studied gravity and capillary cross-flow in fractured porous media, and showed that the contribution of capillary cross-flow from the side faces of the matrix rock increased with the tilt angle. Gu and Yang (2003) used numerical modeling to study the interfacial profile between two immiscible fluids in a reservoir with a fracture with random orientation, and found that the equilibrium shape of the interfacial profile depends on the ratio of gravity and capillary forces.

When a wetting fluid flows through a fracture, capillarity may drive the wetting fluid from the fracture into the matrix, while viscous forces propel the fluid to flow through the fracture with less resistance. Rangel-German, *et al.* (2006) used stacked Boise sandstone blocks to study multi-phase flow in a fractured system. They concluded that both capillary and viscous forces control the flow in the fracture and that capillary continuity can occur in any direction, depending on the relative strengths of the capillary and Darcy (viscous) terms in the flow equations. These forces dominate fluid migration in fractured formations and strongly influence fracture-matrix transfer mechanisms. In addition, Babadagli (1994) found that as the injection rate is increased, fracture pattern becomes an important parameter controlling the saturation distribution in the rock matrix. As the rate is lowered, however, the system begins to behave like a homogenous system showing a frontal displacement regardless of the fracture configuration. Although these and other studies (Prodanovic, *et al.* (2008), Hoteit and Firoozabadi (2008), Donato, *et al.* (2007), and Rangel-German, *et al.* (2006)) have contributed to current understanding of multiphase flow in fractured systems, there is still limited understanding of the relative impact of variables affecting two-phase displacement mechanisms in fractured rocks. Analyzing how injection rate affects fracture-matrix flow, especially under capillary dominated conditions, remains largely unexplored and it is the main goal of this paper. The present work is one component of a multi-variable, experimental analysis of fracture-matrix flow including the effects of fracture orientation, bedding plane orientation relative to the fractures, flow direction, and fluid type.

A laboratory flow apparatus was designed specifically for this set of experiments, in which saturation maps are monitored as a function of time for two injection flow rates. X-ray computed tomography (CT) was used to record these saturation maps as a function of time. Continuous CT scanning allowed us to track capillary imbibition into a fractured Berea sandstone sample originally saturated with Kerosene. Results from this work help visualize the impact of injection flow rate on the dynamics of fracture-matrix transport and, at the same time, provide detailed quantitative information for the validation of representative numerical models of fractured permeable geomaterials.

2. Experiment design

The purpose of this design was to create a flow system and holder for a two-dimensional fractured sample, to allow complete sample monitoring with single-slice CT scanning. By making the sample a thin disk, we eliminate one flow direction, the one orthogonal to the disk. Therefore, a single slice is sufficient to capture the entire fracture and the surrounding rock matrix, thereby allowing us to keep track of saturation changes at small time intervals. The rock sample used in this study was Berea Sandstone. Each Sample disk has a diameter of 102 mm and thickness of 10 mm. A single tensile fracture was created artificially on each disk. All fractures are aligned with the centre of the sample and perpendicular to the bedding layers. Fracture apertures are around 0.5 mm. Pore volume of the matrix and fracture are 18.51 mL and 0.51 mL respectively.

The experimental apparatus includes three major portions: sample holder, fluid supply system and x-ray CT scanner. A medical HD350 scanner with a detection limit of 25 microns was used in this study. Each CT image produced a matrix of 512 by 512 pixels covering the entire sample. The voxel size was $0.20 \times 0.20 \times 5.00$ mm.

A cake-shaped sample holder was fabricated according to the diagram shown in Figure 1. This sample holder is made of Teflon to avoid chemical reaction with fluids, and only Teflon and the rock sample are intercepted by the scanning plane. Viton rubber sheets are used to seal the gap between the sample and the walls of the holder, thus blocking potential pathways around the sample and allowing fluid flow through the fracture alone. The supporting rod, also shown in the top-right insert in Figure 1 can rotate in 45 degree increments, and it is attached in such a way that the cell can rotate on its horizontal axis. Although the core holder is designed to allow different fracture inclinations and flow directions with this rotation system, this paper focuses on experiments using a horizontal fracture.

A schematic representation of fluid circulation system is presented in Figure 2. Water and kerosene represent the wetting and non-wetting fluid phases, respectively. Water was tagged with 15% by weight of sodium iodide (NaI) to increase its CT registration and provide a high contrast between the two phases. A vacuum pump enables the sample holder reach 250 microns vacuum condition. This vacuum state is used to pre-saturate the sample with kerosene. A syringe pump (LC-5000) delivers the tagged water through the fracture. To guarantee a predominantly capillary-driven displacement, injection flow rates are low, in the order of 40 mL/hr and 4 mL/hr, which correspond to capillary numbers in the order of 8.6×10^{-6} and 8.6×10^{-7}, respectively. The capillary number represents the relative control of viscous over capillary forces. For capillary numbers below 10^{-5}, flow in porous media is considered to be dominated by capillary forces.

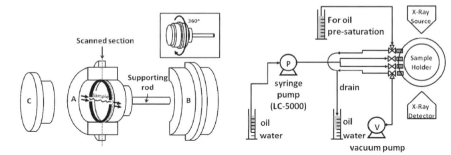

| **Figure 1.** *Schematic drawing of sample holder design* | **Figure 2.** *Schematic drawing of fluid circulation system* |

3. Experimental procedure

A schematic representation of the experimental procedure is presented in Figure 3. The dry fractured sample is packed and vacuumed in the sample holder and scanned to observe its heterogeneity and its layered structure (stage 1 in Figure 3). In stage 2, the sample is saturated with kerosene (non-wetting phase) and scanned. The image difference between these two stages is used for porosity calculations. In stages 3 and 4, the sample was flooded by injecting water. Fluid saturations were continuously monitored by scanning at specific time interval until residual oil saturation was reached. At the same time, kerosene recovery is recorded as a function of time at the outlet.

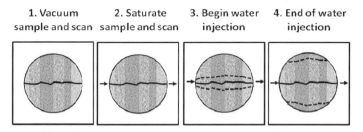

Figure 3. *Experimental procedure and CT scanning sequence*

4. Determination of porosity and fluid saturation

The average sample porosity (ϕ_{avg}) is 23.27% obtained from the volume of kerosene used in saturating the sample and the bulk volume of the sample. Pixel porosities (ϕ_{pixel}) were obtained from equation [1] using x-ray CT registrations from the vacuum condition in stage 1 (CT_{vacuum}) and the kerosene saturated condition in stage 2 (CT_{wet}) as shown in Figure 3.

$$\phi_{pixel} = \frac{CT_{wet} - CT_{vacuum}}{(CT_{wet} - CT_{vacuum})_{avg}} \phi_{avg} \qquad [1]$$

In situ saturations were also determined using data from the CT scanner. Pixel water saturations ($S_{w,pixel}$) were obtained from equation [2], where ϕ_{pixel} is the pixel porosity from equation [1], $S_{w,avg}$ is the average saturation of water in the sample obtained from the linear correlation between 100% water saturated and 100% kerosene saturated sample CT_{wet}.

$$S_{w,pixel} = \frac{CT_{wet} - CT_f}{(CT_{wet} - CT_f)_{avg}} \left(\frac{\phi_{avg}}{\phi_{pixel}}\right) S_{w,avg} \qquad [2]$$

5. Results and discussion

Average saturation changes as a function of time and pore volume injected are presented in Figure 4. These saturations were averaged over the entire sample, including fracture and rock matrix. In Figure 4-top, water breakthrough for the low-rate curve in red is delayed 10 min with respect to the higher-rate blue curve. In addition, the higher flow rate curve (q=40 mL/hr) reaches higher water saturation, and thus higher oil recovery, sooner than the low-rate case, but it requires more pore volume injected to reach that saturation level, as seen in Figure 4-bottom. After approximately 300 minutes of water injection, oil recovery becomes negligible in both cases, when water saturation reaches 0.56. Under this final saturation conditions, both oil and water are still mobile inside the rock sample, but the increments in water saturation are too small to be appreciated in the lapse of a few days.

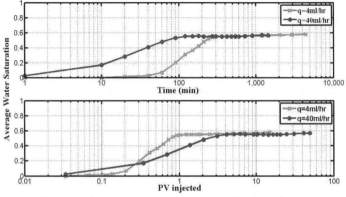

Figure 4. *Average water saturation as a function of time*

Figure 5 and Figure 6 are time progressions of water saturation (Sw) maps corresponding to high and low water injection rate. Dark blue represents regions saturated with kerosene (Sw=0.0), red represents regions saturated with water (Sw=1.0), and intermediate colors represent the co-existence of kerosene and water. For the high rate case (Figure 5) we see a sharp increase in water saturation in the neighborhood of the fracture. Under this flowing condition, the fractures refills with water at a faster rate than it can be transferred through the fracture-matrix interface, confirming similar experimental observations found in the literature (Rangel-German and Kovscek (2002)). Simultaneously, counter-current imbibition is occurring in the water invaded zone as oil is expelled from the matrix into the fracture. As time progresses, the imbibition front moves away from the fracture, and water accumulation becomes evident around the outlet end of the fracture (right side) in red, supporting the fact that the rate of capillary dispersion through the matrix is low compared to the rate of injection.

The rate of injection is also responsible for the shape of the imbibing front, which is farther away from the fracture inlet than the outlet. These mechanistic observations are less pronounced when the rate of injection is reducing. Figure 6 shows an analogous progression of water saturation maps obtained at 4 mL/hr of water injection. The contrast in saturation ahead and behind the water front is not as sharp as that in Figure 5. This is evident in a smoother color transition, passing from dark to light blue, to green, and finally yellow and red. In addition, for approximately the same pore volume injected, that is 0.69 PVI at 20 min high-rate and 180 min low-rate, we observe a much larger imbibed region in the low-rate case. This implies low injection rate (4 mL/hr) allows a more effective spreading of water for the same volume injected. Plots in Figure 5 and Figure 6 also show symmetry of the saturation fronts relative to the fracture location indicates gravitational effects are negligible.

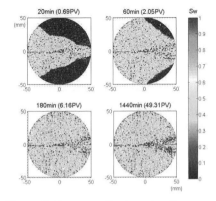

Figure 5. *Sequence of water saturation maps obtained from CT scanning at 40 mL/hr water injection rate*

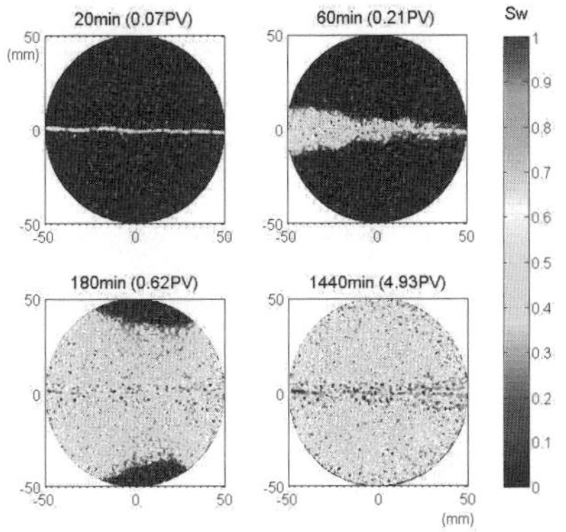

Figure 6. *Sequence of water saturation maps obtained from CT scanning at 4 mL/hr water injection rate*

Further quantitative examination of saturation changes obtained from CT scanning is presented in Figure 7. These vertical saturation profiles averaged over the central 6 mm of each CT slice for the two flow rates under study. These profiles capture saturation changes with time in the direction perpendicular to the fracture. For both experiments, continuous high water saturation is observed in the center of the sample, where the fracture is located. The most salient differences between these two groups of vertical profiles are: (1) the speed at which the water front moves away from the fracture, which was also evident in the saturation maps presented in Figures 5 and 6; and (2) the change in saturation as we move away from the fracture. Figure 7-right shows a gradual saturation change at the front, while there is a drastic drop in saturation across the water front in Figure 7-left. Furthermore, water saturations remain in the 0.50-to-0.55 range within the imbibed zone, which suggests that both fluid phases are under a dynamic equilibrium at that saturation. This is consistent with the knowledge that counter-current flow is the prevalent flow mechanism in the imbibed zone. As the water front progresses, the resident oil is displaced towards the fracture, in a counter-current manner, and replenished by the oil that is sitting ahead of the front, thus maintaining a dynamic equilibrium and a constant saturation in the imbibed zone.

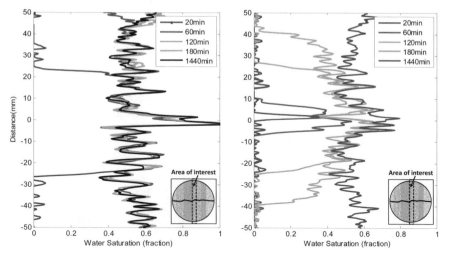

Figure 7. *Vertical saturation profiles perpendicular to the fracture and averaged over the central 6 mm of the sample, q=40mL/hr (left) and q=4mL/hr (right) (see CD for color version)*

6. Conclusions

In this experimental study, we analyze the impact of two different injection rates on the capillary dominated displacement of oil by water in a fractured rock sample, using x-ray computed tomography. The higher injection flow rate (q=40 mL/hr) showed higher oil recovery, although more pore volume was needed to reach the ultimate oil recovery. A sharp imbibing front was observed in the high-rate experiment, while a smoother saturation gradient was observed at low rate. Water saturation in the imbibed zone remains constant at around 0.50-to-0.55, suggesting a dynamic equilibrium in the mobility of oil and water phases. We also describe the counter-current flow mechanisms that are evident from the experimental results, and support our observations on the evolution of saturation maps and profiles obtained in the laboratory. These experiments will be extended in upcoming studies to cover a broader range of flowing conditions and fracture orientations, and served as reference data to test the validity of traditional fracture-matrix transfer function.

7. Acknowledgements

The National Science Foundation, project CMMI-0747585, and the Penn State Center for Quantitative Imaging.

8. References

Babadagli T., "Injection rate controlled capillary imbibition transfer in fractured systems", SPE 28640, *Proceedings of the SPE Annual Technical Conference and Exhibition*, New Orleans, LA, USA, 25-28 September 1994.

Donato G., Lu H. Y., Tavassoli Z., Blunt M. J., "Multirate-transfer dual-porosity modeling of gravity drainage and imbibition", *SPE Journal*, vol. 12 no. 1, 2007, p. 77-88.

Firoozabadi A., Markeset T., "An experimental study of capillary and gravity crossflow fractured porous media", SPE 24918, *Proceedings of the SPE Annual Technical Conference and Exhibition*, Washington, DC, 4-7 October 1992.

Gu Y. G., Yang C. D., "The effects of capillary force and gravity on the interfacial profile in a reservoir fracture or pore", *Journal of Petroleum Science and Engineering*, vol. 40 no. 1-2, 2003, p. 77-87.

Hoteit H., Firoozabadi A., "An efficient numerical model for incompressible two-phase flow in fractured media", *Advances in Water Resources*, vol. 31 no. 6, 2008, p. 891-905.

Karpyn Z. T., Alajmi A., Radaelli F., Halleck P. M., Grader A. S., "X-ray CT and hydraulic evidence for a relationship between fracture conductivity and adjacent matrix porosity", *Engineering Geology*, vol. 103 no. 3-4, 2009, p. 139-145.

Prodanovic M., Bryant S. L., Karpyn Z. T., "Investigating matrix-fracture transfer via a level set method for drainage and imbibition", SPE 116110, *Proceedings of the SPE Annual Technical Conference and Exhibition*, Denver, Colorado, 21-24 September 2008.

Rangel-German E., Akin S., Castanier L., "Multiphase-flow properties of fractured porous media", *Journal of Petroleum Science and Engineering*, vol. 51 no. 3-4, 2006, p. 197-213.

Rangel-German E. R., Kovscek A. R., "Experimental and analytical study of multidimensional imbibition in fractured porous media", *Journal of Petroleum Science and Engineering*, vol. 36 no. 1-2, 2002, p. 45-60.

Micro-Petrophysical Experiments Via Tomography and Simulation

Micro-petrophysical experiments

M. Kumar* — E. Lebedeva* — Y. Melean* — M. Madadi* — A. P. Sheppard* — T. K. Varslot* — A. M. Kingston* — S. J. Latham* — R. M. Sok** — A. Sakellariou* — C. H. Arns*** — T. J. Senden* — M. A. Knackstedt*

* Dept Applied Mathematics
Research School of Physics and Engineering
The Australian National University, Canberra, ACT 0200, Australia
mun110, evg110, ymb110, mah110, aps110, tkv110, amk110, sjl110, asa110, tjs110
mak110@physics.anu.edu.au

** Digital Core Laboratories, Pty Ltd
Innovations Building, Acton, Canberra ACT 0200 Australia
rob.sok@digitalcorelabs.com

*** School of Petroleum Engineering
The University of New South Wales, Sydney NSW 2052 Australia
c.arns@unsw.edu.au

ABSTRACT. *Increasingly, x-ray micro-tomography is being used in the observation and prediction of petrophysical properties. To support this field there is a vital need to have well integrated and parallel research programs in hardware development, structural description and physical property modeling. There is a constant need to validate simulation with physical measurement, and vice versa. Greatly assisting these demands is the ability to perform direct image-based registration in which the tomograms of successively disturbed (dissolved, fractured, cleaned, etc) specimens can be correlated to an original undisturbed state in 3D. In addition, 2D microscopy of prepared sections tie chemical analysis back to the 3D datasets giving a multimodal assessment.*

KEYWORDS: *quantitative analysis, micro-tomography, visualization, petrophysics*

1. Introduction

Over the past 2 decades x-ray tomography has found wide application in the petroleum industry (Withjack, 1988), with particular focus on visualizing potential core damage and the tracking of fluids distribution during flooding experiments. At first, tomography was conducted on the millimetric scale with medical instruments but now micrometric tomography is readily accessible. Ready access to faster computer hardware has provided the opportunity to conduct more complex experiments and the possibility to directly incorporate the 3D tomographic data into modeling/simulation. The increased ease of data collection has not been matched by the improved ability to manage, compare or process this data. Clearly, to take advantage of this "data storm" it is necessary to control the whole data lifecycle, starting with hardware design through to the validation of simulated data. The task is not trivial and must be supported in turn by a holistic view of data and analysis management. In the sections below we will first examine some of the key steps that aid the integration of data on all levels. Finally, several cases studies will be used to illustrate the utility of this approach.

The profile of a tomography facility depends very much on the user-base. The facility described here was specifically assembled to develop the associated hardware to a point where it feeds the highest quality data feasible into a research program in granular and porous materials. Necessarily, there is overlap in skills between experimental, theoretical and computational researchers in the group. In one instance, hardware development is guided by developments in reconstruction theory. Through calibration the comparison of experiment with simulation allows normally inseparable factors to be differentiated.

We will present a range of petrophysical measurements, both virtual and actual, which use tomography as a starting point and matching computer simulation to lab experiments as the goal. Specific systems investigated will include the tracking of acidic dissolution in carbonate sediments, the mapping of residual hydrocarbons after waterflooding into mixed-wet vs. water-wet pore space in reservoir rocks, coincident multi-modal, multiscale imaging of porosity (nm to mm) in carbonate core and monitoring the effects of brine concentration on brine-rock interactions. The key to many of these *in situ* experiments is the ability to perform direct volume registration in which the tomograms of successively disturbed (dissolved, fractured, cleaned, etc.) specimens can be correlated with each other and with the original undisturbed state in 3D.

2. Work flow

The process to prepare, collect and reconstruct a single high quality tomographic dataset is increasingly straightforward but becomes considerably more challenging once multiple physical analyses or complementary techniques are applied to a specimen. The workflow in Figure 1 illustrates a typical workflow for a specimen analyzed at our facility. The need for calibration and validation at all the stages shown as shaded boxes is paramount if a quantitative view is to be gained. Each of the shaded steps contains intrinsic calibration protocols. It is clear that reconstruction accuracy depends on certain critical CT system parameters, such as magnification and detector calibration, however for many commercial systems these are generally determined only at the time of initial instrument installation. Similarly, electron microscopy involves chemical and detector calibration at a level which is normally insufficient to be reliable for quantitative comparison with tomographic data. A regime of monitoring image quality for 3D and 2D datasets must be an on-going part of a workflow.

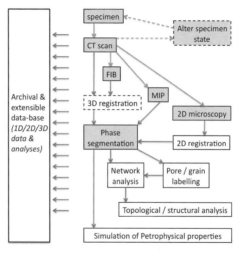

Figure 1. *The workflow from experiment through to analysis and simulation. The red arrows illustrate a path that the material specimen takes, the blue arrows refer to data flow The dashed elements show steps that rely on sequential scanning of the same specimen after alteration. The shaded steps require specific calibration against external references. MIP = Mercury Intrusion Porosimetry.*

There are a number of ancillary tools, experimental and computer-based which a team relies upon such as, porosity and elemental analyses, visualization and data base management. Data visualization is not, generally speaking, a quantitative tool in its own right but more often than not plays an important role in familiarizing the worker with the dataset. Not to be understated, is the importance data visualization plays in the rapid portrayal of a specimen and accompanying simulation to an uninitiated audience. Tools such as *Drishti* (http://en.wikipedia.org/wiki/Drishti) help the worker to rapidly understand the relationships between material density, density gradients and spatial distribution. This is particularly useful when other types of data can be superimposed, such as simulation results or 2D micrographs. The shear visual density of tomographic datasets forces the precise identification of material phases to lie in the realm of computational segmentation. Other imaging modalities invariably give either lower spatial resolution or require a 2D section to be taken destructively through the specimen. Comparison of this complementary image data from techniques such as Scanning Electron Microscopy (SEM), with 3D volumetric data greatly enhances material differentiation.

The task of recording the tomographic workflow, from experimental parameters, and related imaging modalities to an historical account of analyses and results requires an extensible data format, in our case, *NetCDF (http://en.wikipedia.org/wiki/NetCDF)*. For the grander task of inter-collating the entire workflow for each experiment in a manner which permits efficient data mining a purpose-built database was designed, *Plexus* (http://plexus.anu.edu.au/).

3. Material phase segmentation

In order to quantitatively calculate physical properties of a specimen from its tomogram, the first step is to classify image voxels according to physical composition. In an x-ray tomogram, the value at each voxel is related to the average x-ray attenuation and depends on the substance located in each voxel. For microtomography, one can assume the material is homogenous within regions much larger than the voxel size. The assumption that voxels are predominantly composed of a single substance will be examined below. Nonetheless, using this assumption reduces the task of determining composition to *segmentation*: classifying each voxel as a particular component (or phase) according to its associated attenuation value. This task is complicated by the fact that materials are inhomogeneous, they have complex absorption spectra, laboratory-based x-ray sources are strongly polychromatic, and images have noise. Although segmentation is inherently imperfect at component boundaries – where voxels may be composed of several substances – it nonetheless represents the best starting point for quantitative analysis. Once identified, each component is assigned material properties using *a priori* specimen knowledge, sometimes refined with experimental measurements. Prior to segmentation, we have found that image quality can be significantly improved by the application of an edge-preserving de-noising filter, and that best results are obtained with the 3D anisotropic diffusion filter (Perona, 1990), particularly the hybrid method of Frangakis (Frangakis, 2001). To perform the segmentation itself, we have found that classifying voxels purely according to their local intensity values (i.e. thresholding by intensity) is sufficient only for very high quality images containing two substances. Therefore, non-local and gradient information needs to be incorporated into the segmentation process. There are many published image segmentation techniques and invariably most have the same basic principle: identify seed regions, grow the seed regions based on some rules, and finally terminate at a suitable point. To analyze large tomograms routinely, segmentation must be computationally efficient and allow for data-parallel execution. We have developed a region-growing algorithm that uses the fast-marching method (Sethian, 1996) to achieve a good compromise between segmentation quality and computational efficiency (Sheppard, 2004). This algorithm cannot automatically segment a tomogram; on the contrary, some user skill is required and there tends to be some uncertainty in the outcome. To identify multiple components, the algorithm is applied repeatedly by masking all previously identified components. To demonstrate the technique a two-phase segmentation is shown in Figure 2. Figure 3 demonstrates one of the measures used to check the quality of a segmentation. A simulation of a non-wetting fluid ("mercury") intrusion is performed on the segmented image and compared against a lab-based MIP measurement on the actual specimen scanned. Although this is a destructive measurement, half the core may be preserved for other analyses. If the simulated intrusion data correlates well with the laboratory intrusion data then we can be confident that the segmented image contains an accurate representation of the pore-space of the specimen. The segmented image is the first stage of processing for many subsequent petrophysical simulations, so it is critical that the MIP experiment validates the segmentation in order to be confident in the subsequent analyses.

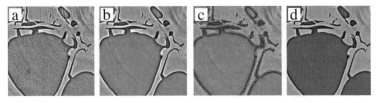

Figure 2. *A sub-region of a slice from a phase-contrast tomogram of wood, with a voxel size of 700 nm. The experimental data (a), filtered with anisotropic diffusion (b), and an initial threshold is chosen to identify seed regions for the segmentation (c). Blue is the void phase seed and red is grain phase seed. The final segmentation (d) is overlaid on the tomogram; regions of wood phase have been colored green. Adapted from Sakellariou, 2007.*

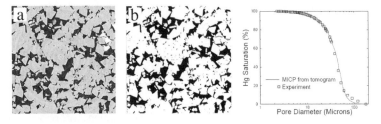

Figure 3. *(a) A 500 μm wide subset of a dolomite, and (b) the corresponding 2-phase segmentation. (c) A plot of simulated vs. measured MIP on same specimen scanned*

4. Registration

A range of 3D image registration techniques have been developed to extend the capabilities of x-ray micro-CT and to allow the calibration of micro-CT data to other microscopic techniques. In contrast to the alignment of 2D images, for which many tools are publicly available, registration of very large 3D images, for which distributed-memory parallel algorithms are required, pose a considerable challenge. As we are not aware of available software which is capable of accomplishing this task the group has implemented registration techniques (Latham, 2008) which combine a low resolution exhaustive search plus a multi-start, multi-resolution local optimization to achieve accurate spatial alignment of an image pair. Registration of a 2D micrograph with a tomogram (Figure 4) takes typically 1 hour on 64 CPUs, whereas a 3D-to-3D registration takes only about 2 hours on the same number of CPUs (SGI Altix BX2 system). The chief requirement is for significant distributed memory, over 100GB being required for the processing of 2048^3 datasets.

Figure 4. *An illustration of the 2D to 3D registration result. The routine orients any arbitrary slice back into the original containing volume. The linearity of the 2D section must be as well matched to the tomogram as possible. This means better than 1 part in 1000 linearity. The process of registration is contingent on good calibration of both tomogram and micrograph.*

The ability to accurately align micro-CT images with images generated from other modalities provides diverse capabilities and the potential to enhance the information present in the micro-CT image. The images in Figure 5 show 2D scanning electron microscopy (SEM) images of polished thin sections extracted from plugs previously imaged by x-ray micro-CT. SEM imaging generates sub-micron resolution images, and can even provide detailed mineralogical maps through energy dispersive x-ray spectroscopy. Accurate registration of the 2D microscopic image within the 3D image volume increases the potential for mapping x-ray gray-levels to mineral phases, reveals information about features too fine for x-ray micro-CT and provides an invaluable quality control on the 3D image data.

To the eye, 2D micrographs seem undistorted, however to register against a 3D dataset the micrograph must be corrected for non-linearities to better than 1 part in a 1000. SEM images can contain distortions of a few percent, optical micrographs can be worse. Tomograms, by design have undistorted cubic elements (voxels) and, if the x-ray optics have been calibrated appropriately, provide a distortion free cubic grid. In Figure 5 we have also registered an optical thin section of the same slice (again warp corrected). Although polarizing information is difficult to incorporate into a 3D dataset it provides an instant visual key for a trained mineralogist. The advantage here is that a specialist can adapt their intuition for making 3D inferences from a tried and tested 2D methodology.

Figure 5. *(a) A small region from a tomographic section (~1 mm wide), (b) the corresponding registered SEM from a polished thin section of the same region, (c) the optical micrograph from the same thin section. All micrographs require linear calibration against an appropriate standard.*

Another application of registration is the alignment of tomographic datasets acquired at different resolutions. Figure 6 illustrates the difficulty in considering the multi-scale heterogeneity in carbonate reservoirs. Accurate spatial alignment provides the ability to reconcile differences in physical porosity measurements against computationally derived values from segmentation. In Figure 7 a tomogram of a 25 mm carbonate core was generated with 15 micron voxel size. Subsequently, a 5mm core was drilled from the 25 mm core and scanned with a 5 micron voxel size.

The higher resolution tomogram was then registered with the lower resolution tomogram. The accurate alignment of the two images then allows one to more easily transfer (upscale) the high-resolution porosity data to the larger-scale, or lower resolution, tomogram.

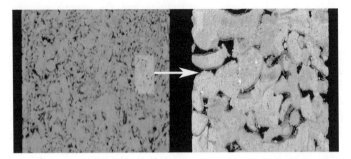

Figure 6. *A cross-section through a 44 mm carbonate core with a tomogram from the 5 mm core physically sub-sampled from the region shown. This result is that the resolution is increased by almost a factor of 10. A quantitative measure of the error in porosity can be made by considering these multiscale measurements.*

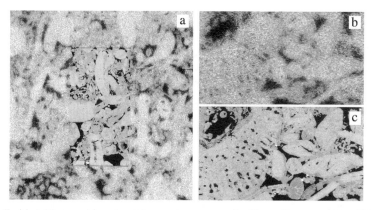

Figure 7. *(a) A 15 mm wide sub-set from a tomogram of a 40 mm limestone core (25 micron voxels. The higher resolution inset is shown overlain centrally and is the corresponding slice from the 5 mm core physically sub-sampled from the 40 mm core, and re-scanned. (b) and (c) show registered slices from the 40 mm and 5 mm cores, respectively.*

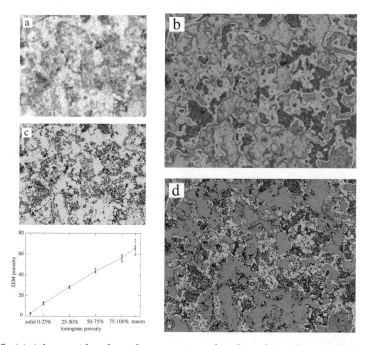

Figure 8. *(a) A 1 mm wide sub-set from a tomographic slice of a carbonate. (b) shows four ranges in apparent porosity from solid (blue), light and dark green (micro-porous, sub-voxel resolution) to macro-pore (red, resolved in the tomogram). (c) The corresponding registered polished section from SEM, with (d) illustrating the levels of micro-porosity not assessed in the tomogram. The plot left shows porosity as a function of percentage of voxel filled. The error bars are given by the segmentation of the SEM.*

In Figure 8 a study is shown demonstrating a situation where the gray values of the tomogram voxels indicate a partial volume, that is, neither fully pore nor fully solid. This is a typical occurrence in specimens where pore sizes are much smaller than the actual voxel size. In the mono-mineralic system shown, it allows porosity to be calibrated based on gray value. For multi-mineralic systems ambiguities with mineral density and porosity can make this approach very challenging. The opportunity that registration with SEM allows is to validate the choice of effective porosity against apparent 2D porosity. In Figure 8 the registered SEM is used as a mask to define the solid phase so thereby adding certainty to a calibration of effective porosity versus voxel intensity.

5. Case studies

In this section we rely on the registration process to quantitatively incorporate other imaging data and to track the micro-structural changes that occur upon dissolution or fluid ingress. The following classes of studies aim to illustrate the importance of data quality both as acquired from tomography and other modalities. For ease of presentation all cases show only small 2D subsets, but all comments have been derived from analysis on the original 3D datasets.

Sub-resolution porosity: Porosity which is below the imaging resolution continues to challenge tomographic modeling, particularly for multi-mineralic systems such as clay-rich siliclastics. Figure 9 shows how important the integration of other imaging modalities can be for tomography-based simulation. Based on CT data alone only a small proportion of the clay phase is correctly identified. Much of the pore-filling material shows too little contrast to be usefully segmented as clay. Comparison with the registered SEM "calibrates" the selection of segmentation parameters and substantially more clay is identified. At this point the simulation of permeability more closely matches the lab-based whole-core measurements. It is highly instructive to see that, apart from the change in porosity, the electrical conductivity is not very sensitive to the refined segmentation.

Fluid mapping and wettability: the migration of fluid, particularly immiscible fluids such as oil, water and gas on the scale where surface tension dominates the fluid distribution is the chief domain for the geological application of micro-tomography today. To track fluid interfaces at the pore-scale is a major technical challenge. Here we look at addressing this challenge by imaging specimens in different saturation states, using registration to accurately align the tomographic images and subsequently subtracting the aligned images to analyse the fluid distributions. Figure 10 illustrates an initially dry, siliclastic which undergoes two sequential floods with water. The registration is so effective in distinguishing liquid from gas that a minimum of X-ray contrasting medium is needed in the fluid phase, improving overall clarity of menisci and thin films. In the experiment shown in

Figure 11 a water solution is allowed to spontaneously imbibe into a dry 5mm core and allow us to equilibrate for 24 hours, Figure 11(b). The trapped gas phase is clearly evident from the image in Fig 11(c). The trapped gas phase saturation is 31%. Imbibition is complex and imaging the resulting pore-scale distribution of fluids can assist in improving our understanding and modeling of the process. The imbibition of a wetting fluid into a porous media is influenced by rate, heterogeneity of the pore space and local pore geometry (Morrow, 1985), which can lead to a wide variety of wetting patterns including site invasion percolation, cluster growth and flat frontal advance (Lenormand, 1984; Blunt, 1995).

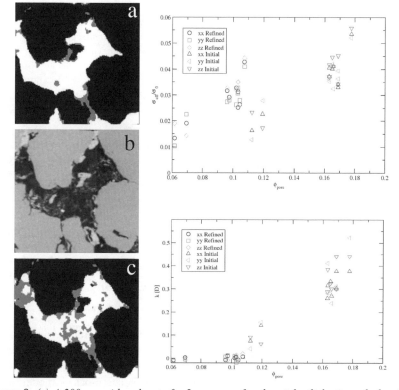

Figure 9. *(a) A 200 μm wide sub-set of a 5mm core of a clay-rich siliclastic rock showing a "naïve" segmentation based on the tomogram alone. Black represents the grain phase, gray the clay phase and white the pore space. (b) The registered SEM of the same region. (c) A refined segmentation based on the SEM. The top plot shows the effect both on estimated porosity and simulated conductivity for naïve and refined segmentations. The conductivity simulation is finite element based and assumes the presence of a clay phase halves the conductance. Surface conductivity is ignored. Permeability was simulated on sub-volumes of 360^3 voxels, and run in three orthogonal directions. The lower plot shows a marked change when clays are more realistically considered. In this case the clay phase is assumed to be essentially non-permeable. The lab-based measurement of permeability on the whole 50 mm core found 0.009 D, consistent with the SEM-informed segmentation.*

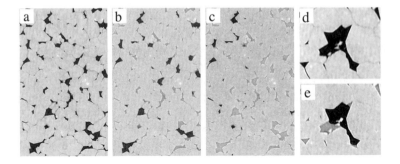

Figure 10. *(a) A 500 μm wide tomographic subset of a dry siliclastic, followed by successive water floods (b), (c). (d), (e) show the level of detail possible in menisci. The next task is to computationally extract interfacial curvatures on the entire dataset*

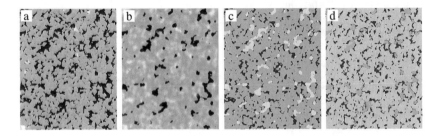

Figure 11. *(a) A 600 μm wide tomographic subset of a dry sucrosic dolomite; (b) registered slice after spontaneous imbibition with 0.1M CsI solution; (c) phase separation of pore space allows separation of water (blue) and residual gas (yellow) in 3D. (d) shows the distribution of the trapped phase (green) based on an ordinary percolation simulation in 3D image having the same water saturation as observed experimentally (Kumar, 2009).*

Notwithstanding the many complexities, the imaged saturation distributions shown in Figure 11(c) show that the trapped gas phase is concentrated in the largest pores. This suggests that a simple percolation type mechanism may be relevant in describing the process (Wilkinson, 1984). In a percolation-based model of an imbibition displacement the residual (trapped) gas saturation is formed when the displaced fluid (gas) stops percolating or becomes disconnected. This ordinary percolation threshold is easily determined on the 3D image data by incrementally removing spheres in the covering radius map from smallest to largest and tracking when the sphere map first disconnects across the entire core, Figure 11(d).

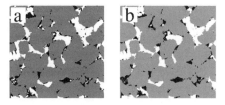

Figure 12. *A 800 μm wide slice from a siliclastic showing dolomitic and anhydrite cement, before (a) and after (b) flushing with low salinity brine (10 milliMolar range). Both tomograms have been registered and 3-phase segmented (black is the pore space, gray is the quartz phase, and white the dolomitic/anhydrite cement phase). CT does not distinguish the different cements well in this case. The red dots in (a) indicate regions where material has mobilized or dissolved.*

Particulate migration: In a geological context, mineral solubility plays a major role in fluid permeability. As part of a broader study into the effects of salinity of the connate brine on clay migration, Figure 12 shows how single dolomite particles can be mobilized with a drop in brine salinity. The registration of the mineralogical map from SEM, Figure 13, with the tomogram implicates the role of sulfate in the surface conversion of dolomite into anhydrite. This leads to the dislodgement of the dolomite which is then caught by the fluid flow (Lebedeva, 2009).

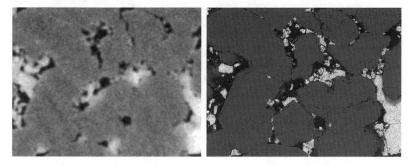

Figure 13. *A 300 μm wide tomographic slice from the siliclastic above registered with a mineralogical map from SEM for dolomite (green) and anhydrite (red). Note that the dolomitic phase on the left is often poorly distinguished from the quartz phase owning to small size. Without registration the difference between surface roughness and a distinctly different phase can be difficult to resolve.*

Reactive Fluids: pore structures of rocks can be strongly altered when subject to reactive fluids. For example, the acid dissolution of carbonates leads to a wide range of alteration of pore morphologies; due to the complexity of the pore structures, experiments are necessary to identify the controlling mechanisms of alteration. Micro-CT experiments coupled with 3D registration allow one to probe alteration at the pore scale. A simple reactive experiment was undertaken; a limestone core was vacuum saturated in a bath of CO_2 saturated water at 10 atm for 96 hours. Imaging was undertaken before and after the saturation experiment. Image registration enabled us to perfectly align the two images and locally mapping the evolution of the porous microstructure of the carbonate during dissolution/precipitation (Figure 14). The dissolution was mapped despite the sample being moved from the beamline for the 96 hour experiment. A number of longer-term experiments can be undertaken using the registration process rather than having to dedicate an entire CT beamline to a slow process.

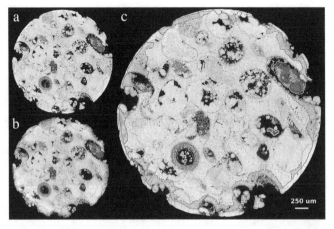

Figure 14. *Slices from two registered tomograms of a 5 mm diameter oolitic carbonate core exposed to carbonic acid (10 atm CO_2 for 96 hours). (a) The initial state, and (b) after dissolution. (c) The red line superimposed on the initial state corresponds to the new boundary after dissolution and the shared green area shows the consequential increase in porosity of the microporous regions. Particles that have been displaced are marked with a yellow line, while particles that are a result of reprecipitation have been indicated with blue.*

Mechanical modeling: The morphology of fractured porous media is very complex and involves heterogeneities at several length scales. Analysis of fractures as with sub-resolution porosity relies upon an accurate determination of X-ray optical and reconstruction parameters. The blurring in tomographic data when these parameters are not well defined will readily cause loss of feature definition, but critically loss of phase (fracture) connectivity.

Finite Element Modeling (FEM) is a numerical technique which can be used for solving static elastic equations over a discretized volume and we can simulate the elastic deformation of porous materials and calculate the acoustic response of arbitrary complex media (within resolution limits). More often than not, the computed results are in good agreement with the experiments (Roberts, 2002; Arns, 2002, Gerchka, 2006). Bohn and Garboczi (Bohn, 2003) present a comprehensive description of the FEM's numerical implementation as applied to 3D digitized images, using a conjugate-gradient algorithm to solve the discretized set of the equations. Although we can use the FEM to determine the numerical solution of any partial differential equation, we consider only the equation of linear elasticity here. Figure 15 shows a 2D slice of a real 3D-fractured medium and the analytical approximation for estimating its properties. For the actual calculations on the fractured samples, we assume that the sample is dry and assign phase moduli of dolomite to the solid phase (K = 69.4 GPa and G = 51.6 GPa). Figure 15 (c),(d),(e) shows slices through the displacement vector fields for three component fields and the fractured sample after the energy is minimized. These results show that displacements on the fracture have been reflected in other directions, indicating that they occurred because of the fracture (which induced a shear-like effect).

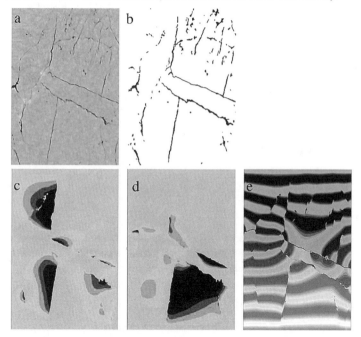

Figure 15. *(a) A single slice through the fractured sample after anisotropic filtering, (b) the corresponding segmented dataset. (c, d, e) show unit displacements in the three orthogonal directions (see Madadi, 2009 for details)*

6. Conclusions

While the estimation of micro-petrophysical properties in 3D still provides enormous opportunities much of the future utility of micro-CT lies in the ability to quantitatively correlate other measurements with this technique, and over a greater range of length scales. Due to the need for up-scaling micro-CT both mineralogy and material moduli mapping will receive attention in the near future. Clearly the management of hydrocarbon reserves is one area to benefit from micro-CT, but studies on regolith and groundwater, mineralization and hydrothermal synthesis, and microfossil identification are just a few to see the benefits of quantitative 3D imaging.

7. Acknowledgements

The authors acknowledge the member companies of the Digital Core Consortium for providing funding support and the Australian Partnership for Advanced Computation for supplying computing resources. We thank also the Australian Research Council for ongoing research support.

8. References

Arns, C.H. *et al.*, "Computation of Linear Elastic Properties from Microtomographic Images: Methodology and Agreement Theory and Experiment," *Geophysics,* vol. 67, no. 5, 2002, pp. 1396–1405.

Blunt, M.J. and Scher, H., "Pore level modeling of wetting", *Phys. Rev. E*, 52, 6387-6403 (1995)

Bohn, R.B. and Garboczi, E.J. User Manual for Finite Element Difference Programs: A Parallel Version of NIST IR 6269, tech. report 6997, US Nat'l Inst. Standards and Tech., 2003.

Frangakis, AS, Stoschek, A, Hegerl, R. "Wavelet transform filtering and nonlinear anisotropic diffusion assessed for signal reconstruction performance on multidimensional biomedical data", *IEEE Trans. Biomed. Eng. 48*, 213 (2001)

Gerchka, V. and Kachanov, M. "Effective Elasticity of Rocks with Closely Spaced and Intersecting Cracks," *Geophysics*, vol. 71, no. 3, 2006, pp. D85–D91.

Kumar M, Senden TJ, Knackstedt MA, Latham SJ, Pinczewski, V, Sok RM, Sheppard AP and Turner ML. "Imaging of Pore Scale Distribution of Fluids and Wettability", *Petrophysics,* 50(4) 311-321 (2009)

Latham, S.J., Varslot,T. and Sheppard, A.P. "Automated registration for augmenting micro-CT 3D images", in *Proc. 14th Biennial Computational Techniques and Applications Conf, J. ANZIAM*, vol.50, pp. C534--C548, 2008.

Lebedeva, E., Senden, TJ, Knackstedt, M.A and Morrow, N. "Improved Oil Recovery From Tensleep Sandstone – Studies of Brine-Rock Interactions by Micro-CT and AFM." *15th European Symposium on Improved Oil Recovery*, Paris 29 April 2009

Lenormand, R. and Zarcone, C., "Role of Roughness and edges during imbibition in square capillaries", *SPE 13264, Presented at 59th Annual Technical Conference,* Houston, 1984.

Madadi, M; Jones, AC; Arns, CH, and Knackstedt, MA. "3D Imaging and Simulation of Elastic Properties of Porous Materials", *Computing in Science & Engineering,* 11(4), 65-80, 2009.

Morrow, N.R.: "A Review of the Effects of Initial Saturation, Pore Structure, and Wettabilityon Oil Recovery by Waterflooding," *Proc. North Sea Oil and Gas Reservoirs, Trondheim* (Dec. 2-4, 1985) Graham and Trotman, Ltd., London,, 179-91 1987.

Perona, P., and Malik, J, "Scale-space and edge-detection using anisotropic diffusion" *IEEE Trans. Pattern Anal. Machine Intell.* 12, 629 (1990)

Roberts, A.P. and Garboczi, E.J. "Computation of the Linear Elastic Properties of Random Porous Materials with a Wide Variety of Microstructures," *Proc. Royal Soc.* London, vol. 458, no. 2021, 2002, pp. 1033–1054.

Sakellariou A, Arns CH, Sheppard AP, Sok RM, Averdunk H, Limaye A, Jones AC, Senden TJ and Knackstedt, MA. "Developing a virtual materials laboratory." *Materials Today,* 10, 44-51 (2007)

Sethian, J. A., "A fast marching level set method for monotonically advancing fronts", *Proc. Natl. Acad. Sci.* USA (1996) 93, 1591

Sheppard AP, Sok RM, and Averdunk H "Techniques for image enhancement and segmentation of tomographic images of porous materials", *Physica A (339),* 145-151 (2004)

Wilkinson, D., (1984), "Percolation in immiscible displacement", *Phys. Rev. A,* 34, 1380-1391.

Withjack, E. "Computed tomography for rock-property determination and fluid flow visualization." *SPE Formation Evaluation,* 696–704, (1988).

Segmentation of Low-contrast Three-phase X-ray Computed Tomography Images of Porous Media

P. Bhattad[*] — C. S. Willson[**] — K. E. Thompson[***]

* Louisiana State University
Department of Chemical Engineering
Baton Rouge, LA 70803
* pradeep@lsu.edu, *** karsten@lsu.edu

** Louisiana State University
Department of Civil and Environmental Engineering
Baton Rouge, LA 70803
cwillson@lsu.edu

ABSTRACT. *X-Ray Computed Tomography (XCT) is an important tool to study porous-media microstructure and fluids present within the void space. In the presence of multiple fluid phases (e.g. air-water in soil science or oil-water-gas in petroleum engineering), the contrast between the fluid phases becomes important for accurate image segmentation. In some cases (e.g. a white light source or low flux), it is not possible to illuminate one of the fluid phases. The result is then a single image containing multiple phases that may contain overlapping peaks of the fluid phases due to little difference in the absorption coefficients. Building upon work done in medical-image-processing research, we have adopted a nonlinear anisotropic diffusion technique to remove noise from the XCT image that also leads to improved peak separation in the image histogram. The noise-free image is the then segmented using indicator kriging and the results are compared with segmentation results obtained using absorption-edge imaging.*

KEYWORDS: *porous media, segmentation, three phase, anisotropic diffusion, indicator kriging*

1. Introduction

In 3D imaging of porous media using XCT, attenuation coefficients of the chemical species and phase density dictate the x-ray absorption and result in a grayscale-absorption-image. Solids present in the natural porous media typically contain minerals with high x-ray attenuation resulting in a grayscale image with good contrast between the solid and void phases. For porous media systems where more than one fluid phase is present, a small amount of chemicals with high attenuation coefficients can be added to one of the fluid phases to increase the contrast in the resulting grayscale-absorption-image. Selective illumination of one or more doped fluids is possible using a monochromatic x-ray beam through the selection of specific energies above and below the element-specific x-ray absorption edge(s). However, this is not possible when using a polychromatic x-ray beam resulting in a single image containing two or more phases.. Accurate segmentation of the phases is critical for accurate calculation of interfacial areas, curvatures, etc.

The segmentation of an image is performed to partition the image into regions with similar predefined criteria such as similarity of the intensity values, connected components or bounding edges. Edge-based techniques include level set, active contour and watershed techniques (Sheppard, Sok, and Averdunk, 2004 and Vincent, Soille 1991). Histogram-based techniques such as simple and multiple thresholding, clustering and Indicator Kriging operate based on the discontinuity or abrupt change in the intensity values in the histogram. The accuracy of segmentation techniques is affected by the noise present in the image. Noise in XCT images can arise from various sources, such as scattered x-ray photons impinging on the detector to create speckle noise, imperfections in the detector causing ring artifacts or error in the reconstruction causes blurring (Herman, 1980). Impurities in natural samples and little separation in attenuation values of phases can lead to blurring and merging of two or more peaks in the image histogram. The main challenge for any noise removal algorithm is to be able to distinguish the phase edges and small features in the image and selectively alter only the noisy voxels. The literature contains many methods ranging from simple low pass averaging filters such as the Gaussian filter and the median filter to process-based filters such as anisotropic diffusion and wavelet transform. Here, we demonstrate the application of nonlinear anisotropic diffusion for de-noising of an image of a multiphase porous media system and threshold indicator kriging for segmentation of the phases.

2. Anisotropic diffusion

The diffusion process equilibrates fluctuations in a property to its neighborhood. In image processing, the noise in a grayscale image can be visualized as the fluctuations in the grayscale values. The diffusion equation (equation [1]) is applied to the grayscale (intensity) values $I(x, y, z)$ to equilibrate the noise to its surrounding neighborhood. Perona and Malik (Perona and Malik, 1990) proposed a non-linear isotropic diffusion approach where the diffusion coefficient is an inverse function of the local image gradient (equation [2]). This approach causes the diffusive flux to be minimized across the edges since image gradient is maximum at the phase edges. The tensor form of the diffusion coefficient leads to nonlinear anisotropic diffusion. The anisotropy causes the tangential component of the diffusion coefficient to remove noise around the object edges, while minimizing the diffusion normal to the object boundaries. The diffusion tensor calculation takes into account local features such as the gradient in the grayscale values and directionality of the gradients in the form of eigenvalues and eigenvectors. These local features are calculated using the structural tensor smoothed using the Gaussian filter of width σ. The eigenvector $e_{i,j}$ orients the co-ordinates along the local image structure and the relative magnitude of the eigenvalues λ_i can be used to identify the nature of the feature in the image such as corners, edges, planes and isolated noise voxels. Refer to Spies et $al.$ (Scharr, and Spies, 2005 and Spies, Jahne, and Barron, 2002) for more details on the eigenvalue analysis.

$$\frac{\partial I}{\partial t} = div(\boldsymbol{C} \cdot \nabla I); \; I(x, t = 0) = I_0 \tag{1}$$

$$c(|\lambda_i|) = \frac{1}{\left(1 + \frac{|\lambda_i|}{K}\right)}; \; C_{ij} = e_{ij} \begin{bmatrix} c(\lambda_1) & 0 & 0 \\ 0 & c(\lambda_2) & 0 \\ 0 & 0 & c(\lambda_3) \end{bmatrix} e_{ji} \tag{2}$$

An edge stopping parameter (K) is used to distinguish the noise present in the homogenous region of the image from the edges. This information is difficult to obtain by visual observation of the image histogram. Eigenvalue analysis shows that the smallest eigenvalue averaged over the whole image is a good measure of the uncorrelated noise in the image (Spies, Jahne, and Barron, 2002). Since equation [1] is solved iteratively and there is no definite measure of the enhancement of the image, a pre-defined stopping criterion does not exist for the diffusion application on an image. Here, we use the initial measure of the image noise as the starting point and the iterative process is stopped when the confidence level (Scharr, and Spies, 2005) reaches the pre-specified value of 0.75.

3. Indicator kriging

Kriging (Rossi, Dungan, and Beck, 1994) is a commonly used geo-statistics technique to estimate a value at an un-sampled location based on measured values in the neighborhood locations. However from an image segmentation point of view, the probability of the unknown having a value larger than the threshold is more useful. Indicator kriging provides this probability using linear-weighted combinations of the measured values in the neighborhood. Oh and Lindquist (Oh, and Lindquist, 1999) describe an indicator kriging technique for image thresholding in conjunction with the two point spatial covariance. The algorithm is implemented in two steps: (1) The thresholding step involves *a priori* assignment of a portion of the image voxels into two populations; and (2) the kriging step where the kriging weights are calculated from the two point spatial covariance relation calculated using smoothed indicators. The linear combinations of weights and indicators are then used to calculate the probability of the unknown voxel belonging to either population. The two sets of weights are calculated from the globally averaged covariance relations for each assigned population and hence the weights are globally averaged and need to be calculated only once. Information on balancing of the negative weights and how to treat the voxels near the image boundaries can be found in the original paper (Oh, and Lindquist, 1999). Their algorithm also uses two passes of a majority filter to remove the isolated voxels in the a priori assigned population. In our implementation we do not follow this approach since we have found that it introduces artifacts such as joining of separated particles and loss of thin features from the segmented image.

4. Results and discussion

Here, we will demonstrate our methods on a three phase image of size $350*350*350$ voxels^3 (with a voxel resolution of 9.92 μm) that contains quartz sand (solid), a chlorinated solvent (tetrachloroethylene) as oil phase and water that was doped with 8% KI. The imaging was performed at X-ray energies below (33.07 keV, Figure 1a) and above (33.27 keV, Figure 1b) the iodine absorption k- edge. Visual observation of the below edge image (Figure 1a) show three distinct phases. However, in the image histogram (Figure 1c), the peaks for the oil and water phases are merged due to similarity in absorbance at this energy and large variance in the phase gray scale values. This makes the below-edge image ideal for our study. The above-edge image will be used for comparison of the final segmentation results. The below-edge image was first subjected to the nonlinear anisotropic diffusion process to remove noise from the image. Visual inspection of Figures 2a-2b shows that there is over smoothing of the image at higher user-specified values of K resulting in blurring, while the lower values of K results in lower diffusive flux and require higher numbers of iterations to achieve satisfactory noise removal. The selection of the appropriate K values (faster noise removal while still preserving the features) is an important consideration. After the application of anisotropic diffusion, the image can be segmented using threshold-based techniques. The impact of noise removal on image segmentation can be seen in Figure 3. Simple thresholding of the original image (Figure 3a) shows a significantly larger number of noise voxels than the image that was processed with the nonlinear anisotropic noise removal algorithm.

This negates the need to employ post-segmentation noise removal schemes such as the majority filter.

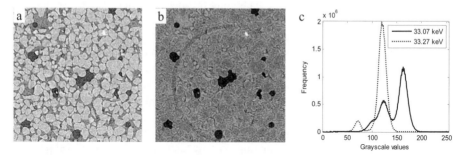

Figure 1. *Section of the image acquired at two energies above and below the iodine absorption K edge. (a) Below edge image at energy 33.07keV, (b) above edge image at energy 33.27keV, (c) histogram of the below edge image. All images are of size 350*350*350*

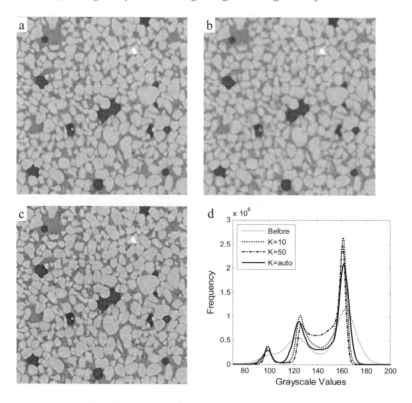

Figure 2 . *Section of the image showing the effect of edge stopping parameter (K) using EED filter. (a) K=10 after 10 iterations; (b) K=50 after 10 iterations and (c) K selected automatically using the average of the lowest eigenvalue, and stopping criteria 0.75 was also applied. Results achieved in 3 iterations and (d) histogram showing peak separation.*

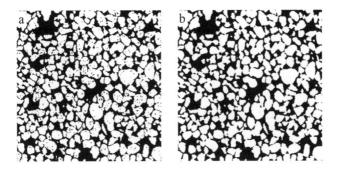

Figure 3. *Section of the image showing the results of simple thresholding test: (a) simple threshold results showing misclassified voxels in the original image; (b) simple threshold results showing the smoothing of noise voxels using anisotropic diffusion*

From the histogram of the filtered image (Figure 2d) we can see that there is a range of values between two peaks, and in the filtered image these values lie mostly at the phase boundaries. Hence a single threshold value cannot be used for the segmentation of phases.

Here, we use an indicator kriging-based algorithm, which uses two threshold values for the partial segmentation of the image into two phases and grayscale values between the two thresholds are assigned to either of the phases using the kriging statistics. Indicator kriging is designed mainly for data with two univariate peaks (e.g. two phases in a grayscale image). In our segmentation scheme for the three-phase-image, we choose to first segment the phase that results in a smaller standard deviation on either side of the chosen threshold values. For example, thresholding the solid phase (the far right peak in Figure 2d) results in a maximum standard deviation of 11.98 in the *a priori* assigned voxels and the segmentation of the oil phase (the far left peak in Figure 2d) results in a maximum standard deviation of 16.78.

Thus, we segment the solid phase on the first pass (Figure 4c). In the second pass, the segmented image is subtracted from the clean gray scale image so that the range of grayscale values in the image represents only the two fluid phases (Figure 4d). During the next step, the segmented solid phase is not used to calculate the two point covariance correlation, and during the probability calculation, the segmented part is treated as the boundary with an indicator value of 0.5. The resulting segmented image (Figure 4e) contains all three phases: oil, water and solids.

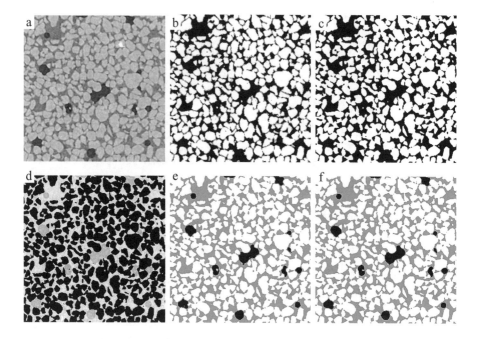

Figure 4. *Section of the image showing the Indicator Kriging segmentation for 3 phase image. (a) Noise free below-edge image, (b)Partially thresholded image: White-solid, Black-Rest of the image containing Oil and Water, Gray-Unassigned voxels, (c) Segmented image showing solid (White) and rest of the image (Black), (d) Solid subtracted from the clean grayscale image, (e) Fully segmented three phase image using below edge image only white-Solid (57.508%), Gray-Water (35.436%) and Black-Oil phase (7.056%) and (f)Fully segmented three phase image using both above and below edge images, white-Solid (57.508%), Gray-Water (35.395%) and Black-Oil phase (7.097%).*

Following the method described above, the above-edge image is also subjected to nonlinear anisotropic diffusion (with the same parameters) and the indicator kriging program is used for segmentation of the oil phase from the rest of the image. The solid phase segmentation from the below edge image is the same as shown in Figure 4c. The solid-segmented image is combined with the oil-segmented image to obtain a three-phase image (Figure 4f). Comparison of the two images (Figures 4e and 4f) shows a nearly identical segmentation.

5. Conclusion

Nonlinear anisotropic diffusion, aided by automatic eigenvalue analysis of the edge stopping parameter, effectively removes noise from porous media images while preserving the local structure in the data and not blurring the edges. The peak separation and reduction in the grayscale variance of each peak increases the accuracy of the image segmentation. Comparison of the segmentation of a single image segmented to separate three phases with a segmented image obtained from above- and below-edge images show that this approach is effective and should prove useful for separating phases in XCT images where absorption edge imaging is not an option.

6. Acknowledgements

The authors acknowledge Keegan L. Roberts, Civil and Environmental Engineering, Louisiana State University (LSU) for providing us with the XCT images. The images were collected at GeoSoilEnviroCARS (Sector 13), Advanced Photon Source (APS), Argonne National Laboratory. GeoSoilEnviroCARS is supported by the National Science Foundation - Earth Sciences (EAR-0622171) and Department of Energy - Geosciences (DE-FG02-94ER14466). Use of the Advanced Photon Source was supported by the U. S. Department of Energy, Office of Science, Office of Basic Energy Sciences, under Contract No. DE-AC02-06CH11357. The authors acknowledge the members of the PoreSim Research Consortium at LSU (BP America, Millipore, Schlumberger) for financial support.

7. References

Sheppard, A.P., R.M. Sok, and H. Averdunk, "Techniques for image enhancement and segmentation of tomographic images of porous materials", *Physica A*, 2004, 339, 145-151.

Vincent, L. and P. Soille, "Watersheds in digital spaces: an efficient algorithm based on immersion simulations", *IEEE Trans. Pattern. Anal. Machine Intell.*, 1991, 13, 583-598.

Herman, G.T., "On the noise in images produced by computed tomography", *Computer Graphics and Image Processing*, 1980, 12, 271-285.

Perona, P. and J. Malik, "Scale-space and edge detection using anisotropic diffusion", *IEEE Trans. Pattern Anal. Machine Intell.*, 1990, 12, 629-639.

Scharr, H. and H. Spies, "Accurate optical flow in noisy image sequences using flow adapted anisotropic diffusion", *Signal Process.Image Comm.*, 2005, 20, 537-553.

Spies, H., B. Jahne, and J.L. Barron, Range flow estimation, *Computer Vision and Image Understanding*, 2002, 85, 209-231.

Rossi, R.E., J.L. Dungan, and L.R. Beck, Kriging in the Shadows - Geostatistical Interpolation for Remote-Sensing, *Remote Sensing of Environment*, 1994, 49, 32-40.

Oh, W. and W.B. Lindquist, Image thresholding by indicator kriging, *IEEE Trans. Pattern Anal. Machine Intell.*, 1999, 21, 590-602.

X-ray Imaging of Fluid Flow in Capillary Imbibition Experiments

Influence of compaction and localized deformation

C. David* — L. Louis* — B. Menéndez* — A. Pons** — J. Fortin** — S. Stanchits*** — J.M. Mengus****

*Université de Cergy-Pontoise – Laboratoire Géosciences et Environnement Cergy
5 mail Gay-Lussac, F-95031 Cergy-Pontoise, France
christian.david@u-cergy.fr, laurent.louis@u-cergy.fr, beatriz.menendez@u-cergy.fr

**Ecole Normale Supérieure – Laboratoire de Géologie
24 rue Lhomond, F-75252 Paris Cedex 05, France
fortin@geologie.ens.fr, apons@clipper.ens.fr

***GeoForschungsZentrum
Telegrafenberg, D-14473 Potsdam, Germany
stanch@gfz-potsdam.de

****Institut Français du Pétrole
1-4 avenue de Bois-Préau, F-92852 Rueil-Malmaison, France
j-marie.mengus@ifp.fr

ABSTRACT. *We used an industrial scanner to image capillary imbibition processes in order to get some insight into fluid motion processes in several porous rocks. Images obtained at different stages in standard capillary rise experiments on intact and damaged rock samples are analyzed. The geometry of the water front depends on the rock microstructure and its curvature changes as the water rises up to the top of the samples. The mechanical damage was induced either during creep experiments at increasing stress levels or in standard triaxial tests. The continuous recording and localization of acoustic emissions was extremely useful in identifying clusters where damage was concentrated. We show that the velocity and the geometry of the water front are strongly disturbed by localized or distributed damage.*

KEYWORDS: *x-ray imaging, capillary imbibition, reservoir rocks, image analysis*

1. Introduction

Imaging techniques have been widely used to investigate microstructural features in porous rocks. Among them x-ray CT scanning methods are of common use in geosciences (Van Geet *et al.*, 2000). In order to study hydromechanical coupling in reservoir rocks, we developed a three stage methodology combining x-ray CT scanning on intact rock samples, mechanical tests with acoustic emissions recording to investigate the localization of induced damage (Fortin *et al.*, 2006), and again x-ray CT scanning on the damaged samples after unloading. Some of our results have been published recently (David *et al.*, 2008); we present here a complementary data set on a larger number of porous rocks, focusing on the influence of rock microstructure, compaction processes and compaction bands on fluid flow patterns.

2. Description of the selected rocks and methodology

Four rocks were selected for the present study: two sandstones – Bentheim sandstone (Romberg quarry, Nothwestern Germany) and Vosges sandstone (Eastern France), as well as two calcareous rocks – Majella grainstone (central Apennines, Italy) and Saint-Maximin limestone (Paris basin, France). Both calcareous rocks have been extensively studied by Baud *et al.* (2009). Table 1 gives the rock composition and some relevant attributes like porosity, mean grain diameter, permeability and peak diameter on mercury porosimetry spectrum.

	LIMESTONES		SANDSTONES	
	Majella	*Saint-Maximin*	*Vosges*	*Bentheim*
porosity (%)	30	37	25	22
mean grain diameter (μm)	54	140	110	200
major composition*	CA(100%)	CA(61%) QZ(39%)	QZ(50%) FE(30%) CL-OX(20%)	QZ(95%) CL(5%)
peak diameter on Hg porosimetry spectrum (μm)	16.2	26.4	28.7	28.8
permeability (D)	0.145	0.97	0.165	1.1

* QZ=quartz; FE=feldspar; CA=calcite; CL=clays; OX=oxydes

Table 1. *Microstructural attributes of the studied rocks*

The methodology has been described in detail by David *et al.* (2008). In few words, dry cylindrical rock samples (diameter 40 mm, length 80 mm) are put on a stand under an industrial scanner available at Institut Français du Pétrole, France. Our experimental device allows for water to be brought into contact with the bottom part of the sample: at this point capillary rise starts and one image of the central cross-section is taken every 2.5 seconds by the x-ray CT scanner, until the water reaches the top of the core sample. Figure 1 shows a picture and a sketch of the experimental device.

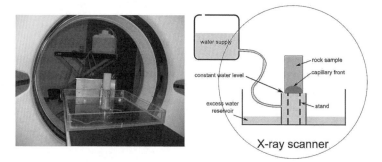

Figure 1. *Experimental device used for our experiments*

In Figure 2, we show an example of a snapshot taken during capillary rise in each rock sample. The "processed" images show how image processing techniques described by David *et al.* (2008) make the analysis of water invasion easier thanks to an enhancement of density contrast between dry (dark) and wet (bright) zones.

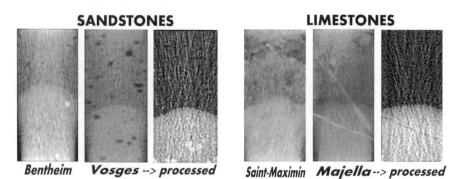

Figure 2. *Example of raw images obtained and image processing applied*

An automated procedure allows us to obtain from each series of enhanced images the height of the water front in the center and at the border, from which the velocity of capillary rise and the mean radius of curvature of the water front are calculated (David *et al.*, 2008). Complementary experiments were performed without imaging, to measure the mass of water entering the rock samples during capillary rise.

3. Comparison of capillary parameters and microstructural interpretation

As can be seen in Figure 2, the density maps reveal heterogeneous features in every rock sample: white (i.e. dense) spots or lines in Bentheim and Majella respectively, black spots (i.e. more porous zones) in Vosges and Saint-Maximin. Significant differences appear also in the geometry of the water front. A quantitative analysis of our capillary experiments is shown in Figure 3.

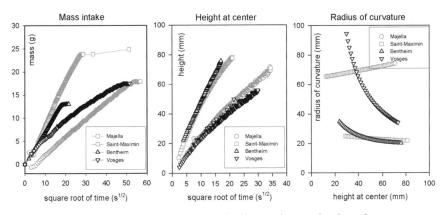

Figure 3. *Evolution of mass intake, height of water front and radius of curvature*

The geometry of the water front changes during capillary rise, and is rounder in Bentheim and Saint-Maximin compared to Majella and Vosges. The mass and height are usually plotted vs. the square root of time. Two parameters can be inferred from these plots: firstly the "A" parameter (cm.s$^{-1/2}$) which is the slope of the height plot (assumed to be linear) and secondly the "B" parameter (g.cm^{-2}.s$^{-1/2}$) which is the slope of the mass plot normalized to the surface in contact with water (also assumed to be linear, at least until water reaches the top). Considering imbibition into a single pipe, the "A" parameter which describes the kinetics of water imbibition should be a constant depending on the pipe diameter. A third parameter is calculated, the Hirschwald coefficient, which is the water saturation when the water front reaches the top of the sample, derived from the mass at the plateau in Figure 3 (Zinszner *et al.*, 2007). The results are summarized in Table 2.

	LIMESTONES		SANDSTONES	
	Majella	*Saint-Maximin*	*Vosges*	*Bentheim*
Hirschwald coefficient	70%	70%	68%	60%
capillary parameter A (cm^{-1}.s$^{-1/2}$)	0.185	0.332	0.181	0.375
capillary parameter B (g.cm^{-2}.s$^{-1/2}$)	0.0335	0.0702	0.0244	0.0477
radius of curvature at mid-height	69	25	65	25

Table 2. *Capillary parameters of the studied rocks*

To go further, we intend to interpret these data in terms of rock microstructure. Figure 4 shows how the capillary parameters scale with microstructural parameters.

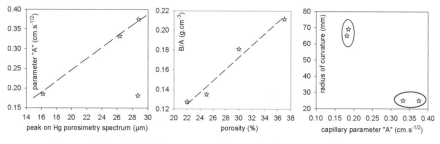

Figure 4. *Correlation between capillary and microstructural parameters*

Except for one point, there is a clear correlation between capillary parameter "A" and the peak diameter on the mercury porosimetry spectrum. This shows that the relevant length scale for capillary processes is given by the mercury injection test (the so-called pore entry diameter), in agreement with our previous results (David *et al.*, 2008). Notice however that according to Washburn's equation in the single pipe model (Zinszner *et al.*, 2007), one would expect a linear relationship between capillary parameter "A" and the square root of pore diameter, rather than diameter. The data point clearly off the trend is the one corresponding to the Vosges sandstone. This discrepancy might be explained by the large clay content in this rock, as it is well known that clay coating or filling significantly modifies capillary processes in porous rocks. The middle plot shows a clear correlation between the ratio of capillary parameters "B/A" and the rock porosity. This ratio is related to the mass intake per unit volume during capillary imbibition, and is clearly controlled by the bulk porosity. Finally on the last plot we found for all rocks that the smaller the kinetics of water imbibition (low A parameter), the flatter the water-air interface (large radius of curvature at mid height), or equivalently the faster the capillary rise, the smaller (i.e. the rounder) the capillary front.

4. Influence of mechanical compaction on capillary processes

In our previous study, we have conducted mechanical tests in a triaxial setup on Bentheim sandstone samples in the brittle regime. It was clearly shown that mechanical damage strongly modifies fluid flow patterns when comparing CT scan images during capillary imbibition tests on intact and deformed rock samples (David *et al.*, 2008). Here we present our results for Saint-Maximin limestone, a much weaker rock than Bentheim sandstone. In Figure 5 we present the mechanical data (i.e. axial/volumetric strain vs. differential stress) for the creep tests conducted at increasing stress levels at constant confining (3 MPa) and pore (1 MPa) pressures as well as x-ray density maps for the tested rock sample before and after deformation respectively.

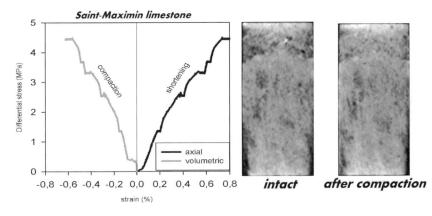

Figure 5. *Mechanical data for creep test and comparison of density maps for the tested Saint-Maximin limestone sample*

During compaction, the intensity of damage was recorded by one single acoustic emission transducer. The sample was unloaded before macroscopic failure occurred, and no fracture was observed after the test. No dilatancy was observed during the loading stage. The comparison of the density maps before and after deformation is not very easy because of the poor resolution of the CT scanner (about 400 µm). Nevertheless the deformed sample looks more homogeneous, as many dark spots have disappeared, probably due to pore collapse.

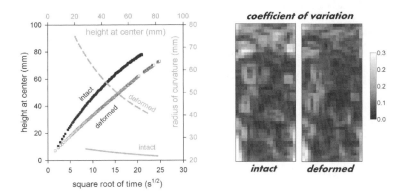

Figure 6. *Left: comparison of results for capillary tests before and after deformation. Right: evolution of the coefficient of variation on 3 x 3 pixel blocks in coarsened images*

Figure 6 shows the results for the capillary tests as well as for the analysis of deformation by image processing. On the capillary plots, there is a clear difference between the intact and the deformed sample. The height evolution vs. square root of time is much more linear for the deformed sample, and the kinetics of capillary rise is reduced. Indeed the "A' parameter decreases from 0.332 cm.s$^{-1/2}$ for the intact sample to 0.281 cm.s$^{-1/2}$ for the deformed sample, a mild but significant decrease. Moreover for the deformed sample, the geometry of the capillary front is nearly flat, with a much larger radius of curvature. The images on the right in Figure 6 show a map of the values corresponding to the coefficient of variation (COV = standard error normalized to mean value) calculated on 3 x 3 pixel blocks after decreasing 8 times the image resolution in order to capture higher scale features on the x-ray density maps. Lower values for the COV are obtained for the deformed sample showing that the heterogeneity has been reduced by mechanical compaction. This is in agreement with the visual inspection of the raw density maps in Figure 5.

5. Influence of stress-induced compaction bands

Following the experimental procedure described by Stanchits *et al.* (2009), a set of Bentheim sandstone samples with diameter 50 mm and length 105 mm were deformed in classical triaxial tests at 195 MPa confining pressure and 10 MPa pore pressure. Circumferential notches with 0.8 mm width and 5 mm depth were machined in order to guide the compaction bands in the central part of the samples.

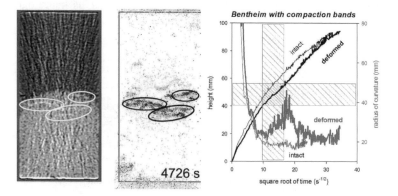

Figure 7. *Left: enhanced image at intermediate stage of capillary rise. Middle: location of acoustic emissions during triaxial test. Right: capillary plots and inferred parameters*

Figure 7 presents an image from CT scanning of the capillary test on one deformed sample, the location of acoustic emissions generated during triaxial loading and the capillary plots for the intact and deformed samples. Acoustic emissions clearly image the presence of compaction bands, which appear on the CT scan image as white linear sections: this is so because the density of the compaction bands increases due to porosity reduction induced by pore collapse and grain crushing. We calculated a reduction of about 15% in porosity within the bands compared to the intact rock. The capillary plots show a significant difference in the capillary rise kinetics, with a lower "A" parameter on the average for the deformed sample. Furthermore when the capillary front sweeps the compaction bands region (shaded area in Figure 7), a sharp increase in the radius of curvature is observed which shows that the water front flattens in the region where compactant deformation is concentrated.

6. Conclusion

We have investigated the influence of mechanical deformation on fluid flow using a combination of different techniques, among them x-ray CT scanning. Different porous rocks have been studied, and the results have been interpreted on a microstructural basis. It is shown that conducting capillary imbibition experiments under CT scanning is a powerful technique to image fluid flow patterns in reservoir rocks and how they are affected by deformation.

7. References

Baud, P., Vinciguerra, S., David, C., Cavallo, A., Walker, E., Reuschlé, T., "Compaction and failure in high porosity carbonates: mechanical data and microstructural observations", *Pure Appl. Geophys.*, vol. 166, 2009, p. 869-898.

David, C., Menéndez, B., Mengus, J.M., "Influence of mechanical damage on fluid flow patterns investigated using CT scanning imaging and acoustic emissions techniques", *Geophys. Res. Lett.*, vol. 35, 2008, doi:10.1029/2008GL034879.

Fortin, J., Stanchits, S., Dresen, G., Guéguen, Y., "Acoustic emission and velocities associated with the formation of compaction bands in sandstone", *J. Geophys. Res.*, vol. 111, 2006, doi:10.1029/2005JB003854.

Stanchits, S., Fortin, J., Guéguen, Y., Dresen, G., "Initiation and propagation of compaction bands in dry and wet Bentheim sandstone", *Pure Appl. Geophys.*, vol. 166, 2009, p. 843-868.

Van Geet, M., Swennen, R., Wevers, M., "Quantitative analysis of reservoir rocks by microfocus X-ray computerised tomography", *Sediment. Geol.*, vol. 132, 2000, p. 25-36.

Zinszner, B., Pellerin, F.M., *A Geoscientist Guide to Petrophysics*, Paris, Technip, 2007.

Evaluating the Influence of Wall-Roughness on Fracture Transmissivity with CT Scanning and Flow Simulations

D. Crandall — G. Bromhal — D. McIntyre

National Energy Technology Laboratory
3610 Collins Ferry Road
Morgantown, WV 26507-0880
USA
Dustin.Crandall@pp.netl.doe.gov
Grant.Bromhal@netl.doe.gov
Dustin.McIntyre@netl.doe.gov

ABSTRACT. *Combining CT imaging of geomaterials with computational fluid dynamics provides substantial benefits to researchers. With simulations, geometric parameters can be varied in systematic ways that are not possible in the lab. This paper details the conversion of micro-CT images of a physical fracture in Berea sandstone to several tractable finite volume meshes. By computationally varying the level of detail captured from the scans we produced several realistic fracture geometries with different degrees of wall-roughness and various geometric properties. Simulations were performed and it was noted that increasing roughness increased the resistance to fluid flow. Also, as the distance between walls was increased, the mean aperture approached the effective aperture.*

KEYWORDS: *fracture flow, finite volume CFD, Navier-Stokes, permeability*

1. Introduction

Fractures often provide primary flow pathways within low-permeability geological media. For decades researchers have examined the effects of fracture wall-roughness on fluid transport by separating fractures, measuring the fracture surfaces with a profilometer, reconstructing analogue geometric models, and performing simulations and experiments within the reconstructed geometries (e.g. Brown, 1989; National Research Council, 1996; Konzuk and Kueper, 2004). Micro-CT scanning of rough-walled fractures has been shown to be an accurate method by which fracture geometries can be acquired without using destructive techniques (Karpyn et al., 2007). Computational fluid dynamics (CFD) have been used to model fluid transport in rough-walled fracture geometries (e.g. Eker and Akin, 2006; Hughes and Blunt, 2001; Crandall et al., 2009), but there is scant literature on methods to convert CT images into tractable CFD meshes. Likewise there are few studies that attempt to determine relationships between fracture geometric properties and fracture transmissivity. In this paper both topics are discussed.

The original fracture used to generate the computational domains for this study was created experimentally by Karpyn et al. (2007) within a 101.4 mm long, 25.4 mm diameter core of Berea sandstone. Prior to fracturing, the core had 18% porosity and a permeability of 200 mD. Micro-CT scanning was performed on the fractured sample with a voxel resolution of 27.3 μm × 27.3 μm × 32.5 μm, where 32.5 μm is the slice thickness measured along the length of the rock core. A total of 3116 slices were obtained along the core length. Image processing was conducted to isolate the fracture and identify the inner structure of the crack. Examination of CT registrations allowed the identification of voxels that were characteristic of the fracture void structure and voxels of surrounding rock were removed. This method extracts an approximation of the void structure due to partial volume effects. The differences in reconstructed void space, and its effect on fluid flow, is the main driver behind this investigation. This fracture data set was converted to a smaller data set, with 767 slices along the length and a voxel resolution of 109 μm × 109 μm × 132 μm, to reduce computational memory requirements when converting the data set to CFD tractable geometries.

2. Conversion of CT data to CFD models

The micro-CT scanned fracture was converted into several three-dimensional (3D) meshes using three software packages; Amira™ (Visage Imaging Inc.), GAMBIT® (Ansys Inc.), and TGrid™ (Ansys Inc.). The resulting meshes were used in the finite-volume CFD solver, FLUENT® (Ansys Inc.) to obtain the flow solutions. Three different fracture roughnesses were obtained by varying the amount of detail retained from the scanned data. The rough fracture walls were separated so that the mean fracture aperture (b_{avg}) was 1, 0.8, 0.6, 0.5, and 0.4 mm. Nine fracture meshes are shown in Figure 1 with two-dimensional (2D) profiles along the fracture lengths. Grayscale aperture plots are shown to illustrate the aperture variability within the geometries, and that the overall aperture distribution is maintained between the different models. In addition, 2D profiles along the fracture lengths and through the center of the width are shown, with the height increased three-fold to illustrate the fracture-wall roughness.

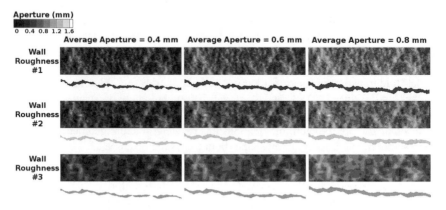

Figure 1. *Nine representative rough fracture meshes with 2D profiles*

The CT data set was initially constructed into a workable 3D volume by importing the data set into the visualization software Amira and exporting the ASCII geometry. A minor amount of data manipulation was performed in this step to remove isolated voxels and ensure connectivity of the bulk fracture void space. This ASCII file was read into the pre-processing software TGrid. Within TGrid the fracture walls were "wrapped", creating a completely enclosed 3D geometry that was then meshed using tetrahedral cells. Three volume meshes were generated with different fracture wall-roughnesses by varying the size of the triangles the original ASCII volume was wrapped with. These three meshes were then imported into GAMBIT and the fracture walls were separated by set amounts, resulting in the 15 fracture geometries used for this study. These fracture geometries were then remeshed within GAMBIT, refined within TGrid, and exported to FLUENT for CFD simulations. This procedure is shown graphically in Figure 2. The items to the left in Figure 2 involve capturing the original data and reconstructing an adequate representation of the volume while the items on the right are required to create a tractable computational mesh for detailed CFD simulations. The entire process, from data acquisition to mesh construction, is important for obtaining useful and reliable CFD results.

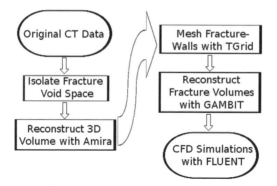

Figure 2. *Construction flow chart*

2.1. Roughness properties of fracture meshes

The Joint Roughness Coefficient (JRC) is an index from 0 to 20 that was proposed by Barton and Choubey (1977) as a method of describing the shear strength of rock fractures. Tse and Cruden (1979) developed an accurate empirical strength of rock fractures.

	Z_2	JRC	H	D_f	b_{orig} (mm)
#1	0.213	10.35	0.64	1.36	0.708
#2	0.171	7.24	0.71	1.29	0.796
#3	0.134	3.72	0.72	1.28	0.883

Table 1. *Fracture mesh roughness properties*

Tse and Cruden (1979) developed an accurate empirical relationship between the root-mean-square of the fracture profile wall slope, Z_2, and the JRC. To determine the Z_2 of a fracture profile, M measurements of the fracture surface height, y, some small distance apart, Δx, can be used,

$$Z_2 = \sqrt{\frac{1}{M(\Delta x)^2} \sum_{i=1}^{M} (y_{i+1} - y_i)^2} .$$ [1]

This was determined to fit the empirical relationship by Tse and Cruden (1979).

$$JRC = 32.2 + 32.47 \log Z_2$$ [2]

A more common descriptor of the fracture wall-roughness is the fractal dimension, D_f. The Hurst exponent, H, is related to the D_f of 2D fracture profiles by $D_f = 2 - H$. The H of a profile can be determined by using the variable-bandwidth technique (Dougan *et al.*, 2000), where the difference in height, Δy, between two points a distance 's' apart are measured and the standard deviation, σ_s, of this displacement is calculated,

$$\sigma_s = \sqrt{\frac{1}{M-s}\sum_{i=1}^{M-s}\left(y_{i+s} - y_i\right)^2} \ . \qquad\qquad [3]$$

This process is repeated for a range of s. If the fracture profile is well described by fractional Brownian motion, a non-Fickian diffusive scaling will exist between the σ_s and the H, such that $\sigma_s \propto s^H$. By plotting σ_s and s on a log-log plot the H can be determined from the slope of straight lines that exhibit this linear behavior. The JRC and the D_f are listed in Table 1 for the three different wall-roughnesses studied.

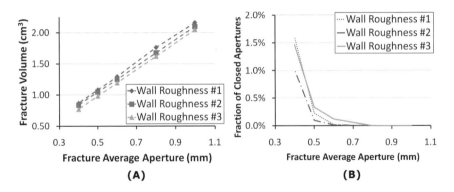

Figure 3. *(A) fracture mesh volume (B) obstructed areal fraction of the fracture plane*

2.2. *Varying apertures fracture meshes*

It has been shown by previous experimental studies that the fracture wall-roughness has an effect on the resistance of fluids flowing through a fracture (e.g. Brown, 1989; National Research Council, 1996). Wall-roughness also creates asperities (i.e. zero-aperture regions) when the fracture walls are close. While it has been shown that these asperities will deform under high confining pressures (Gangi *et al.* 1996) in practice it is rather difficult to perform flow experiments in the lab with these high pressures on individual rock fractures. In addition, these high pressures will tend to permanently deform the rock surfaces. By isolating the individual fracture walls from the three previously described computational meshes we reconstructed 15 separate numerical models, with three different wall roughnesses and mean apertures of 1, 0.8, 0.6, 0.5, and 0.4 mm. The fracture volume and the areal fraction of the fracture plane that is occupied by obstructions for all 15 meshes are plotted in Figure 3. The obstructed apertures have been shown to restrict flow in fractures by previous researchers (e.g. Brown, 1989; Konzuk and Kueper, 2004) so it is worth noting that all fracture meshes with mean apertures greater than 0.5 mm had open flow domains, with less than 0.25% of the region closed to fluid flow. The volume of the fractures was observed to decrease linearly as the fracture aperture is reduced, with a slope of approximately 2.

2.3. *Modeling parameters*

For all geometries the computational mesh was refined to capture the details of flow. Refinement was performed until the change in mean velocity of the flow solution varied less than 1% between successive mesh refinements. The final refined cell count in each 3D fracture model varied from 2 to 5 million cells. Numerical models and methods used are identified in Crandall *et al.* (2009). The steady solution was deemed adequate when both the velocity residuals were less than 10^{-4} and less than 0.1% of the mass flux was lost due to numerical approximations.

3. Results and discussion

Water was modeled as moving through the fracture geometries due to imposed pressure gradients, ∇P. The transmissivity, T, of the fracture was measured with the following relationship,

$$Q = T \frac{\nabla P}{\mu}, \qquad [4]$$

where Q is the mass flow rate calculated through the fracture and μ is the viscosity of water (10^{-3} Pa*s). In a similar fashion the effective aperture, δ, of the fractures were determined with the following "cubic-law" relationship,

$$Q = -\delta^3 \frac{W \cdot \nabla P}{12\mu}, \qquad [5]$$

with the fracture width, W = 2.1 cm. These two fracture flow resistance values were used to quantify the changes in the fluid transport within the reconstructed volumes as functions of the wall roughness and the mean apertures.

3.1. *Flow changes with fracture wall roughness*

The changes in the δ and T with JRC and D_f, are shown in Figure 4(A) and 4(B), respectively for three different wall roughnesses and a mean aperture of 0.6 mm. As can be seen from these plots, with increases in the roughness both T and δ decreased. When T and δ decrease, the resistance to fluid transport increases. Interestingly, the large change in T and δ between the two smoothest fractures, Wall Roughness #2 and #3, corresponds to a relatively large change in the JRC, unlike the small change of D_f shown in Figure 4(B)

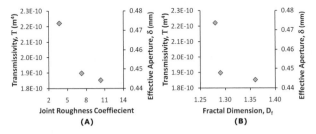

Figure 4. *T and δ of fractures with an aperture of 0.6 mm as functions of (A) JRC and (B) D_f.*

3.2. *Flow changes with fracture aperture*

The changes in the transmissivity, T, and the ratio of the effective aperture to the mean aperture, δ/b, as a function of the mean aperture are shown in Figure 5. Ideally, T would be increasing as the cube of the mean aperture, and the ratio of the effective and mean aperture would be 1. Figure 5(A) shows that T does increase with increasing mean aperture. However, the fit to the ideal line becomes increasingly worse with smaller fracture apertures. This is even clearer in Figure 5(B), where the ratio of δ/b is far from 1 below apertures of 0.7 mm and gets closer to 1 (i.e. the cubic law is more applicable) as b increases and the effects of micro-scale fracture wall roughness on the fluid flow become smaller.

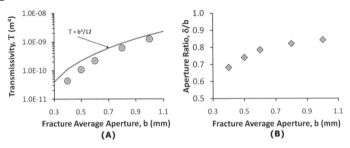

Figure 5. *Calculated transmissivity (A) and δ/b (B) as functions of the mean aperture for fractures with a Wall Roughness #3*

4. Conclusions

The following conclusions can be inferred from these studies:

– reconstruction of natural geometries from CT scans to computational volumes can alter the micro-scale properties significantly,

– alteration of micro-scale wall-roughness of rock fractures has a significant effect on fluid transport through a fracture,

– the Joint Roughness Coefficient appears to characterize small changes in micro-scale wall roughness in a way that correlates well to fluid resistance parameters,

– as the mean distance between fracture walls increases the fracture wall roughness becomes less important, and

– the transmissivity of fractures decreases more rapidly when fracture walls touch at discrete locations.

5. Acknowledgements

This research was performed while Dustin Crandall held a National Research Council Research Associateship Award at the National Energy Technology Laboratory in Morgantown, WV USA. The authors also wish to thank Zuliema T. Karpyn at Penn State University for providing the CT scanned fracture data.

6. References

Barton, N., Choubey, V., "The Shear Strength of Rock Joints in Theory and Practice", vol.10, No.1-2, pp.1-54, *Rock Mechanics, 1977.*

Brown, S.R., "Transport of Fluid and Electric Current through a Single Fracture", vol.94, No.B7, pp.9429–9438, *Journal of Geophysical Research*, 1989.

Crandall, D., Bromhal, G., Karpyn, Z.T., "Numerical Simulations Examining the Relationship between Wall-Roughness and Fluid Flow in Fractures", submitted, *Journal of Contaminant Hydrology*, 2009.

Dougan, L.T., Addison, P.S., McKenzie, W.M.C., "Fractal Analysis of Fracture: A Comparison of Dimension Estimates", vol.24, No.4, pp.383-392, *Mechanics Research Communications*, 2000.

Eker, E., Akin, S., "Lattice Boltzmann Simulation of Fluid Flow in Synthetic Fractures", vol.65, pp.363-384, *Transport in Porous Media*, 2006.

Hughes, R.G., Blunt, M.J., "Network Modeling of Multiphase Flow in Fractures", vol.24, pp.409-421, *Advances in Water Resources*, 2001.

Karpyn, Z.T., Grader, A.S., Halleck, P.M., "Visualization of Fluid Occupancy in a Rough Fracture Using Micro-Tomography", vol.307, No.1, pp.181–187, *Journal of Colloid and Interface Science*, 2007.

Konzuk, J.S., Kueper, B.H., "Evaluation of Cubic Law Based Models Describing Single-Phase Flow through a Rough-Walled Fracture", vol.40, pp.W02402, *Water Resources Research*, 2004.

National Research Council, Committee on Fracture Characterization and Fluid Flow, *Rock Fractures and Fluid Flow: Contemporary Understanding and Applications,* National Academy Press, Washington DC, 1996.

Tse, R., Cruden, D.M., "Estimating Joint Roughness Coefficients", vol.16, pp.303-307, *International Journal of Rock Mechanics and Mining Sciences Geomechanical Abstracts,* 1979.

In Situ Permeability Measurements inside Compaction Bands Using X-ray CT and Lattice Boltzmann Calculations

N. Lenoir — J. E. Andrade — W. C. Sun — J. W. Rudnicki

Northwestern University
Department of Civil and Environmental Engineering
2145 Sheridan road
Evanston, IL 60202
USA
n-lenoir@northwestern.edu
j-andrade@northwestern.edu
steve.sun@u.northwestern.edu
jwrudn@northwestern.edu

ABSTRACT. *The paper presents some results of the characterization of microstuctural differences inside and outside compaction bands in natural sandstone. Specimens of Aztec sandstone were scanned at the synchrotron APS, at Argonne National Laboratory (USA). Porosity measurements obtained by image analysis show a sharp transition inside/outside the compaction band. Finally, preliminary permeability measurements were conducted via the lattice Boltzmann method and demonstrate the dramatic decrease in permeability caused by the reduced porosity inside the compaction band relative to the host rock.*

KEYWORDS: *compaction band, sandstone, permeability, lattice Boltzmann, synchrotron*

1. Introduction

The study of deformation bands including compaction bands (CB) in granular porous rocks is an important topic for many applications ranging from oil and gas reservoirs to CO_2 sequestration to geologic repositories for nuclear waste. For instance, compaction bands have been found in sandstones typical of possible repository host rock for CO_2 sequestration. There is the question of how formation, presence or extension of compaction affects the injection of CO_2 into reservoirs. Due to the observed dramatic decrease in permeability relative to the host rock (see for a review, Holcomb et al., 2007), CB could play the important role of flow barriers and trap the CO_2 injected or, on the other hand, adversely affect injection.

Over the last decade, observations and permeability characterizations in geomaterials have been made at different scales. In particular, lattice Boltzmann (LB) calculations applied to 3D images obtained from X-ray CT have proven to be an efficient method in this area (e.g. Arns et al., 2004, White et al., 2006). However, the method has only been used for characterizing permeability in pristine sandstones without the presence of a CB (e.g. Fredrich et al., 2006). The only LB computation inside of CB was conducted on a simulated 3D image created from a 2D SEM image with a geostatistical model (Keehm et al., 2006). Thus, to our knowledge, this is the first instance of application of these methods to a sample obtained from a CB in the field.

The purpose of this work is to use direct experimental observations and LB calculations to characterize the difference in permeability inside and outside CB obtained from field cores of sandstone. Small specimens of Aztec sandstone cored from the inside and outside of a CB were scanned at a synchrotron in order to characterize the 3D change in microstructure. Porosity was then measured from the images, allowing us to obtain a clear profile of the void-space and observe the transition from the host rock into the CB. From the 3D images, some preliminary permeability measurements were performed using the LB method and show the dramatic effect of CB on the rock's flow properties.

2. Material and methods

The material studied in this work is Aztec sandstone from the Valley of Fire State park (Nevada, USA) where Sternlof (2006) had conducted a field study of the geometry and physical characterization of the bands. This rock is an eolian sandstone with a mean grain size of about 0.25 mm and a porosity ranging from 15 to 25% (Antonellini et al., 1994). Blocks were taken in the field and small cylindrical specimens were cored with a diamond core drill inside and outside of a CB. In order to prevent disturbance of the microstructure due to the coring process, the block was first impregnated with a low viscosity epoxy. The left part of Figure 1 shows a small block of sandstone with the compaction band appearing in lighter color and on the right, the block used in this study, first impregnated with epoxy. The two white rectangles indicate the location of the cored specimens presented here. Specimen no. 1 was cored inside the CB with the inside/outside CB transition included at its end. The specimen no. 2 was cored just straight ahead of no. 1 outside of the CB. Both were cored in a direction perpendicular to the CB. The specimen diameter is about 4.2 mm. The heights are about 20 and 11 mm for no. 1 and no. 2 respectively.

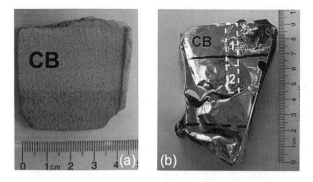

Figure 1. *(a) Piece of Aztec sandstone obtained from the field. The compaction band CB appears in lighter color here; (b) core impregnated of epoxy with the coring location of the 2 specimens presented here*

For scanning the specimens, a synchrotron was chosen as it provides the possibility of obtaining a monochromatic beam. This avoids the artefact of beam hardening that can occur with images obtained with laboratory x-ray scanners. Thus, the image segmentation is simplified and makes it possible to obtain a high quality binary image of the microstructure. The specimens were scanned on the DND-CAT 5-BM-C beamline at the Advanced Photon Source, Argonne National Laboratory (USA). The specimens were mounted on a rotating stage by using wax in order to avoid any motion of the sample during the scanning process. The data were collected with a scintillator associated with a CCD camera. The energy used was 25 keV and 1200 radiographies were recorded over half rotation of the specimen, with an exposure time of 725 ms. The spatial resolution was a voxel size of 6x6x6 μm^3. The specimens were scanned in different overlapping sequential sections equally spaced (every 3 mm) in order to obtain a full image of the cylinders. The images are 16bit and the volume size for a section was 1299x1299x600 voxels which corresponds to a height of 3.6 mm. The images were then stacked and the final volumes were 800x800x3360 and 800x800x1860 voxels for specimen no. 1 and no. 2 respectively.

3. X-ray CT images and porosity measurements

The CT images clearly show the microstructure of the sandstone and the difference in the microstructure inside and outside the CB. Figure 2 presents a horizontal slice obtained from specimen no. 2 on the left and one from no. 1 on the right. Aztec sandstone is mainly composed of round quartz grains (99%) appearing in light gray, and some hematite appearing in white. The porous network mainly filled of epoxy appears in dark gray and black. A packing phenomenon of the grains inside the CB is clearly observed by simple comparison of the images as previously noted by Sternlof (2006) in his field study of CB in Aztec sandstone.

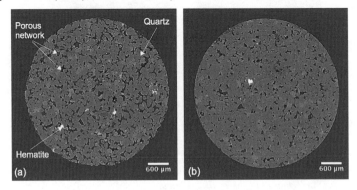

Figure 2. *(a) Horizontal CT slice from specimen no. 2.*
(b) Horizontal CT slice from specimen no. 1

Porosity measurements were performed through image analysis by using the open source software ImageJ (http://rsbweb.nih.gov/ij/). The 16 bit images were first cropped to keep an inner core of 600 voxels width and then converted to binary form in order to measure the permeability. As the grains, especially the boundary grain/pore space, are well defined due to the high spatial resolution, the segmentation was done by a "manual" threshold method. The choice of the threshold value was made as follows. A wide range of gray level values was used and the corresponding porosities were calculated. The obtained profile consists of a linear increase in porosity while within the porous area and a sudden change of slope once the grain boundaries are reached. The onset of this sudden increase was chosen as the threshold value. An example of the obtained 3D porous network after segmentation is shown in Figure 3 on a small volume (200x200x200 voxels) obtained from outside the CB. The mean porosity obtained is about 21% outside and 14.5% inside the CB which is in accordance with the literature (Antonellini *et al.*, 1994; Holcomb *et al.*, 2007). Study of the transition from the CB to the outside material in more detail is possible in the images from specimen no. 1. Figure 3 shows the evolution of the porosity along the continuous specimens no. 1 and no. 2.

The transition is very sharp. The porosity decreases from 21% to 14.5% within a zone of approximately 2 mm. Note, however, that within this transition zone is a region of slightly lower porosity (12%). We will perform further studies to determine whether this lower porosity (than with the band itself) is a feature of this particular specimen is more prevalent. If the latter, it may give some insight into factors controlling the width of the band (since it seems to bear no simple relation to the grain size).

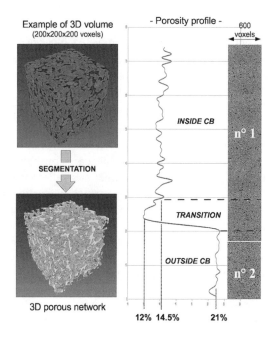

Figure 3. *Image segmentation and porosity profile inside/outside compaction band*

4. Permeability measurements with lattice Boltzmann method

The lattice Boltzmann (LB) method is employed to estimate the permeability of selected sub-domains of the specimen inside and outside the CB. Known for its ability to handle complex geometry, LB method takes advantage of the kinetic theory to seek probability density functions that leads to velocity and pressure fields satisfying momentum and mass balance principles without solving the Navier-Stokes equation directly at the continuum level (Succi, 2001). In this study, the effective permeability is estimated via the procedures described as follows. First, a hydraulic gradient is applied along a basis direction. Then, LB simulations are performed until the pore-fluid velocity field reaches its steady state. Since LB flow is by nature compressible, the permeability corresponding to the basis direction is obtained via Darcy's law augmented with the continuity equation. Some preliminary simulations were conducted on two small sub-domains (50x50x50 voxels), one inside and one outside the CB, with a measured porosity about 15% and 20% respectively. Figure 4 depicts the obtained simulated flows within the 3D porous network. The permeability along the vertical direction of the sub-domain, i.e. perpendicular to the CB, outside and inside are estimated as $k_{in}=1\times10^{-12}$ m^2 and $k_{out}=6\times10^{-19}$ m^2 respectively. Such a dramatic change on permeability reflects the geometrical difference of pore space inside and outside the band. Figure 4 suggests that due to the packing phenomenon inside the CB, pore-scale flow channels are likely to be either severely narrowed or completed blocked when a CB forms. Although the difference in porosity between the two sub-domains is about 25%, the results of the LB simulation show that the geometric changes of flow channels triggered by the formation of CB can significantly reduce the flow velocity and cause a low effective permeability.

Note that the obtained ratio in permeabilities k_{in}/k_{out} is about 10^{-7} which is higher than the one reported in the literature (10^{-2}-10^{-3}) (Sternlof *et al.*, 2004). The smaller ratio obtained here occurs because the calculated permeability is an averaged property having a spatial fluctuation that is larger at smaller scales. LB simulations at a sufficiently large scale would suppress spatial fluctuation but have a high computational cost. The choice for the size of the samples used for permeability calculations is bounded by the minimum size necessary for continuum to make sense and a maximum size imposed by our computational capability. Work is ongoing to parallelize the code and to overcome that limitation of computational cost. Thus, LB simulations will be performed on a large range of volumes in order to determine the elementary volume that will give more representative results of the macroscopic permeability. Moreover, those preliminary results were obtained in a single direction. The anisotropy in permeability will be investigated in a forthcoming publication.

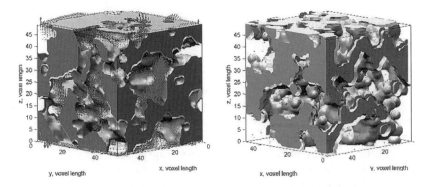

Figure 4. *Fluid flow (represented by arrows) within the 3D porous network obtained with LB computation outside (on the left) and inside the CB (on the right)*

5. Conclusions and perspectives

We have used synchrotron measurements to obtain detailed 3D images of the microstructure outside and inside a CB formed in the field (Valley of Fire, Nevada). These images are consistent with previous observations of increased grain packing within the band. Porosity measurements made it possible to study the transition from outside of the compaction band to inside in more detail than has been done previously. The porosity decreases sharply, over about 2 mm. A new feature observed here is that within the transition zone, there is a region with porosity slightly lower than within the CB. Further investigation is needed to determine whether this feature is particular to this specimen or more typical of CBs formed in the field. Preliminary permeability measurements obtained using LB method on small volumes outside and inside the CB gave promising results. As anticipated, the reduced porosity in the band results in a dramatic decrease in permeability. Work is ongoing to reduce the computation time in order to perform computations in larger volumes and more samples, which will give more representative results. Moreover, it should be emphasized that this study concerns a single specimen and, though the results are promising, more analysis is necessary to get a faithful picture of the overall hydraulic properties in the context of compaction bands.

6. Acknowledgements

The authors would like to thank P. Eichhubl, University of Texas, for providing the sandstone blocks and D.T. Keane, APS-Argonne Laboratory, for his guidance and help on the scanning process. This work is funded by the U. S. Department of Energy, Office of Basic Energy Sciences, under the project no. DE-FG02-08ER15980. Use of the Advanced Photon Source was supported by the U. S. Department of Energy, Office of Science, Office of Basic Energy Sciences, under Contract No. DE-AC02-06CH11357.

7. References

Antonellini, M.A., Aydin, A., Pollard, A.A. and D'Onfro, P., "Petrophysical study of faults in sandstone using petrophic image analysis and x-ray computerized tomography", *Pure and Applied Geophysics*, 1994, vol.143, No.1-3, pp.181–201.

Arns, C.H., Knackstedt, M.A., Val Pinczewski, W. and Martys, N.S., "Virtual permeametry on microtomographic images", *Journal of Petroleum Science and Engineering*, 2004. vol.45, No.1-2, pp.41–46.

Fredrich, J.T., Di Giovanni, A.A. and Noble, D.R., "Predicting macroscopic transport properties using microscopic image data", *Journal of Geophysical Research*, 2006, vol.111, B03201.

Holcomb, D., Rudnicki, J.W., Issen, K.A. and Sternlof, K., "Compaction localization in the Earth and the laboratory: state of the research and research directions", *Acta Geotechnica*, 2007 vol.2, No.1, pp.1–15.

Keehm, Y., Sternlof, K. and Mukerji, T., "Computational estimation of compaction band permeability in sandstone", *Geosciences Journal*, 2006 vol.10, No.4, pp.499–505.

Sternlof, K.R., Chapin, J.R., Pollard, D.D. and Durlofsky, L.J., "Permeability effects of deformation band arrays in sandstone", *AAPG Bulletin*, 2004, vol.88, No.9, pp.1315-1329,

Sternlof, K.R., Structural geology, propagation mechanics and hydraulic effects of compaction bands in sandstone, PhD Thesis, Stanford University, 2006.

Succi, S., *The Lattice Boltzmann Equations*, Oxford Science Publications, 2001.

White, J.A., Borja, R.I. and Fredrich, J.T., "Calculating the effective permeability of sandstone with multiscale lattice Boltzmann/finite element simulations", *Acta Geotechnica*, 2006, vol.1, No.4, pp.195–209.

Evaluation of Porosity in Geomaterials Treated with Biogrout Considering Partial Volume Effect

Y. Kobayashi* — S. Kawasaki* — M. Kato* — T. Mukunoki** — K. Kaneko*

*Hokkaido University
Kita 13 Nishi 8, Kita-ku, Sapporo 060-8628
Japan
ykobayashi@geo-er.eng.hokudai.ac.jp
kawasaki@geo-er.eng.hokudai.ac.jp
kato@geo-er.eng.hokudai.ac.jp
kaneko@geo-er.eng.hokudai.ac.jp

** Kumamoto University
2-39-1, Kurokami, Kumamoto 860-8555
Japan
mukunoki@kumamoto-u.ac.jp

ABSTRACT. This study evaluates the change in porosity of a biogrout specimen using micro-focus x-ray CT scanner and image processing. First, we selected the suitable geomaterials for evaluation of porosity. The results showed that glass beads whose particle size was 0.50-0.71mm were suitable. Second, we calculated the porosity of the glass beads using a method for evaluation of porosity based on a maximum likelihood thresholding method considering the effect of mixels, which are pixels having a two-phase structure. The result showed that the ratio of the porosity of grouted geomaterial to that of ungrouted one was 0.98-0.99, whereas the value estimated by the measurement of changes in concentration of calcium ion was 0.98. Thus, both values closely agreed. Therefore, this study clarifies that the method enables us to evaluate small changes in porosity with great accuracy.

KEYWORDS: porosity, threshold, mixel, biogrout, micro-focus x-ray CT

1. Introduction

Grouting is a technology to inject cementitious and non-cementitious materials into the ground to enhance the strength and/or to reduce the permeability of sands and sandy soils. Recently Portland cement and sodium silicate whose pH is relatively high (e.g. approximately pH=11-12) have been used as the most common compounds in grouting. In contrast, biogrout (Kawasaki et al., 2006) is an eco-friendly grouting technology. It uses microbial metabolism and consists primarily of calcium carbonate ($CaCO_3$), which fills the voids and cracks in soils and rocks. The avoidance of high pH ingredients enables surrounding soils not to be contaminated in the process.

According to the past study, the strength of the biogrouted specimens is about one-tenth part of the expected value (Shoji, 2008). To increase the strength, we needed to increase $CaCO_3$ precipitation. This means, first of all, that we had to quantitatively understand $CaCO_3$ precipitation. That's why, in this study, we evaluated the change of the *porosity* caused by $CaCO_3$ precipitation with micro-focus X-ray computed tomography (CT) system and image processing. Additionally, we considered effectiveness of the method for evaluation of porosity.

The biogrouted specimen contains two materials: *solid* consisting of geomaterial and precipitated $CaCO_3$ and *liquid* consisting of grout solution. In order to evaluate the porosity, we need the geomaterials whose CT number histogram gained with x-ray CT is bimodal. Based on partial volume effect, the important parameters for the geomaterials having bimodal CT number histogram are as follows: the density and the particle size of geomaterial. Thus, we first selected the geomaterials having bimodal CT number histogram based on this standpoint. Second, we calculated the porosity of the geomaterials by using a method for evaluation of porosity.

As a result, the change of porosity evaluated in this study agreed with the estimated value by the measurement of concentration of calcium ion. Thus, we showed the effectiveness of the method for evaluation of porosity.

2. Methodology

2.1. *Selection of the suitable geomaterials for evaluation of porosity*

We selected the geomaterials whose CT number histogram gained with x-ray CT is bimodal so that we can set a threshold between liquid and solid and divide these to evaluate the porosity.

Table 1 shows the properties of the geomaterials tested in this study. Specimen preparation is as follows: (1) mix the materials shown in Table 2 to make grout solution; and (2) put this grout solution and geomaterial into a polystyrene container (diameter = 21.0 mm, height = 54.8 mm).

Then, we measured the specimen with the micro-focus x-ray CT scanner (TOSCANER 30900 μhd) made by Toshiba IT & Control Systems Corporation. The scan mode was single slice, and the size of a voxel was 5 μm×5 μm×16 μm. The x-ray tube voltage was 130 kV and x-ray tube current was 62 μA.

Geomaterials	Density (g/cm^3)	Particle size (mm)
Souma sand	2.65	0.1 (50% particle size)
Gum tips	1.15	2
Plastic pellets	1.53	φ3×h4
Steel balls	7.87	1.0
Glass beads 1	2.50	0.105-0.125
Glass beads 2	2.50	0.50-0.71

Table 1. *Properties of geomaterials*

1M Tris-HCl	100 (ml)
Glucose	3.0 (g)
Calcium nitrate tetrahydrate	23.6 (g)

Table 2. *Constituents of grout solution*

2.2. Data collection for evaluation of porosity

We collected the data for evaluation of porosity with respect to the geomaterials selected in the preceding section. Though it was desirable to use single grouted and ungrouted specimens, we used two different ones to prevent yeast from becoming extinct caused by the first x-ray irradiation of the ungrouted specimen. The grouted specimen was made by mixing yeast into the specimen made in the same way as the preceding section, and then left to rest in an incubator at 25°C. On the other hand, the ungrouted specimen was made in the same way as the preceding section. We mixed no yeast into the ungrouted specimen to prevent the reaction caused by yeast from proceeding. After making the specimens, we imaged them with the micro-focus x-ray CT scanner. The scan mode, the size of voxel and x-ray tube voltage and current were the same as shown in the preceding section. To enhance the reliability, we imaged a total of 12 slices for each specimen.

3. Results

3.1. Selection of the suitable geomaterials for evaluation of porosity

Table 3 shows the histogram type and the suitability for evaluation of porosity of each geomaterial. As shown in Table 3, steel balls and glass beads 2 had a bimodal CT number histogram. Therefore, these geomaterials could be suitable for evaluation of porosity. However, an effect known as beam hardening occurred on the X-ray image of steel balls. Because beam hardening can significantly decrease the accuracy of quantitative results, steel balls were unsuitable for evaluation of porosity. Therefore we selected only glass beads 2 as the suitable geomaterials for evaluation of porosity. Glass beads 2 will hereinafter simply called glass beads.

Geomaterials	Histogram type	Suitability for evaluation of porosity
Souma sand	Monomodal	×
Gum tips	Monomodal	×
Plastic pellets	Monomodal	×
Steel balls	Bimodal	×
Glass beads 1	Monomodal	×
Glass beads 2	Bimodal	○

Table 3. *Histogram type and suitability for evaluation of porosity of geomaterials*

3.2. Data collection for evaluation of porosity

Figure 1 shows examples of x-ray image of grouted and ungrouted specimen. The range of white to light-gray color shows glass beads, while the range of black to dark gray color shows solution.

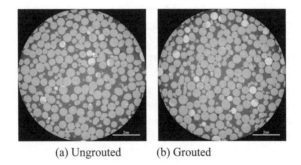

(a) Ungrouted   (b) Grouted

Figure 1. *X-ray CT images for each specimen of glass beads*

Figure 2 shows the average CT number histogram of measured 12 slices. By a difference in density between glass beads and solution, it can be identified that the left peak in Figure 2 indicates solution and the right peak indicates glass beads.

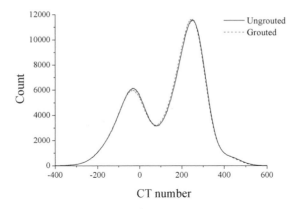

Figure 2. *CT number histogram for each specimen of glass beads*

4. Evaluation of porosity

4.1. *Thresholding method*

To calculate porosity, we needed to set a threshold dividing solid and liquid on the CT number histogram. In this study, we used a thresholding method called "*a maximum likelihood thresholding method considering the effect of mixel*" (Kitamoto, 1999) to set a threshold properly. A general description of this method is as follows.

Figure 3 shows the imaging process of a real image. Phase 1 and Phase 2 in Figure 3(a) are different materials. In this study, we can regard liquid as Phase 1 and solid as Phase 2. In the sampling and quantization process, the voxels containing two materials occur as shown in Figure 3(b). Then, the CT number of these voxels has an averaged value based on *partial volume effect*. Furthermore, in gray level conversion, these voxels have a pixel level that is intermediate between the other two pixel levels (see Figure 3(c)). These pixels, Class 3 shown in Figure 3 (c), are called *mixed pixels* (*mixels*) (Kitamoto and Takagi, 1998). In contrast, the others, Class 1 and Class 2 shown in Figure 3 (c), are called *pure pixels*.

Depending on subjectivity of threshold setter, Class 3 can be divided into both Class 1 and Class 2. Therefore, it is important to take mixel into consideration and not to depend on subjectivity of threshold setter when setting threshold. From these perspectives, this method considers the effect of mixel. This means that it can also take quantization error into account in the imaging process. Additionally, this method calculates threshold from a maximum likelihood. Hence, it can calculate the threshold by a method independent of subjectivity of the threshold setter.

According to this method, superposition shown in Figure 4 measured with x-ray CT can divided into Class 1, Class 2 and Class 3. To divide these three classes, we need two thresholds: t_1 and t_2. Figure 5 shows thresholds calculated by this method with a measured CT number histogram of a slice of Figure 1(a).

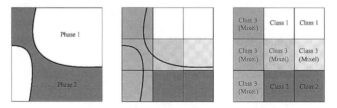

(a) Real image   (b) Sampling and quantization   (c) Image

Figure 3. *Imaging process of real image*

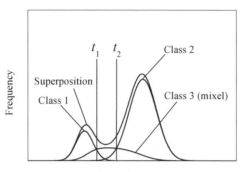

Figure 4. *Concept of threshold setting process*

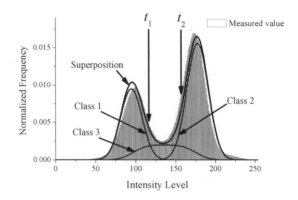

Figure 5. *A result of threshold setting*

4.2. *Calculation of porosity and discussion*

We define pure pixels of liquid as Class 1, pure pixels of solid as Class 2 and mixed pixels (mixels) as Class 3. The proportion of liquid and solid in Class 3 stochastically agreed with the proportional of liquid and solid in the entire region of the image. Thus, when α_1, α_2 and α_3 show area occupancy of Class 1, Class 2 and Class 3 respectively, Class 3 can be divided into Class 1 in the proportion of $\alpha_1\alpha_3$ / $(\alpha_1+\alpha_2)$ and into Class 2 in the proportion of $\alpha_2\alpha_3$ / $(\alpha_1+\alpha_2)$. Therefore, porosity can be calculated from the following equation:

$$\varphi = \frac{\alpha_1 + \dfrac{\alpha_1\alpha_3}{\alpha_1 + \alpha_2}}{\alpha_1 + \alpha_2 + \alpha_3} = \frac{\alpha_1}{\alpha_1 + \alpha_2} \quad [1]$$

where φ is porosity.

We calculated area occupancy of each class by integration of frequency of each class using the estimated thresholds (see Table 4). Then, we calculated porosity by substitution of calculated area occupancy into equation [1]. Moreover, we calculated average porosity of 12 slices of both grouted and ungrouted specimen (see Table 4). The ratio of average porosity of the grouted specimen (φ_2) to that of the ungrouted specimen (φ_1) was as follows:

$$\frac{\varphi_2}{\varphi_1} = 0.98$$

On the other hand, the ratio of porosity of the grouted specimen (φ_2) to that of the ungrouted specimen (φ_1) was estimated by the measurement of concentration of calcium ion as follows:

$$\frac{\varphi_2}{\varphi_1} = 0.98 - 0.99$$

As shown above, these two values closely agreed. Thus, it was clarified that the method using x-ray CT and a maximum likelihood thresholding method considering the effect of mixel as threshold setting enables us to evaluate the change in porosity with great accuracy. Moreover, from the fact that the method made it possible to evaluate such a small change such as 1-2%, we can expect that it can also evaluate a larger change in porosity.

	Area occupancy			Porosity
	Class 1 (α_1)	Class 2 (α_2)	Class 3 (α_3)	
Ungrouted	0.295	0.551	0.154	0.349
Grouted	0.288	0.557	0.155	0.340

Table 4. *Average area occupancies and porosities for each specimen*

5. Conclusion

We have evaluated porosity of biogrouted specimen by using X-ray CT and a thresholding method. Moreover we selected the suitable geomaterials for evaluation of porosity. In conclusion, the following points were made.

- Glass beads whose particle size was 0.50-0.71mm were suitable for evaluation of porosity.

- The method enables us to evaluate small change in porosity with great accuracy.

6. References

Kawasaki, S., Murao, A., Hiroyoshi, N., Tsunekawa M., and Kaneko K., 2006. "Fundamental Study on Novel Grout Cementing Due to Microbial Metabolism", *Jour. Japan Soc. Eng. Geol.*, vol. 47, no. 1, p.2-12 (in Japanese with English abstract).

Kitamoto, A. and Takagi, M., 1998. "Image Classification Method Using Area Proportion Density that Reflects the Internal Structure of Mixels", *Trans. IEICE*, vol. J81-D-2, no.6, p.2582-2597 (in Japanese).

Kitamoto, A., 1999. "A Maximum Likelihood Thresholding Methods Considering the Effect of Mixels", *Technical Report of IEICE*, vol. PRMU99-166, p.7-14 (in Japanese with English abstract).

Shoji, H., 2008. Mechanical properties of soil specimens improved by biogrout, Graduation thesis, Hokkaido University (in Japanese with English abstract).

Image-Based Pore-Scale Modeling Using the Finite Element Method

N. Lane — K. E. Thompson

Cain Department of Chemical Engineering
Louisiana State University
Baton Rouge, LA 70803
USA
nlane4@lsu.edu
karsten@lsu.edu

ABSTRACT. Image-based pore-scale modeling is a powerful computational tool for investigating transport in porous geologic materials. Network modeling has been the main tool for pore-scale modeling in the past, with the lattice-Boltzmann method becoming prevalent more recently. The finite element method offers significant advantages because of its ability to operate in multiscale and multiphysics frameworks. We perform image-based modeling of single phase flow in a Berea sandstone to begin understanding issues associated with domain size and mesh refinement. Results suggest a characteristic scale for permeability of approximately 1 mm for the Berea sample. Mesh coarsening is shown to affect both porosity and permeability. However, the flow patterns remain qualitatively consistent (with the same pores taking the dominant flow) between the meshes.

KEYWORDS: finite element, Berea, permeability, image-based, meshing, refinement

1. Introduction

Modeling of porous geologic materials such as reservoir rocks, marine sands, soils, etc. can be performed at two scales: the pore scale or the continuum scale. Pore scale models distinguish the void and solid phases and represent a more fundamental approach to modeling transport in porous media. Alternatively, continuum-scale models treat the porous material as a continuum phase and employ spatially averaged parameters such as porosity and permeability to describe transport processes. This approach is necessary in many applications because of lengths scales and other issues of practicality. However, pore-scale modeling is essential for understanding fundamental behavior and a number of techniques show promise as predictive tools that might replace certain time-consuming laboratory tests.

Computed tomography (CT) techniques have promoted significant advances in the area of pore-scale modeling because microCT and nanoCT techniques allow for non-destructive, quantitative imaging of the interior of porous rocks, soils, and sands. The use of digital images for computational modeling helps ensure that the pore morphology (e.g. pore size distributions, pore-scale heterogeneity, and spatial correlations) found in real materials are captured by the models.

In this paper, we discuss image-based modeling techniques that operate on data obtained from x-ray microtomography. Most image-based modeling techniques consist of three main steps:

1. Segmentation of the CT data: the initial grayscale image is transformed to a binary image with voxels assigned to either the solid or void phase.
2. Numerical discretization: the void space is discretized using nodes, elements, finite volumes, or pores and pore throats, depending on the modeling approach.
3. Numerical modeling: the relevant equations of motion are solved numerically to obtain velocity and/or pressure fields inside the pore space.

2. Pore-scale modeling techniques

During the past 60 years, network modeling has been the main technique for pore-scale modeling. Network modeling was originated by Fatt (1956), and is predicated on discretization of the pore space into pore and pore throats, followed by imposition of mass conservation at each pore. Flow within each pore throat is approximated by a Poiseuille-type relationship, the exact form of which depends on assumptions about the pore geometry. Early modeling employed lattice-based networks, but modern techniques include image-based models that are mapped directly from 3D digital images of porous media (Thompson et al., 2008).

The main advantage of network modeling is its outstanding computational efficiency. However, due to the severe approximations that are employed, it is not as rigorous as methods that approximate the equations of motion directly, which we refer to here as streamline-scale models. This terminology emphasizes their ability to reproduce the vector velocity field within the pore space (and hence fluid streamlines), which contrasts with network models that can be solved for pore-throat flowrates but not sub-pore-scale velocities.

In principle, streamline-scale modeling is simply the application of computational fluid dynamics (CFD) techniques within the void space of porous media. However, this is challenging because of the spatial complexity, heterogeneity, and hierarchy of scales that are present in geologic materials, and the need to incorporate boundary conditions using multiscale and/or multiphysics techniques.

The lattice-Boltzmann method (LBM) has become the method of choice for streamline-scale modeling in the geosciences (Fredrich *et al.*, 2006). Its advantages include the ability to operate directly on the voxel structure obtained from segmented XMCT images (thus eliminating the need for meshing), the easily parallelizable structure of the code, and the relative ease of simulating finite Reynolds number flows.

3. Image-based FEM modeling

Despite the popularity of LBM methods, there are applications for which other CFD techniques will prove advantageous. In this paper, we focus on preliminary results using the finite element method (FEM) for image-based modeling. The main challenge associated with the FEM for this application (and many others) is the issue of meshing. However, if effective image-based mesh generation techniques are available, then the FEM offers significant advantages in modeling hierarchical structures (which require flow to be resolved at multiple scales) and for phenomena that involve solid mechanics.

3.1. *Materials*

Current simulations were performed using three-dimensional CT images of a Berea Sandstone sample. Data used in the current simulations were obtained at the Center for Advanced Microstructures and Devices (CAMD) at LSU (Baton Rouge, Louisiana, USA). Resolution in this image was 9 microns (on the size of a voxel). The tomography image was segmented using anisotropic diffusion and indicator-kriging thresholding (described in Bhattad *et al.*, "Segmentation of low-contrast three-phase X-Ray Computed Tomography images of porous media", a separate proceedings paper at GeoX 2010).

Figure 1a shows a subsection of the segmented voxel image, which was the domain size used for the B.120 mesh. The center slice of this domain is shown in Figure 1b. For consistency, this same slice is used to plot flow results and mesh structures below.

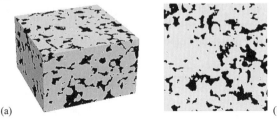

(a)   (b)

Figure 1. *Image of CT volume (a) and the slice used for presenting results (b)*

3.2. *Meshing*

Meshing was performed using an in-house mesh generation algorithm that operates directly on the segmented voxel image and offers significant flexibility in the resulting mesh. Parameters that can be used to control the mesh structure include the overall mesh resolution, levels of mesh refinement near solid/void interfaces, and the incorporation of refinement at specific locations of interest. The meshing algorithm is described in detail in a forthcoming paper.

In the current investigation, subregions were extracted from the Berea image and tetrahedral meshes were constructed using these small subsections to study effects of domain size, mesh resolution, mesh refinement, mesh quality and other factors. The work shown here is part of a larger study to understand the impact of mesh structure and mesh coarsening on pore-scale flow simulations. Discussion below is focused on only two of the relevant issues: the impact of the domain thickness in the direction of flow and the impact of mesh refinement at the solid-void interfaces (which can improve how well the mesh conforms to the pore structure).

Table 1. *Mesh properties and permeability values for the simulations*

Mesh Code	Domain Size (voxels)	Mesh Res. (voxels)	Number Elements	Number Nodes	Porosity (%)	Permeability (cm^2)
B.030	200×200×030	5,10	163,021	326,827	18.60	$3.88×10^{-8}$
B.060	200×200×060	5,10	304,367	599,216	17.25	$1.60×10^{-8}$
B.120	200×200×120	5,10	654,929	1,289,072	17.79	$1.13×10^{-8}$
B.180	200×200×180	5,10	984,665	1,932,991	18.08	$1.16×10^{-8}$
B.240	200×200×240	5,10	1,297,492	2,544,834	18.05	$1.20×10^{-8}$
C.R10	200×200×120	10,10	403,549	779,908	19.31	$2.02×10^{-8}$
C.R07	200×200×120	7,10	478,497	934,178	18.39	$1.43×10^{-8}$
C.R05	200×200×120	5,10	654,929	1,289,072	17.79	$1.13×10^{-8}$
C.R03	200×200×120	3,10	911,213	1,790,579	17.69	$1.03×10^{-8}$

Table 1 contains mesh parameters for the set of meshes used in this study. All the meshes were created using the 200×200 voxel cross section shown in Figure 1b (1.8mm×1.8mm). The series-B meshes vary in thickness (30 to 240 voxels, as indicated by the name of each mesh. The series C.R10 through C.R03 has increasing mesh refinement near the solid-void interface. For both the B- and C-series meshes, the interior tetrahedral elements have linear dimension of order 10 voxels (90 micrometers). The B series meshes contain smaller elements near the solid/void surfaces: order 2.5 voxels in linear dimension. The C-series of meshes have the same overall domain size as the B.120 mesh. However, they vary in the level of refinement at the solid/void interfaces, from approximately 5 voxels in linear dimension (C.R10) down to approximately 1.5 voxels in linear dimension (C.R03).

Figure 2 shows a comparison of the C.R10 versus the C.R03 meshes to illustrate the corresponding change in structure. These images represent a thin slice of the 3D mesh in the immediate region of the Figure 1b slice. Note that the improved resolution of the C.R03 mesh requires a more than doubling of the number of elements (see Table 1).

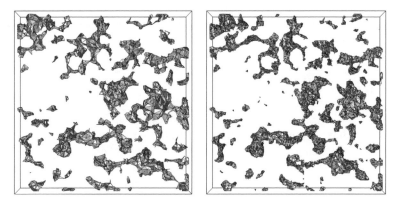

Figure 2. *Thin sections of two meshes: C.R10 (left) and C.R03 (right)*

3.3. Permeability and porosity values

Because the meshes are created using fairly coarse tetrahedral elements (relative to the voxel resolution), the void space in the segmented image is not captured exactly; this is a necessary consequence of attempting to work at coarser resolutions to improve computational performance. A qualitative assessment of the conformance can be made by comparing Figures 1 and 2. The mesh porosity in Table 1 helps to quantify the effect. The porosity of the voxel image used for the largest mesh (B.240) is 18.94%, and this value varies by approximately 0.7% for different subsections. The porosity of the B-series meshes are approximately 0.9%

smaller than the corresponding CT data. The porosity variation for the C-series meshes is more significant due to different levels of mesh refinement near the solid/void interface. There are methods to force the mesh porosity to match the CT data, and these are being examined as part of the larger study.

Permeability values are computed directly from the flow simulations by substituting the applied pressure gradient and the computed bulk flowrate into Darcy's law. Permeability was not measured on the actual Berea samples that were imaged, so an experimental permeability is not available for comparison. However, the permeability of 1.20×10^{-8} cm^2 obtained from the largest mesh (B.240) is significantly larger than what is expected for Berea samples (more typically 0.2×10^{-8} cm^2). The FEM Stokes-flow algorithm used here has been validated using both analytical results in simple geometries and pulse-echo NMR experimental velocities in porous media, so it is not currently clear what is causing this possible discrepancy. The most likely reasons would be one or more of the following: 1. The overall computational domain is not sufficiently large to generate a representative permeability. 2. The image segmentation process created some connectivity that was not present in the real material. 3. Mesh resolution was too coarse, particularly in the tighter constrictions that are most responsible for the resistance to flow. 4. The particular Berea sample that was imaged had a higher-than-expected permeability. These issues will continue to be studied.

3.4. *Effect of domain thickness*

Meshes B.030-B.240 are all centered (with respect to z) around slice 140, thus providing a point of comparison for the different-sized domains. Figure 3 contains a color map of the z component of velocity in this plane obtained using mesh B.240, which we will treat as the most reliable flow field out of the current set. The highest velocity is not in the largest pores on this plane (as it would be in a bundle-of-tubes geometry), but rather in the small circular pore toward the left of the domain. This is typical of flow in a 3D porous material. Specific pathways end up carrying larger fractions of the total flow because of structure at a multiple-pore scale. The highest velocities are found at smaller constrictions along these high-flow paths where the fluid had to accelerate through the constriction.

As the thickness of the mesh is decreased symmetrically around this central slice, the flow distribution will change because the z dimension of the domain becomes too small compared to a characteristic scale for permeability. This causes larger pores that were previously not in high-flow pathways to become more easily connected to the inlet/outlet of the computational domain. The B.120 mesh has essentially the same flow pattern as the larger B.240 mesh. The B.060 mesh begins to exhibit significant differences from the larger domains. Finally, the smallest mesh tested (B.030) exhibits a dramatically different flow pattern on the central slice

(Figure 3b), with a few of the larger pores conducting the bulk of the total flow. This behavior is also reflected in the permeability values, which are essentially constant for meshes B.240, B.180, and B.120, but are larger for B.060 and B.030. These results suggest that the Berea sample exhibits a characteristic scale for permeability of approximately 1 mm (in the direction of flow).

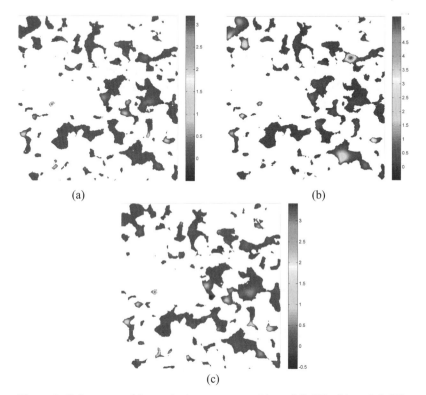

(a)

(b)

(c)

Figure 3. *Color maps of the z velocity component: (a) mesh B.120; (b) mesh B.030; (c) mesh C.R10. The color bars give velocity in units of 10^{-4} cm/s*

3.5. *Effects of mesh coarsening*

The C-series meshes were created by coarsening/refining the elements at the void-solid interface (while leaving interior elements the same size). In principle, this procedure should allow one to improve how the mesh conforms to the pore structure while maintaining computational efficiency. The four C-series results show increasing porosity and permeability as the near-surface elements become coarser, which was an unintended consequence of the surface refinement and will be examined further in the future to ensure better control of this parameter.

In moving from the more refined C.R03 mesh to the coarser C.R10 mesh, the permeability increases by a factor of two. However, the reason for the permeability change is different than what was observed for the B-series meshes. When the domain size was decreased, permeability increased because the flow pattern was dramatically altered, with the highest flows occurring in a completely different set of pores. In contrast, the flow patterns remain similar across the C-series meshes in the sense that the same pores carry the larger flow fractions (i.e. the hot-spots in the velocity plots remain at the same locations). The reason for the increase in permeability is that the velocity becomes proportionally larger over most pores, which is attributable to a combination of the higher porosity meshes and/or the coarser meshes (the latter effect limiting the ability to resolve flow in the tighter constrictions, where the bulk of the frictional losses occur).

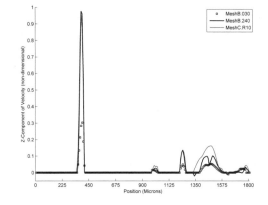

Figure 4. *Velocity (z-component) along a horizontal line for the plots shown in Figure 3*

Figure 4 helps to quantify this effect. It shows non-dimensional z-components of velocity plotted along a horizontal line that passes through the small hot spot in the B.240 mesh. The B.030 velocities along this cut are significantly smaller, despite the fact that the overall flowrate for the B.030 results is 4× higher (for the same pressure gradient), emphasizing the shift in high flows to other pores. The C.R10 velocities remain more or less in proportion compared to the B.240 results, but are larger (and less detailed) in most of the pores, which is attributed to the coarser, higher porosity mesh.

4. Conclusions

The FEM has attributes that will make it a powerful image-based modeling technique. Results shown here emphasize the importance of the mesh generation step: changes in porosity, permeability, and flow patterns accompany strategic coarsening of the mesh. However, mesh coarsening will remain an essential tool for image-based FEM modeling to help optimize computational performance and the ability to perform multiscale modeling. Further research is essential to develop strategies to perform mesh refinement and mesh coarsening without compromising the computational results obtained.

5. Acknowledgements

The authors would like to thank Pradeep Bhattad and Clint Willson for providing the segmented Berea image. We acknowledge Schlumberger Corp. and BP America for supporting this work. N. Lane acknowledges the CFD IGERT at LSU for their support.

6. References

Fatt, I., "The network model of porous media. 1. Capillary pressure characteristics," *Transctions of AIMME,* vol. 207, no. 7, 1956, p. 144-159.

Fredrich, J.T., A.A. Digiovanni, and D.R. Noble, "Predicting macroscopic transport properties using microscopic image data," *Journal of Geophysical Research – Solid Earth,* vol 111, no. B3, 2006.

Thompson, K.E., C.S. Willson, C.D. White, S.L. Nyman, J.P. Bhattacharya, and A.H. Reed, "Application of a new grain-based reconstruction algorithm to microtomography images for quantitative characterization and flow modeling," *SPE Journal,* vol. 13, no. 2, 2008, p. 164-176.

Numerical Modeling of Complex Porous Media For Borehole Applications

NMR-response and Transport in Carbonate and Sandstone Rocks

Seungoh Ryu[*] — **Weishu Zhao**[*] — **Gabriela Leu**[*] — **Philip M. Singer**[**] — **Hyung Joon Cho**[***] — **Youngseuk Keehm**[****]

[*] *Schlumberger Doll Research*
Cambridge, MA, 02139, USA

[**] *Schlumberger Dhahran Carbonate Research*
Al-Khobar 31952, Kingdom of Saudi Arabia

[***] *Schlumberger Product Center*
Sugarland, TX, 77478, USA

[****] *Dept. of Geoenvironmental Sciences*
Kongju National University, Kongju, South Korea

ABSTRACT. *The diffusion/relaxation behavior of polarized spins of pore filling fluid, as often probed by NMR relaxometry, is widely used to extract information on the pore-geometry. Such information is further interpreted as an indicator of the key transport property of the formation in the oil industry. As the importance of reservoirs with complex pore geometry grows, so does the need for deeper understanding of how these properties are inter-related. Numerical modeling of relevant physical processes using a known pore geometry promises to be an effective tool in such endeavor. Using a suite of numerical techniques based on random-walk (RW) and Lattice-Boltzmann (LB) algorithms, we compare sandstone and carbonate pore geometries in their impact on NMR and flow properties. For NMR relaxometry, both laboratory measurement and simulation were done on the same source to address some of the long-standing issues in its borehole applications. Through a series of "numerical experiments" in which the interfacial relaxation properties of the pore matrix is varied systematically, we study the effect of a variable surface relaxivity while fully incorporating the complexity of the pore geometry. From combined RW and LB simulations, we also obtain diffusion-convection propagators and compare the result with experimental and network-simulation counterparts.*

KEYWORDS: *NMR, Lattice-Boltzmann, Randomwalk, Tomogram, Carbonate, Sandstone Rocks*

1. Introduction

Despite the long history of research on porous media, there remain open issues which critically affect various industrial endeavor such as oil/gas exploration, CO_2 sequestration, water management, and storage and migration of toxic waste. A wide range of scales permeates through these disciplines, but pore-scale physical processes remain their common denominator. In this work, we report our recent effort on the pore-level modeling based on micro-tomograms in a borehole application. Several techniques (e.g. (Auzerais et al., 1996)) utilizing detailed 3D pore geometry have reached maturity in recent decades thanks to the affordable computing resource, imaging techniques, and parallelized simulations. These numerical results are in good standing for a class of porous media such as bead packs and sandstones. Here, we focus on aspects of more challenging situations involving carbonates in the oil field.

2. Pore geometry and open issues for the carbonate

Figure 1 shows two contrasting images of the pore-grain interface for a Fontainebleu sandstone and a carbonate rock. While quasi-periodicity is clearly visible in the former, the latter displays pronounced heterogeneity. This is displayed in the right column where the radial pore-to-pore autocorrelation function $g(|\mathbf{r}_2 - \mathbf{r}_1|) \equiv \; < \phi(\mathbf{r}_1)\phi(\mathbf{r}_2) >$ ($\phi(\mathbf{r}) = 1$ in pore, 0 in grains) is plotted for different values of porosity. Note that in both cases, the curves collapses to a generic form (insets) $h_p(r) \equiv (g(r) - \phi^2)/(\phi(1 - \phi))$ with $\phi = < \phi(\mathbf{r}) >$, average porosity. Quasi-periodicity of the sandstone is manifest in the form of mild bumps in $g(r)$, reminiscent of the classic density correlation in a simple liquid, while the carbonate sample displays a monotonic, featureless profile, which actually arises from preponderance of many competing length scales. In many carbonate rocks, the complexity extends even further than suggested in the figure, as there exist structures on finer scales beyond the tomographic resolution, as well as on scales beyond the typical sample sizes. Quantitative elucidation of these aspects remain an open challenge.

3. Numerical NMR relaxometry

Simulations based on realistic 3D pore space allows us to address some of the long-standing issues in probing its geometry. One surrounds the validity of the conventional mapping between the NMR relaxation spectrum and the pore-size distribution(Kleinberg, 1999; Grebenkov, 2007) of carbonate rocks in the possible presence of haphazard heterogeneity in their interfacial properties.(Ryu and Johnson, 2009) The *low field* NMR data of the carbonate sample of Fig.1 (black zagged line in Fig.2 for which bulk relaxation rate of $1/T_b = 0.67\text{s}^{-1}$ was used and 5-points moving average was taken), shows the typical multi-exponential characteristics. This often invites an interpretation in terms of a broad pore size (more precisely, the surface-to-volume (S-V) ratios) distribution. In this scenario, the pore space is approximated as a collection

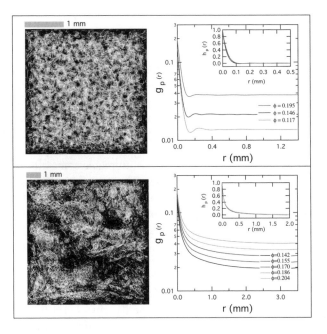

Figure 1. *Pore-grain interface morphology for a typical sandstone (top) and a carbonate rock(bottom panel). Also shown are the radial pore-to-pore auto-correlation function $g_p(r)$ at various porosity values (top:three distinct sandstones; bottom:different thresholding of the original carbonate tomogram). The insets show scaled correlation functions $h_p(r)$ which largely collapse into a universal form.*

of *isolated* pores of varying sizes, all in the so-called *fast diffusion* limit(Brownstein and Tarr, 1979). From the tomogram of the same piece, we obtain the distribution $\mathcal{P}$ of local $\{S_i/V_i\}$ and porosity $\{V_i\}$ (shown in the inset of Figure 2) at various stages of coarse-graining (i labeling sub-blocks of linear dimension L). Attempts to fit experimental data using such recipes neglect the diffusive coupling between pores as shown by series of curves in the figure with a range of controlling surface relaxivity parameter, ρ, values (Kleinberg, 1999) (broken lines with $\rho = 20, 60, 120\mu m/s$) and they fail to work: if ρ is chosen to match the initial slope (red curves), it fails to match the experimental data at long times, and vice versa (blue lines). This suggests that the model neglecting the extended nature of the pore and the heterogeneous diffusive-coupling among its constituents has limited validity in these types of rocks. The role of the latter had been previously considered in simple 1D models.(Ramakrishnan et al., 1999; Zielinski et al., 2002) Random-walk based simulation(Ryu, 2008) on the 3D tomographic pore yields an excellent overall agreement (solid black curve) with the choice of $500\mu m/s$ for the single parameter. The large ρ value thus inferred partly accounts for the fact that at the resolution of $17\mu m$ per voxel, the digital representation of the interface significantly under-estimates the surface area. The effective $\tilde{\rho}$, which represents the combination of raw ρ value and the actual surface-area, dictates

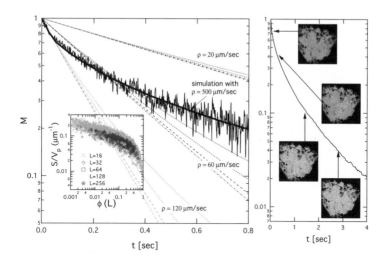

Figure 2. *Low-field experimental data (solid zagged curve) and the best simulation match (solid curve) with $\rho = 500\mu$m/s. Simplistic model results for NMR response at various strengths of surface relaxivity ρ and pore coarsening (colored broken lines, four curves for each value of ρ) are also shown for comparison. The inset shows the porosity and S/V_p distribution at the five stages of coarse graining. Shown on the right panel are the local magnetization density evolution at various stages of the simulation for $\rho = 500\mu$m/s which yields results matching the experiment.*

Figure 3. *The internal magnetic field of a carbonate packstone calculated for the entire to-mogram (1.5^3 cm^3) of the carbonate rock. Shown on the right are the slice-cut views of the field and the probability distribution of its strength for a sub-block of 512^3 voxels at the center rendered separately for the pore (filled black) and grain (gray) space. The curves represent the best Lorentzian (broken lines)(Chen et al., 2005) and Gaussian (solid lines) fits. The latter works better for the pore portion, while neither of the methods successfully fits the grain portion.*

the long time scale dynamics responsible for the good agreement with the entirety of data. These random-walk simulations can be extended for sophisticated NMR probes (Song et al., 2008). The internal field arising from the weak susceptibility contrast between the rock matrix and fluid, often a nuisance for NMR probes, offers a way to probe length scales thanks to the close geometrical correlation between its spatial profile and the geometry of the matrix.(Song et al., 2000) Figure 3 shows an example of the internal field profile in the carbonate rock calculated using a weak dipole field *ansatz*, a method verified via direct imaging on a pack of cylindrical tubes.(Cho et al., 2009) We further derive the local field gradients that play a critical role in stimulated-echo probes(Song et al., 2000) as well as NMR at high fields.(Anand and Hirasaki, 2007; Ryu, 2009) We find that the field inside the 3D pore space can be better approximated by a Gaussian distribution rather than a Lorenzian as reported by Chen *et al.*(Chen et al., 2005) A detailed study on the internal field and their gradients inside various types of rocks is under way, as well as their effect on the NMR response.

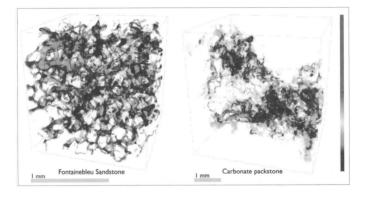

Figure 4. *Local flow speed through the pore matrix (Fontainebleu sandstone (left) and a carbonate packstone (right)), driven from left to right. The color represents the local speed and its scheme was chosen to enhance its features. In both cases, the voxels with less than 2 % of the maximum speed are omitted from view for enhanced views.*

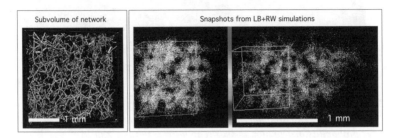

Figure 5. *The first panel shows the carbonate rock as used in the network simulation. The second panel shows the RW-LB based propagation on the sub volume of the same rock at two different stages. The color represents the extent of accumulated displacement for each walker.*

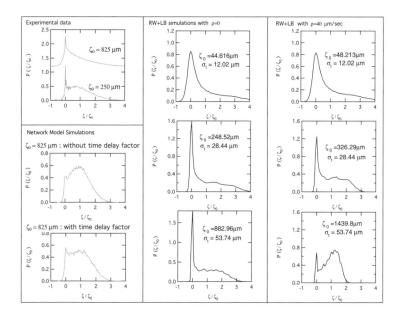

Figure 6. *Top panel on the first column shows experimental $P(\zeta, \zeta_0)$ at two different stages. The lower panels in the first column show the result of a network model simulation with and without the ad hoc time dealy. The middle column shows the combined RW-LB simulated propagator with $\rho = 0$ at three different stages (parametrized by the average displacement ζ_0). The third column shows the same but with $\rho = 40 \mu m/s$. The actual ρ, from more recent NMR relaxometry, is estimated to be about $20 \mu m/s$. Also indicated for reference is the diffusion length $\sigma_t \equiv \sqrt{2Dt}$ for each stage.*

4. Diffusion-Flow propagators

One of the objectives of our numerical modeling is to improve the link between the static pore geometry (as inferred *in situ* from borehole measurements) and its transport, since the latter ultimately controls viability of a hydrocarbon reservoir. The stark contrast between the sandstone and the carbonate pore geometry (Fig.1) underlies the distinct way the flow is distributed. Figure 4 shows the local flow speed distribution for the sub-volume of the rocks obtained from the Lattice Boltzmann simulations.(Succi, 2001) Clearly, the inhomogeneous flux distribution and extreme tortuosity as apparent in the carbonate sample hints at why any framework based soley on parallel channels of effective hydraulic radii should fail.

The flow propagator $P(\zeta, \zeta_0)$, the probability distribution of tagged particle displacements at two different wait-times as indicated by the average displacement ζ_0, (de Gennes, 1983; Stanley and Coniglio, 1984; Aronovitz and Nelson, 1984) provides details on the interplay between the geometrical restriction and the flow, and can be measured by NMR(Hulin et al., 1991; Lebon et al., 1996; Scheven et al., 2004). We simulate the combined diffusion-convection of the fluid molecules using two complementary techniques: one based on network reduction, the other through a combination of random walk and lattice-Boltzmann. (See Figure 5) The former addresses much bigger volumes, while the latter incorporates detailed diffusion/fluid dynamics at finer length scales. The top-left panel of Figure 6 shows the experimental data on a *clean* dolomite rock which has a simpler pore geometry than the carbonate rock used in Figs.1-4. From the tomogram of the same source rock, we first ran a network-model simulation.(Zhao et al., 2009) The result (lower-left panels) under-represents the sharp peak near the origin present in the experimental data. An *ad-hoc* time delay factor was introduced, which to a certain degree enhances the weight near zero displacement(Zhao et al., 2009). Questions arise as to what degree one requires the presence of sub-micron-pores which lie beyond the resolution of the tomogram, and how large-scale heterogeneity and a finite ρ affect $P(\zeta, \zeta_0)$. To clarify these, we developed an alternative method by allowing the random walkers to ride on the background flow from an LB run. (2nd panel of Fig.5) Results (the 2nd and 3rd columns for $\rho = 0$ and $\rho = 40 \mu m/s$) based on the resolution $\sim 3 \mu m/vox$ seem to capture the most salient features of the data. This improvement provides guidance on whether one should attribute such experimental features to the heterogeneity at extreme length scales.

In this work, we demonstrated that various numerical simulations, combined with high quality tomograms, shed valuable insight on complex porous media. Issues due to their large-scale heterogeneity remain to be further investigated.

The authors thank Drs. J. Dunsmuir (Brookhaven NL), J. Goebbels (BAM) and M. Knackstedt (ANU) for microtomograms of various rocks.

5. References

Anand, V. and Hirasaki, G. (2007). Paramagnetic relaxation in sandstones. *SPWLA 48th Annual Logging Symposium*, 446359V.

Aronovitz, J. and Nelson, D. (1984). Anomalous diffusion in steady fluid flow through a porous medium. *Phys. Rev. A*, 30(4):1948.

Auzerais, F., Dunsmuir, J., Ferreol, B., Martys, N., Olson, J., Ramakrishnan, T., Rothman, D., and Schwarts, L. (1996). Transport in sandstone: A study based on three dimensional microtomography. *Geo. Phys. Lett.*, 23:705.

Brownstein, K. and Tarr, C. (1979). Importance of classical diffusion in NMR studies of water in biological cells. *Phys. Rev. A*, 19:2446.

Chen, Q., Marble, A., Colpitts, B., and Balcom, B. (2005). The internal magnetic field distribution, and single exponential magnetic resonance free induction decay, in rocks. *Journal of Magnetic Resonance*, 175:300.

Cho, H., Ryu, S., Ackerman, J., and Song, Y.-Q. (2009). Visualization of inhomogeneous local magnetic field gradient due to susceptibility contrast. *J. Mag. Res.*, 198:88.

de Gennes, P. (1983). Hydrodynamic dispersion in unsaturated porous media. *Journal of Fluid Mechanics Digital Archive*, 136:189.

Grebenkov, D. (2007). NMR survey of reflected brownian motion. *Rev. Mod. Phys.*, 79:1077.

Hulin, J., Guyon, E., Charlaix, E., Leroy, C., and Magnico, P. (1991). Abnormal diffusion and dispersion in porous media. *Physica Scripta*, 1991:26.

Kleinberg, R. (1999). Methods in the physics of porous media. In Wong, P., editor, *Nuclear Magnetic Resonance*, volume 35. Academic Press.

Lebon, L., Oger, L., Leblond, J., Hulin, J., Martys, N., and Schwartz, L. (1996). Pulsed gradient NMR measurements and numerical simulation of flow velocity distribution in sphere packings. *Physics of Fluids Physics of Fluids Phys. Fluids*, 8:293.

Ramakrishnan, T. S., Schwartz, L. M., Fordham, E. J., Kenyon, W. E., and Wilkinson, D. J. (1999). Forward models for nuclear magnetic resonance in carbonate rocks. *The Log Analysist*, 40:260.

Ryu, S. (2008). Effects of spatially varying surface relaxivity and pore shape on NMR logging. *Proceedings of the 49th Annual Logging Symposium*, page 737008 BB.

Ryu, S. (2009). Effect of inhomogeneous surface relaxivity, pore geometry and internal field gradient on NMR logging. *SPWLA 2009*, page JJJJ.

Ryu, S. and Johnson, D. (2009). Aspects of diffusive-relaxation dynamics with a *nonuniform*, partially absorbing boundary in general porous media. *Phys. Rev. Lett.*, 103:118701.

Scheven, U., Seland, J., and Cory, D. (2004). NMR propagator measurements on flow through a random pack of porous glass beads and how they are affected by dispersion, relaxation, and internal field inhomogeneities. *Phys. Rev. E*, 69:021201.

Song, Y.-Q., Ryu, S., and Sen, P. (2000). Determining multiple length scales in rocks. *Nature*, 406:178.

Song, Y.-Q., Zielinski, L., and Ryu, S. (2008). Two-dimensional NMR of diffusion systems. *Phys. Rev. Lett.*, 100:248002.

Stanley, H. E. and Coniglio, A. (1984). Flow in porous media: The 'backbone' fractal at the percolation threshold. *Physical Review B*, 29:522.

Succi, S. (2001). *The Lattice Boltzmann Equation for Fluid Dynamics and Beyond*. Oxford University Press.

Zhao, W., Picard, G., Leu, G., and Singer, P. (2009). Characterization of single phase flow through carbonate rocks. *Trans. Porous. Media.* in press.

Zielinski, L., Song, Y.-Q., Ryu, S., and Sen, P. (2002). Characterization of coupled pore systems from the diffusion eigenspectrum. *J. of Chem. Phys.*, 117:5361.

Characterization of Soil Erosion due to Infiltration into Capping Layers in Landfill

Use of industrial x-ray CT scanner for visualizing soil erosion

T. Mukunoki* — Y. Karasaki* — N. Taniguchi**

* X-Earth Center, Graduate School of Science and Technology
Kumamoto University
1-39-2 Kurokami
Kumamoto, 860-0862
Japan
Kumamoto University
mukunoki@kumamoto-u.ac.jp

** MAEDA Corporation
26-10-2 Fujimi
Chiyoda-ku, Tokyo Metropolitan Government, 102-8151
Japan
MAEDA Corporation
taniguchi.n@jcity.maeda.co.jp

ABSTRACT. *The chemical stabilization of burned ash would be retarded by the local infiltration due to soil erosion and partial variability of density in the capping layer. In order to improve the local infiltration in the final cover-treatment soil, it is important to understand the generation of local preferential flow paths due to rainfall in the cover soil. In this paper, x-ray Computed Tomography (CT) images were used to analyze the density change of the model cover soil after watering test with several test conditions.*

KEYWORDS: *landfill, soil erosion, unsaturated soil, x-ray CT scanner, cover-soil system*

1. Introduction

The long-term infiltration of rainfall into a landfill can cause two potential issues: dilution of leachate concentration and, erosion of fine grains in the final cover-treatment soil. The first factor called Washing Out Effect, would bring the contribution of early chemical stabilization of a burned ash (Higuchi 2005 and McDougall *et al.* 2001). Meanwhile, the second factor would cause negative issues due to increased preferential flow path. Preferential flow path is locally generated in the final cover-treatment soil and waste (Koener and Daniel, 1997 and Ishibashi *et al.* 2008). Local generation of preferential flow path would make the flow rate increase in the cover soil; and then, the harmlessness process of the waste promotes heterogenous stabilization in the landfill. Hence, it is significant to reduce the generation of preferential flow path in the final cover-treatment soil. Authors have focused that the factor of generating local-preferential flow path would be caused by heterogeneity of density in the cover soil (Mukunoki *et al.* 2008). Figure 1 presents photographs of gas release pipe and drainage layer around the leachate collection pipe in a landfill. Both of these are designed to create high hydraulic conductivity paths; hence, the gravels and bolder were installed as shown in Figures 1(a) and (b). In order to collect methane gas and leachate in the landfill, the drainage layer was large voids and they would become enough space for movement of soil particles.

The objective of this study is to observe the condition of soil erosion in the final cover-treatment soil and to understand the mechanism of generating preferential flow paths. This paper will discuss the result of model tests using decomposed granite soil with different density condition and particle size as a final cover-treatment soil using x-ray Computed Tomography (CT) images.

(a) Pipe of gas release   (b) Drainage layer

Figure 1. *Photograph of gas release part and drainage layer in a landfill*

2. Experimental overview

2.1. *Development of test apparatus*

Figure 2 shows an entire system of sprinkling test apparatus. Figure 3 is a picture of the surface of model soil tested. The model ground was 200 mm in depth and

200 mm in diameter. Gravels in the center of the model soil were placed at 150 mm depth from the surface of model soil and 50 mm in diameter. The gravel has a mean diameter (D_{50}) of 20 mm. The gravel part models the drainage layer for gas collection layer. During the rainfall test, water migrated through the model ground and drained from the bottom part of the soil box. Authors developed the system of water circulation. The capacity of water tank is 400 L. The soil box was placed in the large box to capture the drained water, which was the pumped back to the water. This system can use the water of 400 L repeatedly.

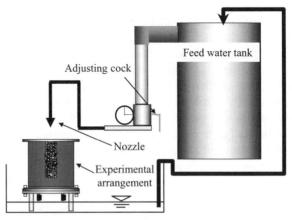

Figure 2. *Entire system of test apparatus*

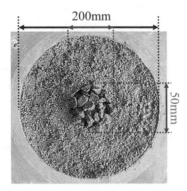

Figure 3. *Photograph of the surface on the model ground*

2.2. *Test condition*

Figure 4 illustrates the cross-sectional view of model ground. In this study, we are interested the density changes from soil erosion due to rainfall near the gas collection layer. Hence, the gravels were installed at the center of the model ground as Case 1 (Figure 3). The mean diameter (D_{50}) of soil used was 0.85 mm and the uniformity coefficient was 22.4.

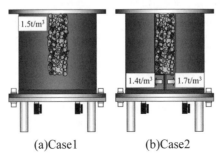

(a)Case1   (b)Case2

Figure 4. *The cross-sectional view of the model ground for Case 1 and Case 2*

In Case 1, the bulk dry density was 1.5 t/m³ and the initial saturation degree was 58%. Meanwhile for Case 2, two different density conditions were established around the gravels. In Case 2, the two different bulk dry densities prepared were 1.4 t/m³, and 1.7 t/m³. The initial saturation of model ground with the density of 1.4 t/m³ was 58% and with 1.7 t/m³ was 91%. The reason why the initial saturation degree was not 100% was because the model test was performed on the outside of the CT room so the model ground had to move to the CT room. During this process, there was a concern that water leaked from the bottom part of the model ground. The extra mass loss of pore water in the model ground does not allow us to quantitatively evaluate density change due to movement of soil particle because of the change of wet density. Therefore, authors confirmed the residual saturation degree of the model soil after the test and its saturation degree was given as the initial saturation degree.

3. Results and discussion

3.1. *Calibration of CT-value, density and hydraulic conductivity*

In general, CT-value is proportional to density so that the correlation between CT-value and hydraulic conductivity could be obtained based on the hydraulic conductivity tests. Figure 5 shows the correlation of CT-value, dry density and hydraulic conductivity of decomposed granite soil tested for each density condition. The solid line and equations showed linear approximation by air (0.0 t/m³, CT-value: -997), water (1.0 t/m³, CT-value: 9) and variant density of model grounds (1.4 t/m³, CT-value: 640 or 1.5 t/m³, CT-value: 663 or 1.7 t/m³, CT-value: 745). Density and hydraulic conductivity of model soil were deduced from each CT-value with Figure 4.

3.2. X-ray CT image of Case 1 (1.5 t/m³)

Figure 6 shows the cross-sectional CT image of the model ground for Case 1 before and after the sprinkling test. After the sprinkling test, soil erosion around gravels can be observed in images at 150 mm height within the white line. The soil erosion progresses due to sprinkling at upper grounds. The white dot line points out the low density area in dry such as 0.36 t/m³ and 0.74 t/m³.

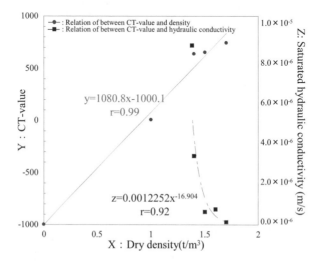

Figure 5. *The correlation of CT-value between CT-value – dry density and hydraulic conductivity*

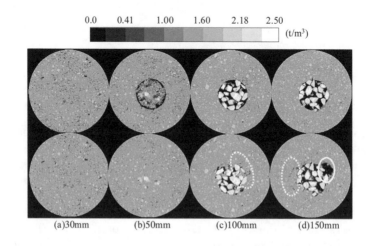

Figure 6. *X-ray CT images for Case 1 before and after the sprinkling test*

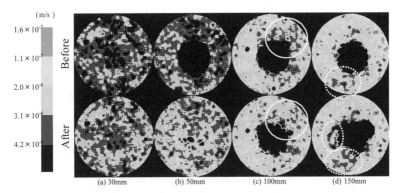

Figure 7. *Contour map of hydraulic conductivity in the model ground for Case 1*

It can be observed that soil particles were moved into the large pores in the gravel layer. The soil erosion would be generated in a continuous flow and presented model ground to make a complicated structure.

Figure 7 shows the color contour map of saturated hydraulic conductivity deduced from the CT-value (Figure 6) and the relations shown in Figure 5. The black area in Figure 7 indicates that the hydraulic conductivity in each contour map is the greatest area in the gravel part. After sprinkling test, the area of hydraulic conductivity between 3.10×10^{-6} m/s and 1.60×10^{-7} m/s increases from 7% to 9% at 30 mm and 50 mm height from the bottom of soil box. The migration of soil particles would make progress with soil erosion, so the soil particles were washed away with seepage flow. On the other hand, the hydraulic conductivity dropped down in white dot-line at 100 mm height. That means the soil particles were drastically washed away from 150 mm high and were clogged in 100 mm height. In contrast, the permeable area would be risen up in white dot line at 150 mm.

3.3. X-ray CT image of Case 2 (1.4 t/m³ and 1.7 t/m³)

Figure 8 shows the cross-sectional image of the model soil for Case 2 before and after the sprinkling test. The white dot line is boundary line between density of 1.4 t/m³ and 1.7 t/m³ in this figure. After the sprinkling test, the area with 1.7 t/m³ density was eroded as shown in Figure 8. By contrast, the area of 1.4 t/m³ has a higher porosity than the area of 1.7 t/m³ so it should be more permeable. Then, it was observed that the eroded area was formed with migration of soil particles around gravels at 100 mm and 150 mm.

Figure 9 shows the distribution of hydraulic conductivity deduced from the CT-value (Figure 8) and the relations shown in Figure 5. In the area of 1.4 t/m³, the hydraulic conductivity increased at each height. So it was considered that migration of soil particles were generated with weak interaction between soil particles.

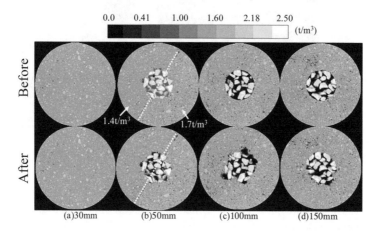

Figure 8. *X-ray CT images for Case 2 before and after the sprinkling test*

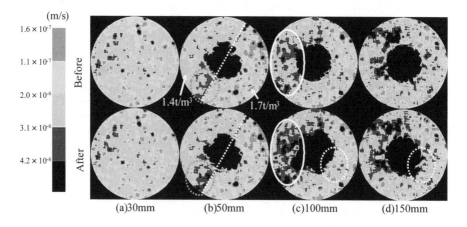

Figure 9. *Contour map of hydraulic conductivity in the model ground for Case 2*

After the sprinkling test, the area with a conductivity of 4.2×10^{-6} m/s was widely distributed. The dot line in white shows soil erosion in the area of 1.7 t/m^3. It cannot be observed that local water permeability area at 1.7 t/m^3 was generated. Then, boundary area between 1.4 t/m^3 and 1.7 t/m^3 was soil erosion area.

3.4. *Factor of generating local preferential flow path in the cover soil*

The factors leading to the generation of preferential flow paths in the cover soil were 1) the density change of soil layer, 2) different particle size and 3) loss of the matrix suction due to increase in the saturation degree in the soil. Fine grain soil eroded first with water seepage and, in addition, the case that there is enough space where coarse grains also could move would cause the final preferential flow path. Simultaneously, the increase of water saturation brought a reduction in the matrix suction between soil particles. As the results have shown, the local-preferential flow path created near the part with the density change and different particle size. In order to have more quantitative discussion, the numerical analysis of unsaturated flow concerned about the boundary condition of cover soil is ongoing.

4. Conclusions

The sprinkling test was performed and the inner condition of model soils at each level was scanned by x-ray CT scanner. Preferential flow path could be visualized for each case and the causes were discussed in this paper. The major conclusions of this study are as follows:

− x-ray CT scanner is effective application to determine distribution of density and hydraulic conductivity in soils;

− large voids in the drainage layer were clogged due to soil erosion, and;

− factor causing soil erosion would be density change of soil layer, difference of particle size and the lost of the matrix suction in the soil.

5. References

Higuchi, S. 2005. "Early stabilization for the landfill", *Proc. of 17th Annual Conference of The Japan Society of Waste Management Experts,* JSWME:959 (in Japanese).

Ishibashi, T., Komiya, T., Nakayama, H. and Shimaoka, T. 2008. "Restraint of Rainwater Infiltration within a Landfill Site Capping Layer Using Geosynthetics for Drainage", *Journal of the Japan Society of Waste Management Experts*, Vol. 19(2), 101-108 (in Japanese).

Koerner, R. and Daniel, D. 1997. "Final covers for solid waste landfills and abandoned dumps", *American Society of Civil Engineering, Press*, Reston, VA, USA.

McDougall, F., White, P., Frank, M. and Hindle P. 2001. *Integrated Solid Waste Management: Life Cycle Inventory*, Blackwell Publishing Ltd, Oxford.

Mukunoki, T., Taniguchi, N., Matsumoto, H. & Murakami, Y. 2008. "Visualization of soil erosion under a geotextile due to infiltration using X-ray CT", *Procs. of the fourth European Geosynthetics Conference*, Euro Geo2008, CD-R.

On Pore Space Partitioning in Relation to Network Model Building for Fluid Flow Computation in Porous Media

From microtomography to pore networks

E. Plougonven — D. Bernard — N. Combaret

CNRS, Université de Bordeaux
ICMCB, Institut de Chimie de la Matière Condensée de Bordeaux
87 avenue Dr A. Schweitzer
33600 Pessac cedex
France
{plougonv, bernard, combaret}@icmcb-bordeaux.cnrs.fr

ABSTRACT. *We present an improved method for generating a representative network model (NM) of a porous medium from the structure of its pore space. We differentiate the problem of positioning pore bodies (PB) and delimiting them via pore throats (PT). From a binary 3D microtomographic reconstruction of a porous material, we use the common watershed approach on the Euclidean distance map, examining several aspects in detail: the skeleton and the necessary pre-processing to remove digitization artefacts, the conversion to a graph that will support the NM, along with the post-processing needed to account for geometric features, delimitation of the PB by a watershed method and comparison to direct PT construction, and finally evaluation of the robustness of the NM generation by application on synthesized images of varying resolutions.*

KEYWORDS: *porous media, pore networks, skeletonization, x-ray microtomography.*

1. Introduction

Porous materials are studied in a wide variety of fields, such as Earth sciences, petrophysics, paper industry, metallurgy, etc. In these applications, pore scale fluid flow is of prime interest, but direct computation of pressure and velocity fields are extremely time and memory-consuming. Network models (NM) constitute a widely accepted alternative: in NM, the complex structure of pore space is represented by an equivalent network of pore bodies (PB) connected via pore throats (PT). Most of the initial works found in the literature were done on synthetic networks arranged in regular grids with simple geometric shapes of pores and throats, their size distribution adjusted by fitting experimental measurements such as mercury intrusion curves. With x-ray microtomography, direct visualization of the internal 3D structure of the pore space is available, making it possible to generate realistic NM. Usually, when real materials are considered, there is no "natural" partition of the pore space in PB and PT. Consequently, the partitioning process that generates this equivalent network must be defined in relation to its final usage. Starting from a 3D binary image of a porous sample, a skeleton of the pore space is built, using a distance-ordered homotopic thinning algorithm [Bertand and Couprie, 2007]. This skeleton is used to define a graph where nodes correspond to PB and branches to PT. Finally, the partitioning itself is performed to precisely delimit the PB associated to the nodes and the PT to the branches. In this paper different crucial steps of this process are analyzed in depth. Section 2 describes a new method to filter the image before skeletonization to avoid spurious branches and other topological artefacts. Section 3 presents the building of the graph, and section 4 describes PB delimitation. Results are presented for 3D images of synthetic materials.

2. Skeletonisation and digitisation artefacts

It is acknowledged that geometric and topological information are not exclusively sufficient to fully account for local features of pore space [Atwood et al., 2004]. The typical solution is to skeletonize the pore space and integrate additional geometric features to construct the graph. However, a well-known disadvantage of skeletonization is its sensitivity to small features. A cavity in an object will automatically result in a surface in the skeleton, regardless of its size and whether or not it is due to noise. Figure 1, for instance, shows a contact between two overlapping balls. The right image presents the resulting skeleton, joining the two ball centers by more than one branch. The bottom branch is due to digitization effects and should not be considered in the final NM. Methods such as median filtering do not handle all cases and modify an unreasonable amount of irrelevant pixels. We propose to directly identify pixels that are responsible for spurious information in the skeleton assuming it branches originates from features smaller than the pixel resolution, i.e. digitization artefacts.

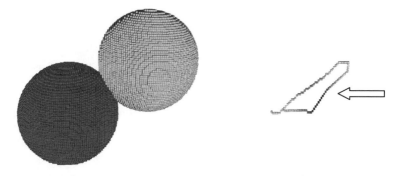

Figure 1. *Two overlapping balls and the resulting skeleton joining their centers. The lower branch (arrow) of the skeleton is obviously unreliable*

As the skeleton of a 3D object is at most 2D, we can classify these artefacts into three categories: 0D (or isolated pixels), 1D (branches), and 2D (surfaces). 0D artefacts are due to misclassification of pixels, typically pixels inside the solid are misclassified during binarization as void, and therefore results in single disconnected pixels in the skeleton. Filling small connected components of the pore space is sufficient to remove these. 2D artefacts are the dual situation from 0D, in the sense that they are caused by misclassifications in the void phase: small components of solid appear *floating* in the porosity. Again, removing small connected components of the solid phase resolves this issue (apart from border cropping effects, there is generally only one connected component for the solid). 1D or branch-type artefacts can be divided into two subtypes, since we consider 26-connectivity for the skeleton: cornerwise and edgewise artefacts. The former occurs when two void pixels are 26-connected and have no other void pixel in their common 26-neighborhoods, as illustrated in Figure 2. Characterization of these pixels can be done locally, using a 2x2x2 configuration given all possible rotations, and generation of the branch in the skeleton is avoided by first reassigning these void pixels to the solid phase. The latter artefacts are more difficult to detect because they are non-local. Edgewise artefacts are caused by two groups of 18-connected void pixels having no other void pixel in their common 26-neighborhood (arrow in Figure 3). In order to detect these pixels, we need to examine the edges of the pixels that are potentially responsible. Several notions are needed: we say an edge is strong if it is contained in at least three pixels of the void phase. An edge is called weak if it is contained in two void pixels that are 18- but not 6-connected. An edge chain is a sequence of edges sharing two by two a common vertex, and the edge chain extremities are the leading and tailing vertices. Finally, we affirm that an edgewise branch artefact is caused by a chain of weak edges having neither extremity connected to a strong edge. This notion is illustrated in Figure 3. With this novel approach to solve the problem of sensitivity of the skeleton to small features, only the pixels causing spurious information in the skeleton are, by construction, modified, therefore producing a minimal number of changes in the original binary image.

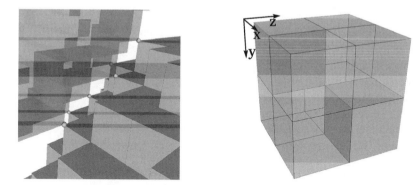

Figure 2. *Left: near contact between two objects, with 6 cornerwise connections, which will result in 6 branches in the skeleton. Right: 2x2x2 configuration to test for each cornerwise branch-type artefact for removal*

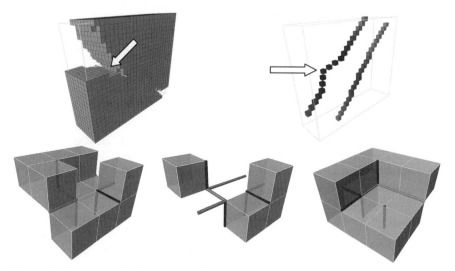

Figure 3. *Illustration of edgewise branch-type artefacts: the top left image shows the pixels (arrow) responsible for the second branch in the skeleton, shown in the top right (arrow). The bottom left image shows the chain of weak edges, in light gray, detected by our method, and the containing pixels removed (bottom centre). The bottom right, however, shows a weak edge chain having an extremity connected to a strong edge, in dark gray, and is therefore part of a relevant connection to be kept in the skeleton.*

3. Graph and post-processing

3.1. *Topological classification of pixels*

From a curvilinear or 1D skeleton, it is fairly straightforward to perform the conversion to a graph, which will be the starting point of the NM, where PB are defined by the vertices of the graph, and PT by the edges. If a pixel of the skeleton is neighbor to at most two other disconnected pixels, then it belongs to an edge of the graph, otherwise it is a vertex. Connected components of node pixels are grouped into a cluster, which identifies one PB. One difficulty is the fact that the skeleton can be 2D, i.e. contain surfaces. Since 2D artefacts were removed previously, the remaining surfaces are caused by cropping effects on the image boundary. Depending on the boundary conditions, they are either open (the exterior is considered as void) or closed (periodic boundary conditions). These surfaces can be removed by simply assigning all but the biggest connected component of the solid to the void phase [Lindquist and Venkatarangan, 1999], although this modification can lead to large void openings that affect flow properties. Instead, we choose to modify the topology of the skeleton by either anchoring surface pixels on the image border for open surfaces, or piercing closed surfaces before performing a second skeletonization. The positions of the anchors or piercings are decided on the distance of the pixels to the solid phase (maxima are used as anchors and holes are pierced at the minima). Of course, this type of modification requires that the surfaces be identified, and for this the topological classification of pixels from [Malandain *et al.*, 1993] was used.

3.2. *Node merging and insertion*

Geometric information needs to be included into the graph so as to account for all the local features in the pore space. Firstly there can exist openings in the void phase that are not marked by a branch intersection in the skeleton, as in a chain of pores [Ioannidis *et al.*, 1997]. Nodes are subsequently inserted along the branches where a significant maximum is found. The significance is determined by performing an H-maximum extraction on the branches. A small value of H (less than 4) is sufficient to ignore surface roughness effects [Soille, 1999]. Another common post-process is the merging of nodes when they identify the same opening in the pore space. Distance criteria are used to determine this. Many methods exist, supposed to be based on the amount of overlapping between the regions each node identifies. The two most common conditions are illustrated in Figure 4 with two nodes, and use the distance between them, and their distance from the solid, illustrated by their maximal balls. The left condition, center inclusion, merges nodes if one is contained in the maximal ball of the other. The right condition merges nodes if their maximal balls intersect.

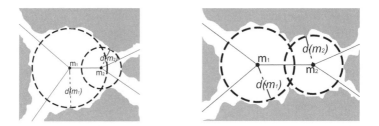

Figure 4. *Two common conditions for merging nodes in the graph.*
Left: center inclusion, and right: ball intersection

If the volumes of each region are sufficiently well represented by the maximal balls centered on the nodes, then we can say that the merging will depend on the amount of overlapping volume $\mathcal{V}$ between these maximal balls $\mathcal{B}(m_i, r_i)$ and $\mathcal{B}(m_j, r_j)$, where $r_x = d(m_x)$. We therefore compute the ratio between this volume and the volume of the smallest maximal ball, and if it exceeds a fixed percentage (30% seems an acceptable value) then the nodes are merged. The following equation defines this condition:

$$\frac{\mathcal{V}(\mathcal{B}(m_i, r_i) \cap \mathcal{B}(m_j, r_j))}{\min_{k \in \{i,j\}} \mathcal{V}(\mathcal{B}(m_k, r_k))} < \alpha \qquad [1]$$

With the skeleton of the filtered binary image converted to a graph, and geometric information integrated in the data structure, the support for the NM is created. The nodes and branches of the NM have been positioned, but we now need to identify the PB and PT in the pore space, respectively.

4. Delimitation and validation

4.1. *Region growing vs. throat construction*

Two types of delimitation methods, PB detection and PT construction, were tested. While the first relies on classical region growing approaches, the second requires new and specific implementations. For the former, we have used the topological watershed algorithm proposed by [Bertrand, 2005]. The inverse of the distance map is used as a relief, and the minima, considered as catchment basins in flooding semantics, are to be separated along the most significant crest lines. Of course, the graph created in the previous section is used to define the basins to separate (all regional minima not identified by a node of the graph will be *filled in* before the watershed). This algorithm creates a pixel-thick watershed, and to obtain a thin separation, these watershed pixels are assigned one of their neighboring PB according to the value in the distance map of the PB pixels. Available work on the second type of delimitation, PT construction, originally strictly applied the definition of Dullien given for a pore: "portions of the void space confined by solid surfaces and planes erected where the hydraulic radius of the pore space exhibits local minima" [Dullien, 1991], but it is clear that planar separations are not adaptable to the complex structures of real materials, and non-planar separations are preferable. The method similar to that of [Shin, 2002] was implemented: at the minimum distance to the solid along each branch of the graph, the closest solid pixels in angular wedges around the branch are determined. Then the 6-connected shortest path of solid pixels on the interface with the void and passing through every pixel detected in each wedge is found. This path defines the border of the PT which is then triangulated. This process is repeated for every branch of the NM, and naturally divides the pore space into PB.

4.2. Evaluating the robustness

These delimitation methods are tested on a random close packing of balls, for which the ball radius is sequentially increased, and at digitizations of varying resolutions. The results, plotted in Figure 4, show that overall the watershed algorithm performs better than the PT construction approach.

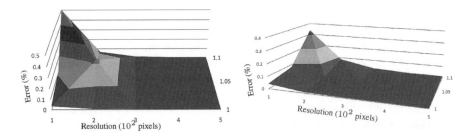

Figure 5. *Percentage of error on the delimitation, as a function of image resolution and ball radius increase (from 0 to 10% increase). On the left, by PT construction, and on the right PB detection*

The error obviously becomes significant when overlapping is so important that individual balls become unidentifiable. In the low resolution image with a 10% radius increase, a PB consisting of 28 balls of the original packing is observed, but a pertinent decomposition because the resulting pore contains neither cavities nor tunnels.

5. Conclusions

We have presented in this paper the necessary stages for obtaining a representative NM from a binary image. An efficient skeletonization algorithm is chosen, along with the exact Euclidean distance for centeredness. Furthermore, we correctly handle the problem of digitization artefacts in the skeleton by a novel prefiltering method. Topological classification of the skeleton pixels is performed, to detect branches and intersection, but also surfaces, which are thinned by a second skeletonization procedure. Geometric information is added to this graph, with a new node merging condition based on maximal ball overlap volume. Finally, pore delimitation procedures are compared, and we show quantitatively that a watershed approach is more efficient than direct PT construction, by application on modified random close ball packings at different resolutions.

6. References

Bertrand G., Couprie M., "Transformations topologiques discrètes", in *Géométrie discrète et images numériques*, pages 187–209, Paris, Hermès-Lavoisier, 2007.

Bertrand G., "On topological watersheds", *J. Math. Imaging Vis.*, vol. 22 no. 2-3, 2005, p. 217–230.

Ioannidis M. A., Kwiecien M. J., Chatzis I., MacDonald I. F., Dullien F. A. L., "Comprehensive pore structure characterization using 3d computer reconstruction and stochastic modelling", *Proceedings of the 1997 SPE Annual Technical Conference and Exhibition*, 1997, Part Omega (pt 1).

Malandain G., Bertrand G., Ayache N., "Topological segmentation of discrete surfaces", *International Journal of Computer Vision*, vol. 10 no. 2, 1993, p. 183–197.

Dullien F.A.L., *Porous Media: Fluid Transport and Pore Structure*, Academic Press, 1991.

Lindquist W.B., Venkatarangan A., "Investigating 3d geometry of porous media from high resolution images", *Physics and Chemistry of the Earth, Part A: Solid Earth and Geodesy*, vol. 24 no. 7, 1999, p. 593–599.

Atwood R.C., Jones J.R., Lee P.D., Hench L.L., "Analysis of pore interconnectivity in bioactive glass foams using x-ray microtomography", *Scripta Materialia*, vol. 51 no. 11, 2004, p. 1029–1033.

Shin H., A throat finding algorithm for medial axis analysis of 3D images of vesiculated basalts, PhD thesis, 2002, State University of New York at Stony Brook.

Soille P., *Morphological Image Analysis, Berlin,* Springer-Verlag, 1999.

3D and Geometric Information of the Pore Structure in Pressurized Clastic Sandstone

M. Takahashi* — M. Kato** — A. Changwan*** — Y. Urushimatsu* — Y. Michiguchi* — H. Park*

*Advanced Industrial Science and Technology
1-1-1, Higashi, Tsukuba, Ibaraki 305-8567
Japan
takahashi-gonsuke@aist.go.jp

**Hokkaido University
Faculty and Graduate School of Engineering
Kita 13, Nishi 8, Kita-ku, Sapporo, Hokkaido, 060-8628
Japan
kato@geo-er.eng.hokudai.ac.jp

***Saitama University
Graduate School of Science and Engineering
255, Shimo-Okubo, Sakura-ku, Saitama, Saitama, 338-8570
Japan
Ahn.changwan@aist.go.jp

ABSTRACT. *The three-dimensional geometry and connectivity of pore space play a fundamental role in governing fluid transport properties of porous media. The spatial and three-dimensional information of pore geometry is difficult to obtain under air and pressurized conditions. To quantify the flow-relevant geometric properties of the pore structure in clastic sandstone, we introduce the three-dimensional medial axis (3DMA) method. We verified the applicability of the 3DMA method for glass beads aggregate as porous material. We present the effect of the hydrostatic pressure on permeability using the transient pulse method and giving a quantitative characterization of pore distributions in intact clastic sandstone and clastic sandstone pressurized to 25 MPa.*

KEYWORDS: *pore, throat, 3-dimentional geometry, permeability, tortuosity, medial axis*

1. Introduction

Sedimentary rocks are named and classified mainly on the basis of their texture and composition. Sedimentary rocks classified as being clastic are composed mainly of broken and worn fragments of preexisting minerals, rock particles that were carried to the site of deposition by moving agencies such as streams, wind, waves, and glaciers, and there cemented. Sandstones are fundamentally clastic in origin.

Porosity and permeability are two primary factors controlling the movement and storage of fluids in rock and sediments. The mechanical and hydraulic properties of porous rocks are a major and common concern in various scientific and engineering fields. The permeability and specific storage of rocks and sedimentary layers deep underground are important parameters for problems such as those related to the evaluation of buried natural gas and oil, CO_2 aquifer storage, and various kinds of waste storage.

The permeability of a geologic material is dependent on the porosity of the material and the degree of connectivity between pores. The specific storage of geologic materials is dependent on the porosity of the material and the compressibility of both the material and the included fluid; accordingly, both the permeability and specific storage of geologic materials vary over wide ranges. For example, the permeability of geologic materials may vary from 0.2 m/s for gravel to less than 10^{-11} m/s for clay and intact rock, representing variation by a factor of 10^{10}. The general equation that describes the transient pulse test method has been used to quantitatively evaluate transient variations in hydraulic head and the transient distributions of hydraulic gradient within a test specimen.

Takahashi *et al.* (2006) measured porosity changes by means of mercury-injection porosimetry and a gas adsorption method for Shirahama sandstones that were deformed under various confining pressures and pore pressures. The brittle and fully ductile deformation regimes are associated with distinct patterns of volumetric strain change measured by displacement transducers and pore volume change evaluated from the pore fluid volume flowing out of or extracted from the specimen. They focused on the correlation of inner structural changes, especially those on a microscopic scale, with mechanical deformation using permeability, total porosity, and BET surface-area measurements. Under higher confining pressures, the specimens behave in a fully ductile manner, and the volumetric strain determined from displacement-transducer measurements and pore volume change shows persistent compaction throughout the experiments. They confirmed that the total porosity determined by means of mercury-intrusion porosimetry increases with increasing confining pressure, and the BET surface area of the pore space obtained using the gas adsorption method supports the porosity increase.

The total porosity and pore size distribution in atmospheric sandstone have been obtained through mercury-injection porosimetry and the gas adsorption method generalized as the BET method. However, three-dimensional information of the pore geometry is difficult to obtain under ambient air and pressurized conditions. Thus, we try to investigate the pressure dependence on geometrical information of pore space in rock specimen by means of micro focus x-ray CT data. In this paper, we use the three-dimensional medial axis (3DMA) method of Lindquist *et al.* (2000) to quantify the flow relevant geometric properties of the pore structure in Shirahama sandstone, clastic sandstone in Japan. At first, we obtain three-dimensional data for different sizes of glass beads by microfocus x-ray CT, and verify 3DMA method for application to glass beads as representative of porous media. In addition, we present the quantitative characterization of each distribution of pore geometry for intact Shirahama sandstone and Shirahama sandstone pressurized to 25 MPa.

2. Inner pore structure and permeability of comparison Berea sandstone and Shirahama sandstone

Most sandstone is composed of quartz and/or feldspar because these are the most common minerals in the Earth's crust. Sandstones are fundamentally clastic in origin. They form from cemented grains that may either be fragments of pre-existing rock or mono-minerallic crystals. The most common cementing materials are silica and calcium carbonate, which are often derived from the dissolution or alteration of sand after it is buried. The environment where it is deposited is crucial in determining the characteristics of the resulting sandstone, which include its grain size, sorting and composition, and the rock geometry and sedimentary structure.

Berea sandstone is fine grained, but the grains are angular rather than rounded, which makes this sandstone an ideal grindstone and abrasive. It has been quarried in many areas as a building stone and was used for foundations, sidewalks, bridge abutments, and buildings. The relatively high porosity of about 18% and permeability of Berea sandstone makes it a good reservoir rock. Berea sandstone is widely used as a standard specimen in oil reservoir engineering. Total porosity is about 18% and rather large pores are dominant. Figure 1(a) shows the permeability evolution for increasing hydrostatic pressure. Permeability was measured with a transient pulse technique to the method by Brace *et al.* (1968), Hsieh *et al.* (1981) and Neuzil *et al.* (1981). The permeability is calculated from the recovered pulse decay across the specimen (Zhang *et al.* 2000a, b). It is seen there is rather high permeability and low dependency on hydrostatic pressure.

The tested sample of Shirahama sandstone was obtained from the Kii Peninsula, Wakayama prefecture, central Japan. The sandstone is Miocene sedimentary rock that consists mainly of quartz grains (44.23%) and rock fragments (46.41%) that are aggregations of various kinds of minerals. The quartz grains have an average size of about 150 μm. The rock fragments have irregular shapes and are relatively easy to deform or crush. The sandstone is rich in rock fragments and has high roundness, low total porosity and rather small pores below a micron scale, and high hydrostatic pressure dependency on permeability reduction (Figure 1(b)). The pre-existing pores are dominantly approximately 1 micron in size, and the total porosity is about 13%. For this clastic sandstone of Shirahama sandstone, we carried out image processing using X-Ray CT volume data and the three-dimensional axis analysis method, and investigated the difficulty of the segmentation treatment for clastic rich sandstone with a rather small pore size and low porosity.

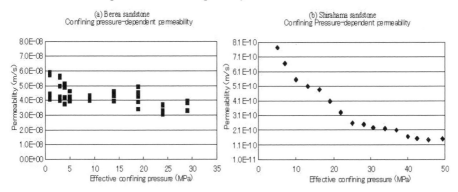

Figure 1. *Permeability as functions of hydrostatic pressure of Berea sandstone (a) and Shirahama sandstone (b). The permeability is measured by the transient pulse method*

3. Outline of the 3DMA method

Lindquist and Venkatarangen (1999) developed the 3DMA computational package as a tool for analyzing the geometry of the pore and grain phases in three-dimensional CT data. Using this package, we can obtain the spatial distributions for pore and throat sizes, coordination number, tortuosity, and other geometric relations between pores and throats. In general, CT images of porous media are grayscale images and have a bimodal population, one mode corresponding to the signal from the pore space and the second to the signal from the grain space. As we are interested in the pore structure, an appropriate segmentation procedure is required to quantify the geometry of the pore space. The most suitable threshold procedure is selected to match a predetermined bulk density and porosity of the porous medium. In the 3DMA package, the kriging-based algorithm is used to segment the voxel images. The medial axis for a sphere is a straight line passing through the center point, and that for a cylinder is the axis of rotational symmetry. The medial axis for any n-dimensional object can be found using an algorithm equivalent to a "burning algorithm" (see Lindquist et al., 1996).

Figure 2 is a two-dimensional illustration of the medial axis behavior of two nodal pores. A throat is defined as a local minimum of the cross-sectional area in a channel.

For an object in a spatial continuum, the medial axis is a network of paths and vertices unless the object contains embedded cavities, in which case the medial axis also contains surface segments. For digitized objects, the medial axis consists of digitized paths corresponding to a line of connected voxels, digitized vertices corresponding to a cluster of one or more voxels, and possibly digitized surfaces corresponding to a sheet of voxels. The usual conceptual model divides the void space of a porous medium into nodal pores connected by pore channels. Nodal pores are located at multigrain junctions. Roughly equidimensional in geometry, these nodal pores are joined to one another by pore channels, which include microcracks situated along interfaces between two neighboring grains and tubular pores along three-grain edges.

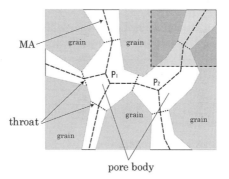

Figure 2. *2-dimensional illustration of Medial Axis behavior*

Figure 3 shows the original CT data and an example of medial axis analysis for the packing of glass beads with 600 μm diameters. We can visually confirm the accuracy of the medial axis analysis. The medial axis is affected by the flow channel, percolating backbone, and flow path in addition to other factors.

Figure 3. *3-dimensional image of Medial Axis in grass beads with 600 μm diameter: (a) 3-dimensional original CT data (b) 3-dimensional pore space image extracted from CT data (c) 3-dimensional medial axis image calculated from pore network data*

4. Microfocus x-ray CT apparatus and geometric information of intact and stressed Shirahama sandstone

A new pressure vessel was developed to simultaneously supply both the confining pressure and pore pressure to a rock specimen with a 10 mm diameter and 20 mm length. To determine a change to the inner structure of a rock specimen under hydrostatic pressure, we took three-dimensional images of stressed Shirahama sandstone by microfocus x-ray CT. The hydrostatic pressure was raised to 25 MPa, which generally corresponds to the stress state at 1,000 m depth.

Figure 4 shows various images for the intact (upper low) and stressed (lower low) samples; (a) (b) 3 dimensional original CT volume data, and (b) (d) the 3DMA network. Figure 5 shows the measured distributions of (a) throats, (b) pores, and (c) coordination numbers for intact and stressed samples. The measured frequencies for throats and pores decreased remarkably under the stressed condition. It is clear that permeability reduction as the hydrostatic pressure increases is attributable to the decrease in the population of pores with a radius of 20–90 µm, decrease in the population of throats with a radius of 10–50 µm, and increase in tortuosity in the X and Z directions. This phenomenon has already been confirmed by theory and many experiments, but the three-dimensional geometry of throats and pores has not been investigated in detail. Ultimately, we can obtain three-dimensional visualizations. Figure 6 shows the changes in tortuosity in three directions for intact and stressed samples. In each direction, the tortuosity was greater under the stressed condition. Hydrostatic pressure decreased the pore size, throat size, and coordination number in Shirahama sandstone. Permeability reduction under pressurization to 25 MPa hydraulic pressure is attributed to these inner structural changes.

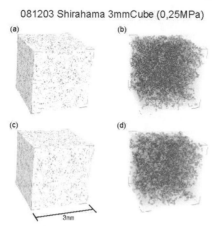

081203 Shirahama 3mmCube (0,25MPa)

Figure 4. *Calculated observations under intact (upper low) and stressed (lower low) state. (a)(c) 3-dimensional original CT data; (b)(d) 3-dimensional medial axis image calculated from pore network data*

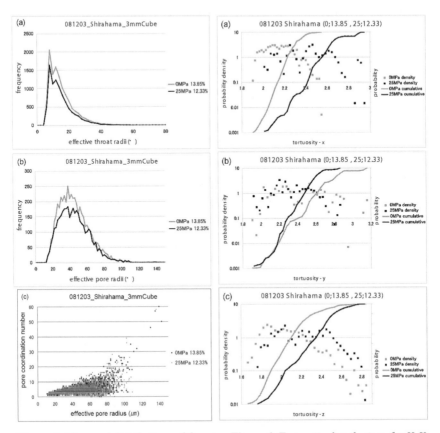

Figure 5. *Measured distributions of throat (a), pore (b) and coordination number (c)*

Figure 6. *Tortuosity distributions for X, Y and Z direction. Z axis corresponds to horizontal direction*

5. Conclusion

Porosity and permeability are two primary factors that control the movement and storage of fluids in rock, especially clastic sandstone in the case of Japanese fields. Permeability experiments were conducted under hydrostatic stress to 25 MPa to investigate geometric differences in Shirahama sandstone. The permeability consistently decreased with increasing hydrostatic pressure and with decreasing volumetric strain. To quantify the flow-relevant geometric properties of the pore structure in Shirahama sandstone, we used the 3DMA method of Lindquist *et al.* (2000) and verified three-dimensional data for different stress conditions obtained by microfocus x-ray CT. We verified the results obtained using the 3DMA method and measured the distributions of pore size, throat size, channel length and coordination number as well as correlations between pore geometry and throat geometry. In addition, we presented the quantitative characterization of each distribution for intact Shirahama sandstone and Shirahama sandstone pressurized to 25 MPa. It is clear that the permeability reduction due to increasing hydrostatic pressure is caused by a decrease in the population of pores with a radius of 20–90 µm, decrease in the population of throats with a radius of 10–50 µm, and increase in the tortuosity in the X and Z directions in stressed Shirahama sandstone.

6. References

Lindquist W. B., Lee S.-M., Coker D. A., Jones K., W., Spanne P., "Medial axis analysis of void structure in three-dimensional tomographic images of porous media", *Journal of Geophysical Research*, vol. 101, no. B4, 1996, p. 8297-8310.

Lindquist W. B., Venkatarangan A., "Investigating 3D geometry of porous media from high resolution images", *Physics and Chemistry of the Earth Part A,* vol. 25, no. 7, 1999, p. 593-599.

Lindquist W. B., Venkatarangan A., Dunsmuir, J., Wong T.-F., "Pore and throat size distributions measured from synchrotron X-ray tomographic images of Fontainebleau sandstones", *Journal of Geophysical Research,* vol. 105, no. B9, 2000, p. 21509-21527.

Brace, W. F., Walsh, J.B., and Frangos, W. T., "Permeability of granite under high pressure", *Journal of Geophysical Research*, vol. 73, no. 6, 1968, p. 2225-2236.

Hsieh, P,A., Tracy, J. V, Neuzil, C, E.,Bredehoeft, J, D., and Silliman, S, E.,"A transient, laboratory method for determing the hydraulic properties of tight rocks – I theory", *International Journal of Rock Mechanics and Minig Sciences & Geomechanics Abstracts*, vol. 18, no. 3, 1981, p. 245-252.

Neuzil, C. E., Cooley, C., Silliman, S. E., Bredehoeft, J. D., and Hsieh, P. A., "A transient laboratory method for determining the hydraulic properties of tight rocks – II. application", *International Journal of Rock Mechanics and Mining Sciences & Geomechanics Abstracts* , vol. 18, no. 3, 1981, p. 253-258.

Zhang, M., Takahashi, M., Morin, R. H., and Esaki, T., "Evaluation and application of the transient-pulse technique for determining the hydraulic projectiles of low-permeability rocks – Part 1:theoretical evaluation", *Geotechnical Testing Journal*, vol. 23, no. 1, 2000a, p. 83-90.

Zhang, M., Takahashi, M., Morin, R. H., and Esaki, T., "Evaluation and application of the transient-pulse technique for determining the hydraulic projecties of low-permeability rocks – Part 2: experimental application", *Geotechnical Testing Journal*, vol. 23, no. 1, 2000b, p. 91-99.

Evaluation of Pressure-dependent Permeability in Rock by Means of the Tracer-aided X-ray CT

D. Fukahori * — K. Sugawara*

Graduate School of Engineering
Kyoto University
Kyoto daigaku-katsura, Nishikyo-ku, Kyoto 615-8540, Japan
d.fukahori@fx8.ecs.kyoto-u.ac.jp

***Professor emeritus, Kumamoto University*

ABSTRACT. *X-ray Computed Tomography is successfully applied to the visualization of water flow in rock, and to the quantitative determination of the intrinsic permeability under various effective confining pressures. In a specific pressure vessel, one-dimensional water permeation test is performed. The transport diffusion phenomena of tracer is visualized and analyzed so as to evaluate the mean pore velocity of water in rock under the saturated condition, by utilizing the subtraction of CT images along with the data stacking technique. The intrinsic permeability is determined, based upon the Darcy's law. Case example clarify that the tracer-aided x-ray CT is a promising tool available for the evaluating the intrinsic permeability influenced by the effective confining pressure.*

KEYWORDS: *rock, visualization, permeability, effective confining pressure, tracer, x-ray CT*

1. Introduction

Knowledge of water flow in rock is important for the design and construction of radioactive waste disposal, since the mass transfer phenomenon of radioactive nuclear matter is deeply associated with water current in rock strata, which is expected to play an important role as a natural barrier. Rock that is deep underground is subjected to high confining pressure. Thus, it is necessary to accurately evaluate the influence of the confining pressure on the intrinsic permeability of rock.

The tracer-aided water permeation test system (Fukahori *et al.* (2006)) is applied for the visualization of one-dimensional horizontal water flow in porous rock, and for the evaluation of the mean pore velocity of water in rock. In this article, the system is briefly described, along with the preparation of high density tracer. Subsequently the image subtraction is proposed for the evaluation of the porosity, along with the image reconstruction of the tracer density, with case examples. Finally the intrinsic permeability is presented and discussed, using the mean pore velocity accurately evaluated from the movement of tracer. Case study makes clear that the present method is a promising tool available for the evaluating the intrinsic permeability influenced by the effective confining pressure.

2. The tracer-aided water permeation test system

2.1. *Principal apparatuses*

The x-ray CT scanner (TOSCANER-20000RE) in Figure 1 is employed for the present test, in the standard configuration of a beam thickness of 2 mm and an image reconstruction of 2048 pixels x 2048 pixels with a pixel size of 73 μm, under an x-ray bulb operation of 300 kV / 2 mA.

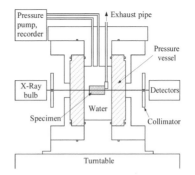

Figure 1. *X-ray CT scanner*   **Figure 2.** *Pressurization system*

The water permeation test is conducted with a cylindrical pressure vessel in Figure 2, under a multi-stage pressurization. The vessel is made from acrylic resin, so as to avoid an attenuation of x-rays. The maximum capacity is 20 MPa, and the inner diameter is 10 cm. A specimen is horizontally placed in the center of the vessel, and the horizontal water flow in the specimen is continuously monitored by the scanner. As shown in Figure 2, at the right-hand end of the specimen, an exhaust pipe is installed to discharge the air involved within a porous specimen.

2.2. Preparation of high density tracer of equivalent with water in viscosity

In order to make no disturbance to the water flow in the specimen, a high density tracer having the same viscosity as water is prepared by mixing the potassium iodide and the urografin used for the human body. As shown in Figure 3, the viscosity of tracer (μ) under a constant temperature is independent of the tracer density (ρ), and equivalent to the viscosity of pure water in a range of $0 < \rho < 400$ kg/m^3.

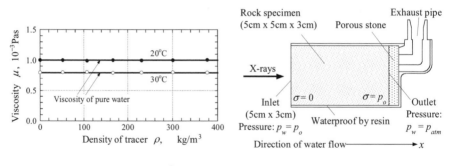

Figure 3. *Viscosity of tracer*   **Figure 4.** *Side view of specimen*

2.3. Rock specimen and procedure of the tracer-aided water permeation test

Kimachi sandstone of 20% porosity was tested. To create a horizontal water flow in the specimen, as shown in Figure 4, it is fully sealed by resin, excepting the inlet. As previously mentioned, the exhaust pipe has been installed at the outlet with a porous stone, and fully sealed by resin. When the specimen is pressurized in the vessel, water permeates into the specimen from the unsealed end and the pressure at the inlet is equivalent to pressure in the vessel (p_o), the pressure at the outlet is equivalent to the atmospheric pressure ($p_{atm} = 0$). Therefore, the effective confining pressure at the inlet is $\sigma = 0$, the effective confining pressure at the outlet is $\sigma = p_o$.

The present tracer-aided water permeation test consists of an initial water injection into a dry specimen, a series of tracer test. Figure 5 shows a schematic pressure-time diagram of the tracer-aided water permeation test.

The test is commenced from taking an image of dry specimen (CT_d). The initial water injection is performed to replace the air in the specimen with water. The higher pressure results in a higher degree of water saturation. This can provide the image of fully-saturation (CT_S) just before the tracer test. The tracer test is performed by controlling the inlet pressure (p_o). Each tracer test is a set of tracer injections over a relatively short time, a subsequent injection of water to transport the tracer into the direction of flow, and finally injection of water under a pressure of 10 MPa so as to remove the tracer and replace with pure water. They can provide the images of tracer migration before the water injection (CT_T), and the images of tracer migration during the water injection (CT_W).

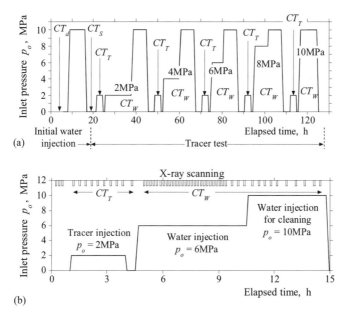

Figure 5. *Schematic pressure-time record of the test: (a) tracer-aided water permeation test; (b) tracer test in case of the inlet pressure p_o= 6 MPa*

2.4. Image processing

The image of porosity is obtained from the images subtraction of $CT_S - CT_d$, and the images of tracer migration at certain step by the images subtraction of $CT_T - CT_S$, and by $CT_W - CT_S$.

3. Theoretical consideration

In general, the intrinsic permeability of rock is determined from the linear form of Darcy's law. In addition, the intrinsic permeability decreases with increasing the effective confining pressure. In this study, the intrinsic permeability k is given by the following equation (Sugawara *et al.* (2004)):

$$k = k_m \cdot (1 - m) \quad \text{at} \quad \sigma = p_o - p_{atm} \qquad [1]$$

where m is a rock constant, k_m is the measurement value given by Darcy's law as follows:

$$k_m = u \cdot \mu \cdot \phi \cdot L/p_o \qquad [2]$$

where u is the mean pore velocity of water, μ is the viscosity of water, ϕ is the porosity of rock, and L is the length of specimen.

Moreover, the rock constant m is obtained by the relation between the measurement value k_m and the effective confining pressure σ. The relation between k_m and σ so as to estimate the rock constant m is approximated by the following equation (Sugawara *et al.* (2004)):

$$k_m = k_{mo} \cdot (\sigma / \sigma_o)^{-m} \qquad [3]$$

where k_{mo} represents the intrinsic permeability at $\sigma = \sigma_o$.

4. Analysis of water flow upon the image processing

4.1. *Precise evaluation of the porosity*

Figure 6 shows the image of the porosity (ϕ) obtained by the CT image subtraction after the local averaging of 51 pixels x 51 pixels. It can be noted from Figure 6 that the continuous plane of high porosity is parallel to the x-axis direction. It is considered that the continuous plan is the bedding plane, and peculiar to Kimachi-sandstone. The image of the porosity is indispensable for understanding the water flow in rock.

4.2. *Non-uniform movement of tracer with water flow*

Figure 7 shows the image of the tracer density (ρ) obtained by the CT images subtraction after the local averaging of 51 pixels x 51 pixels. The increase of the tracer density is represented with white. Then the images describe that the tracer injected from the left hand moves to the right hand in this figure, with water flow. It is noteworthy that the front of the tracer is uneven, and the degree of fluctuation increases with time. The width of the visual tracer zone increases with time, while the maximum density is decreasing. This is deeply related to the heterogenous structure of rock, and the diffusion-dispersion mechanism of tracer. In particular, it can be noted that the movement of tracer near the high porosity region A is small. It is considered that the water flow in rock is dependent not only upon the influence of the porosity but also the pore size distribution.

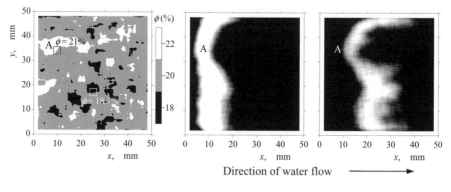

Figure 6. *Image of porosity*   **Figure 7.** *Images of tracer density*

4.3. *Evaluation of the mean pore velocity of water in rock*

The distributions of the tracer density along the x-axis are summarized in Figure 8, which are obtained from the CT images subtraction and stacking procedure (Sugawara *et al.* (2004)). They are used to evaluate the mean pore velocity of water (u). The tracer density-distance curve in Figure 8 has a peak as represented with black circles. The mean pore velocity of u is evaluated from the relation between the distance from the inlet to the point of peak and the elapsed time. The value of u is evaluated as shown in Figure 9.

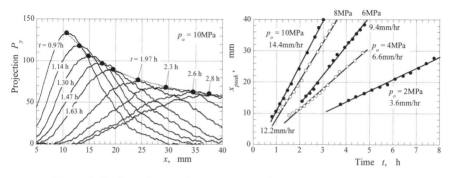

Figure 8. *Profiles of tracer density*   **Figure 9.** *Mean pore velocity obtained*

5. Estimation of the permeability of rock

The intrinsic permeability k evaluated by equation [1] is shown in Figure 10 by black circles. White circles in this figure represent the value of k_m upon the Darcy's law. The value of m was obtained by the relation between the k_m and the effective confining pressure σ. In this figure, results of transient pulse method obtained by using same rock sample are also plotted together. As this figure shows, the computation values in this study are consistent with the results of transient pulse method and it is shown that the intrinsic permeability k decrease with increasing the effective confining pressure.

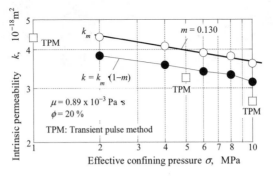

Figure 10. *Evaluation of the intrinsic permeability*

6. Conclusions

The practice of the tracer-aided water permeation test of rock is presented to make clear that tracer-aided x-ray CT is a promising scheme available for minute evaluation of pore velocity in rock and the intrinsic permeability of rock. The results obtained show that despite its simplicity, this suggested method, by using the mean pore velocity and the value of m, gives relatively accurate results.

7. References

Fukahori, D., Saito, Y., Morinaga, D., Ogata, M., Sugawara, K., "Study on water flow in rock by means of the tracer-aided x-rays CT", *Advances in X-ray Tomography for Geomaterials*, 2006, p.287-292.

Sugawara, K., Fukahori, D., Iwatani, T. and Kubota, S., "Analysis of wetting process of rock by means of X-ray CT", *X-ray CT for Geomaterials –Soils, Concrete, Rocks –*, 2004, p.315-334.

Assessment of Time-Space Evolutions of Intertidal Flat Geo-Environments Using an Industrial X-ray CT Scanner

F. Yamada* — A. Tamaki** — Y. Obara***

*Kumamoto University, Graduate School of Science and Technology
2-39-1 Kurokami
Kumamoto 860-8555
Japan
yamada@kumamoto-u.ac.jp

**Nagasaki University, Faculty of Fisheries
1-14 Bunkyo
Nagasaki 852-8521
Japan
tamaki@nagasaki-u.ac.jp

***Kumamoto University, Graduate School of Science and Technology
2-39-1, Kurokami
Kumamoto 860-8555
Japan
obara@kumamoto-u.ac.jp

ABSTRACT. An industrial x-ray computed tomography (CT) scanner was used to investigate the time-space evolutions of vertical sediment structures in an intertidal flat with non-destructive conditions, at the mouth of the Shirakawa river, Japan. Seasonal sediment core samples were collected over one year at two locations. One location was Manila clam (Ruditapes philippinarum)-dominant, and the other clam non-dominant. Reconstructed three-dimensional images using Hounsfield Units (HU) demonstrated that water retention volumes in the intertidal sediment in the clam-dominant area existed up to 4 cm below the sediment surface. The CT results revealed that low-bulk density regions below the water-sediment interface are necessary for the survival of the Manila clam. Moreover, the shell hash composition of the substrate appeared to be related to the clam's distribution.

KEYWORDS: Manila clam, biogenic structure, water retention, shell hash, industrial x-ray CT

1. Introduction

The Manila clam, *Ruditapes philippinarum*, a macrobenthic bivalve, is an important species for commercial fisheries in many countries. In Ariake Sound, Kyushu, Japan, the fishery yield of Manila clams increased considerably in the 1970s, reaching a peak of 65,000 metric tons in 1977, but decreased rapidly throughout the 1980s (Tamaki *et al.* 2008). Ecosystem protection and management in intertidal zones requires an understanding of the influence of geo-environmental properties as well as chemical and biological properties.

Geo-environmental properties including porosity, water retention, and sediment density profiles are known to be modified by the burrowing activity of macrobenthic species (Mazik *et al.*, 2008). Sassa and Watabe (2007) demonstrated that suction, negative pore water pressure relative to atmospheric air pressure, also has a substantial influence on the geo-environmental properties of intertidal zones. Moreover, water retention in the sediments and the burrowing activity of macrobenthic animals were found to be closely linked to suction dynamics. The burrowing activity of macrobenthic species involves the construction of tubes, burrows, and galleries in sediment. These are generally called biogenic structures. Biogenic structures play a key role in the functioning of water-sediment interfaces and are important to consider in functional ecology studies (Rosenberg *et al.*, 2007). Despite the recognized importance of biogenic structures, the visualization and quantification of the space occupied by such structures in the sediment remains difficult using current techniques.

Recently, medical x-ray computed tomography (CT) scanning techniques have been used to visualize biogenic structures and sediment density profiles within marine sediment cores in two and three dimensions. (Mermillod-Blondin *et al.*, 2003; Dufour *et al.*, 2005; Rosenberg *et al.*, 2007; Mazik *et al.* 2008; Bouchet *et al.*, 2009). X-ray CT scanning is one of the best non-destructive techniques for investigating biogenic structures. It has been proposed that this technique could be used in routine environmental analysis (Perez *et al.* 1999). However, all studies using this technique to date have aimed to visualize and quantify the biogenic structures of macrobenthic animals. No study has directly compared the time-space evolutions of geo-environmental properties, including water retention, at multiple locations within an intertidal flat.

We used an industrial x-ray computed tomography (CT) scanner to investigate the time-space evolutions of water retention volumes of an intertidal flat with non-destructive conditions at the mouth of the Shirakawa River in Kyushu, Japan. We measured the water retention volumes of the intertidal flat to clarify the relationship between Manila clam habitats and seasonal variations in water retention volumes. Sediment core sampling was conducted seasonally over one year, at two places: one area was clam-dominant, the other clam non-dominant.

2. Materials and methods

2.1. *Study site*

The study site is located at the center of the eastern coast of Ariake Sound in Japan, which is a closed inner bay as depicted in Figure 1(a). The length, width, and depth of the Sound are approximately 97 km, 20 km, and 20 m, respectively. Figure 1(b) depicts the intertidal flat adjacent to Shirakawa River mouth, where we collected core samples. The tide is semidiurnal, and the mean spring tidal range is 3.86 m (Yamada and Kobayashi, 2004). The average wave height and period are 0.2 m and 3 s, respectively. The characteristics of flow velocity, suspended sediment concentration, salinity, and suspended sediment flux at this study site were described in detail in Yamada *et al.* (2009).

Tamaki *et al*, (2008) described the Manila clam dominant and non-dominant areas within this intertidal flat, based on field observations. In Figure 1(b), the closed red ellipse shows the Manila clam dominant area located in the low-tide zone, and the closed yellow ellipse shows the non-dominant area, located in the mid-tide zone. Black circles show the core sampling points with the points A and B located offshore, at distances of 1,500 and 1,000 m from the seawall.

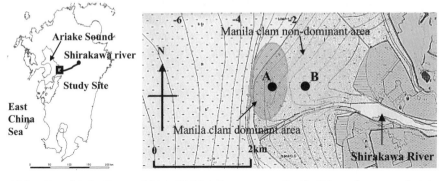

(a) Location of study site   (b) Intertidal flat where core sampling for CT scanning was conducted.

Figure 1. *Study site in Ariake Sound at the mouth of Shirakawa river in Kyushu Island, Japan*

Photo 1 compares the surface topography of both the clam dominant (Point A) and non-dominant (Point B) areas in the intertidal flat during low tide. The sand ripples are more pronounced at Point B, with an average wavelength of 10 cm and wave height of 1 cm. The flat is composed of a mixture of sand and mud. The median diameter at Point A was 0.12 mm, finer than that at Point B. The mud content at Point A was 10 to 20%, higher than that at Point B.

(a) Point A   (b) Point B

Photo 1. *Surface topography of both the Manila clam dominant and non-dominant areas*

2.2. Core sampling

Seasonal sediment core sampling at both points during low tide was conducted in 2007 (Table 1). Sediment sampling was carried out with 400 mm-long cylindrical acrylic tube that was 100 mm in diameter and 5 mm in thickness. The corer was pushed vertically into the sediment, ensuring that no twisting occurred, so as to minimize distortion of biogenic structures. The cores were carefully dug out, again aiming to minimize disturbance. The top of each core was sealed with a paraffin plug to preserve the water-sediment interface and biogenic structures during transportation from the field to the laboratory. The cores were scanned within three days following collection.

No.	Month/Day/Year	No.	Month/Day/Year
1	02/04/2007	5	09/25/2007
2	06/13/2007	6	09/28/2007
3	09/10/2007	7	10/24/2007
4	09/12/2007	8	12/26/2007

Table 1. *Dates of core sampling*

2.3. CT scanning

The x-ray CT scanner uses x-rays to measure the mean absorption coefficient of examined material. The resulting cross-sectional images are displayed in grey-scale; with darker zones representing lower and lighter zones higher x-ray attenuation. X-ray attenuation is defined by the absorption coefficient μ, as expressed in Beer's law. Grey-scale values in numeric files are expressed as *Hounsfield Units (HU)*, obtained by comparing the absorption coefficient for material μ_s to that for water μ_w as follows:

$$HU = \left(\frac{\mu_s}{\mu_w} - 1 \right) \times 1000 \qquad\qquad [1]$$

Equation [1] demonstrates that the *HU* of water is zero. The *HU* of air should be -1,000 because the absorption coefficient for air is zero. Conventional studies have assumed that x-ray attenuation was linearly related to the density of the sample. This is true for biological samples like the human body, but becomes more complex for dense plastic samples (Boespflug *et al.*, 1995). Because medical CT scanners typically use between 40 and 140 kV of power, the two dominant effects characterizing the x-ray interactions with the material are the photo-electron and the Compton effects (Duliu, 1999). Hence, the linear attenuation coefficient depends upon both the effective atomic number Z_{ef} and the density ρ of the examined material. Although Duchesne *et al.* (2009) developed a correction method for the photo-electron effect for x-ray attenuation coefficients, quantitative information on x-ray absorption is sometimes unclear. Therefore, in this study, the *HU* is considered as a specific unit and its variation interpreted empirically.

Scanning of the sediment cores was performed using an industrial x-ray CT scanner (Toshiba, Toscanner-23200) at Kumamoto University, Japan. The scanning system was described in detail by Otani *et al.* (2000). Using a source radiation of 300 kV and 2 mA, cross-sections with a thickness of 1 mm were then made from the water-sediment interface to a depth of 80 mm, giving 80 images per core. A CT transverse section is the uniform distance between sequential slices. We obtained two-dimensional (2D) images composed of pixels on a matrix of 2048×2048 pixels, and three dimensional (3D) images composed of voxels. The pixel resolution was 0.073 mm $\times$ 0.073 mm, and the pixel area was 0.0053 mm^2. The voxel volume was 0.0053 mm^3.

3. Results

A 2D CT image for each cross-section was obtained based on the *HU* distributions. Figure 2 shows 2D images of both the clam dominant and non-dominant areas at a depth of 3 cm from the bottom surface on February 4, 2007. Shell hashes were visible in the clam dominant area, as shown in Figure 2 (a). Live adult specimens of the bivalve *Mactra veneriformis*, were present only in the non-dominant area, as shown in Figure 2 (b). Biogenic structures, such as tubes and burrows are present in both images. In a comparison of the images, the clam dominant area was found to have more low-density regions than the non-dominant area.

The histograms of the *HU* values of all 80 images of both areas were compared, as shown in Figure 3. Although the average and peak *HU* at both locations were similar, a marked difference occurred when *HU* was close to zero. Since the *HU* of water is zero, the water retention volume of the clam dominant area was larger than that of the clam non-dominant area.

To examine the spatial distribution of water retention volume in the sediment core, 3D reconstructions of CT images were obtained using the *HU* corresponding to water (Figure 4). The water retention *HU* range of 0 to 100 was used to estimate the water voxels. As shown in Figures 4(a) and (c), the reconstructed 3D image of the clam dominant area in February 2007 demonstrated water retention in the intertidal sediment up to 4 cm below the sediment surface. Since the shell *HU* range was found to be over 1,200, 3D shell images were also reconstructed, as shown in Figure 4(b). The CT results suggested that the regions of water retention, as well as the low-bulk density regions below the water-sediment interface, are necessary for the survival of the Manila clam.

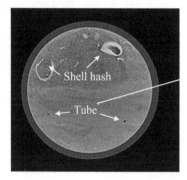
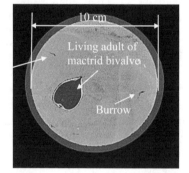

a) Point A   b) Point B

Figure 2. *2D images at 3cm depth from the bottom surface on February 4, 2007*

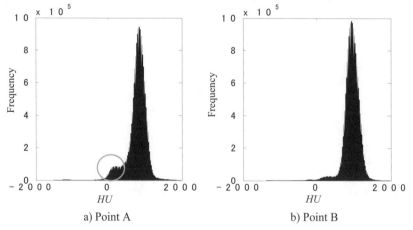

a) Point A   b) Point B

Figure 3. *Histograms of HU values of whole 80 images on February 4, 2007*

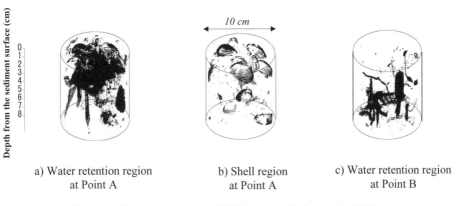

a) Water retention region   b) Shell region   c) Water retention region
at Point A   at Point A   at Point B

Figure 4. *3D reconstructions of CT images on February 4, 2007*

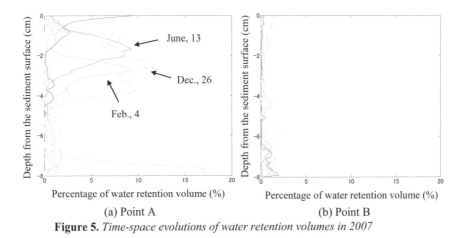

(a) Point A   (b) Point B

Figure 5. *Time-space evolutions of water retention volumes in 2007*

4. Discussion

To examine the time-space evolutions of water retention volumes on a seasonal time scale, we compared reconstructed 3D images using whole observation data from one Manila clam dominant and one non-dominant area (Figure 5). The ratio of water *HU* pixels with a range of 0 to 100 total pixels gave the percentage of water retention volumes in each 2D cross-section. As shown in Figure 5, the water retention regions of the clam dominant area were present in the intertidal sediment up to 4 cm below the sediment surface during the rainy and winter seasons. In summer and autumn, retention of water was almost absent because of high temperatures and sunshine. Sassa and Watabe (2007) proposed that suction dynamics associated with tide-induced ground water level variations have important effects on water retention in intertidal zones. The CT results in the present study lend further support to this notion. Finally, to investigate why water retention was higher in the clam dominant area, after CT scanning each core was sectioned from the top into slices of 50 mm. Each slice was passed through three sieves (mesh size of 4, 8, 16 mm). These results further suggest that shell hash compositions of substrates were related to water retention in the clam dominant area.

5. Conclusions

An industrial x-ray computed tomography (CT) scanner was used to investigate the time-space evolutions of vertical sediment structures in an intertidal flat. Seasonal sediment core samples were collected over one year at two locations, one clam-dominant and one clam non-dominant. 3D images reconstructed using *HU* showed that water retention volumes in the intertidal sediment at the clam-dominant area existed up to 4cm below the sediment surface. The CT results suggested that low-bulk density regions below the water-sediment interface are necessary for the survival of the Manila clam. Moreover, our results suggest that shell hash compositions of substrates are related to the distribution of the Manila clam. Overall, our study demonstrated that the industrial x-ray CT scanner is a useful tool for analyzing the vertical distributions of water retention in the bottom sediment layers, because the cores can be examined under non-destructive conditions.

6. Acknowledgements

This study was supported by a Grant-in-Aid for Scientific Research by the Japan Society for the Promotion of Science and Fisheries Agency of Japan.

7. References

Boespflug, X., Long, B. F. N., and Occhietti, S., "CAT-scan in marine stratigraphy: a quantitative approach", *Marine Geology*, vol. 122, pp. 281-301, 1995.

Duchesne, M. J, Moor, F. Long, B. F., Labrie, J., "A rapid method for converting medical Computed Tomography scanner topogram attenuation scale to Haunsfield unit scale and to obtain relative density values", *Engineering Geology*, vol. 103, pp. 100-105, 2009.

Dufour, S. C., Desrosiers. G. Long, B., Lajeunesse, P. Gagnould, M. Labrie, J., Archambault, P., Stora, G., "A new method for three-dimensional visualization and quantification of biogenic structures in aquatic sediments using axial tomodensitometry", *Limnology and Oceanography: Method*, vol. 3, pp. 372-380, 2005.

Duliu, O. G., "Computer axial tomography in geosciences: an overview", *Earth-Science Review*, vol. 48, pp. 265-281., 1999.

Mazik, K., Curtis, N. Fagen, M. J., Taft, S., Elliot, M., "Accurate quantification of the influence of benthic macro- and meio-fauna on the geometric properties of estuarine muds by micro computer tomography", *J. of Experimental Marine Biology and Ecology*, vol. 354, pp.192-201, 2008.

Mermillod-Blondin, F., Marie, S., Desrosiers, G., Long, B., de Montety, L., Michaud, E., Stora G., "Assessment of the spatial variability of intertidal benthic communities by axial tomodensitometry: importance of fine-scale heterogeneity", *J. of Experimental Marine Biology and Ecology*, vol. 287, pp 193-208, 2003.

Otani, J., Mukunoki, T., Obara, Y., "Application of X-ray CT method for characterization of failure soils", *Soils and Foundations*, 40, pp.111-118, 2000.

Perez, K.T, T. P., Earl, W. D., Richard, H.M., Peter, R.B., Michel, S.R., John, A.C., Roxanne, L.J. and Daniel, N. K., "Application of computer-aided tomography (CT) to the study of estuarine benthic communities", *Ecological Applications*, 9(3), pp. 1050-1058, 1999.

Rosenberg, R., Davey, E., Gunnarsson, Norling, K., and Frank, M., "Application of computer-aided tomography to visualize and quantify biogenic structures in marine sediment", *Marine Ecology Progress Series*, vol. 331, pp. 23-34, 2007.

Sassa, S., Watabe, Y., "Role of suction dynamics in evolution of intertidal sandy flats: Field evidence, experiments, and theoretical model", *J. of Geophysical Research*, vol. 112, F01003, doi:10.1029/2006JF000575, 2007.

Tamaki, A., Nakaoka, A., Maekawa H., Yamada, F., "Spatial partitioning between species of the phytoplankton-feeding guild on an estuarine intertidal sandflat and its implication on habitat carrying capacity", *Estuarine, Coastal and Shelf Science*, vol. 78, pp. 727-738, 2008.

Yamada, F., Kobayashi, N., "Annual tide level and mudflat profile", *J. of Waterway, Port, Coastal and Ocean Engineering*, vol.130, pp.119-126, 2004.

Yamada, F., Kobayashi, N., Sakanishi, Y., Tamaki, A., "Phase averaged suspended sediment fluxes on intertidal mudflat adjacent to river mouth", *J. of Coastal Research*, vol. 25, pp. 350-358, 2009.

Neutron Imaging Methods in Geoscience

A. Kaestner — P. Vontobel — E. Lehmann

Spallation neutron source division
Paul Scherrer Institut
CH-5232 Villigen PSI
Switzerland
anders.kaestner@psi.ch

ABSTRACT. *Neutron imaging is a younger and also less known method than X-ray imaging. Through the different attenuation behavior compared to X-rays this method provides complementary information and may even provide another way to investigate a sample non-destructively. The reasons are due to other contrast relations between different materials and isotopes or that the X-rays are unable to penetrate the sample. The basic principles behind neutron radiography and tomography are explained. Also the main components of a neutron imaging system are described. The two neutron imaging beamlines at Paul Scherrer Institute, Switzerland, are presented to show the infrastructure and instrument palette that can be found at a state of the art neutron imaging beamline. An outlook is given to show future instrument developments like energy selective imaging and phase contrast enhancement. Finally, a collection of examples illustrates applications from soil physics, nuclear waste deposit planning, and mineralized fossils.*

KEYWORDS: *Neutron imaging, tomography, minerals, soil, fluid movements, complementarity*

1. Introduction

Using neutrons for imaging purposes has with the development of advanced detector methods proven to be a successful method of investigation for a wide variety of applications in different fields of research. This imaging method is based on the attenuation of an incident beam similar to X-ray imaging. The difference lies in the attenuation mechanism which provides a completely different set of attenuation coefficients compared to X-rays and the quite different neutron and X-ray source characteristics. Water and its movements is an important actor in many neutron imaging investigations because of its high contrast compared to the medium it wets. Water is however not the only high contrast fluid, oil, for example, also delivers high contrast because of its hydrogen content.

Initial neutron imaging experiments at Paul Scherrer Institute (PSI), Switzerland, were made with films [PLE 95]. Film based imaging is inconvenient since the films have to be developed and later the processing and evaluation of the images is difficult. With the development of digital camera systems it is possible to acquire images with both higher temporal and spatial resolution. This allows real-time investigations of infiltration processes [CAR 07]. Further advantages of digital cameras are linear dynamic range and stationary detector. These features make the acquisition of tomographic data sets and referenced time series possible.

In this paper we will describe the principles of neutron imaging and its applications in geology and soil sciences.

2. Method

2.1. *Image forming process*

Neutron transmission images can be analyzed using the exponential law of radiation attenuation, also known as Beer-Lamberts law. This means that any sample will attenuate the intensity of the illuminating beam (I_0) according the material distribution ($\Sigma(x)$) along the penetration path (L) through the sample. This can be formulated as the following generalized formulation

$$I(E) = I_0(E)e^{-\int_L \Sigma(x,E)\,dx} \qquad [1]$$

This is the same attenuation law that applies for X-ray attenuation. The difference is that the linear X-ray attenuation coefficient is replaced by the macroscopic cross section (Σ) of the investigated material. The attenuation mechanism for neutrons is based on the interaction between the neutron and the nuclei of the atoms. The macroscopic cross section is determined by the neutron and proton configuration in the atomic nucleus [KRA 88]. The type of neutron reaction for many isotopes in the thermal and cold energy region is rather scattering than absorption, see *e.g.* the list of thermal neutron cross sections given by [SEA 92]. As a result Σ is not proportional to the atomic mass as for X-rays which interact with the electrons. Σ can even vary strongly between two isotopes of the same element, *e.g.* ^1H and ^2H (Deuterium). For example, the attenuation of hydrogen is an order of magnitude larger than for lead and many other metals. The practical effect of this fact is that the large cross section of hydrogen gives a high contrast for water relative to many porous media in nature. This makes neutrons the optimal probe for the study of wetting of porous media as will be shown later in the applications section.

Neutrons are neutral particles and as such they are not able to excite atoms. This mean that they have to be detected by a secondary process that generates charged particles which in turn excite a fluorescent material to emit visible light. A review of different neutron scintillation material is found in [EIJ 04]. At PSI mainly two scintillator materials are used; $^6LiF/ZnS(Ag)$ and Gd_2O_2S. The Li based scintillator produces more light which shortens the scan times. The Gd based scintillator on the other hand captures more neutrons which allows thinner layers and higher resolution. In an imaging system this scintillation process takes place on a screen that gives the spatial distribution of the neutron intensity behind the sample. The light produced by the scintillator is mostly acquired by cooled CCD cameras. The produced images represent the shadow or projection of the sample. Figure (1) schematically shows the components of a neutron imaging beamline. They are a neutron source, a collimator to shape the beam, a sample table, and finally a detector camera combination to acquire the projection of the sample.

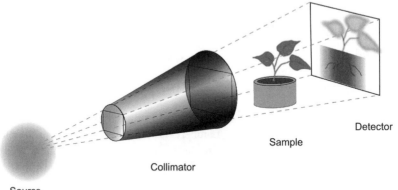

Detector

Sample

Collimator

Source

Figure 1: A schematic illustration of a neutron imaging beamline. Important components are a source, a collimator, the sample, and finally a detector combined with a camera.

2.2. Neutron sources

The practical use of neutron imaging for the investigation of geological samples, where tomography is preferred, requires reasonably performance in beam intensity and collimation. A lower limit in beam intensity can be set to 10^5 neutrons cm^{-2}s^{-1}. Such relatively high neutron fluxes can only be provided by two kinds of sources: nuclear reactors or spallation sources. World-wide, fission research-reactors are more common than the spallation sources. The first one is based on the principle of nuclear fission with U-235 as fissile material. Similarly to nuclear power plants, the chain reaction in fission enables the permanent supply with neutrons. A moderator surrounds the reactor core to moderate the neutron to thermal energies. More recently, spallation was found useful for neutron production. This process is based on the bombardment of high energy particles (in the order of GeV) like protons onto a target of a heavy element, *e.g.* lead. In the spallation act, the lead nuclei are split into parts of lower mass and 10 to 15 neutrons are emitted per proton. Therefore, a spallation source consists of the strong particle accelerator and the target station as the primary neutron source. PSI operates world's strongest spallation source (SINQ) for the moment with 590 MeV protons of a beam with about 1.4 mA, corresponding to about 1 MW thermal power [FIS 97]. Its performance compares to a reactor of about 10 MW. In both type of sources neutrons are then slowed down to thermal (25 meV) or cold energies (several meV) in a neutron moderator system containing heavy water or liquid D_2 at e.g. 25 K. The beam line layout is very advanced in respect to the extraction principle and used components (mirrored guides, cold source, *etc.*).

2.3. *Neutron imaging beamlines*

The configuration and characteristics of neutron imaging beamlines varies from place to place. The reason is that each beamline is optimized for the local conditions of the neutron source. Some of these conditions are the neutron spectrum, distance to the source, beam shape, and space in the experimental area. Here, we describe the two neutron imaging beamlines at PSI; NEUTRA and ICON. NEUTRA is the older beamline at PSI and has been operational since 1998. ICON was taken into operation in 2005, figure (2). They have similar infrastructure in terms of sample positioning and camera systems. The major difference is that NEUTRA is attached to a thermal neutron source whereas ICON is attached to a cold neutron source which operates at 25 K. The spectra of NEUTRA and ICON are plotted in figure (3). The macroscopic cross section for materials is larger for cold neutrons for many isotopes. In general the neutron caption reactions are related to the inverse of the neutron velocity. For cold energies the effect of scattering dominates and the valleys of the Bragg cross sections in polycrystaline materials induce reverse behavior. Large cross section has the effect that smaller amounts of an element can be detected with cold neutrons, *i.e.* higher contrast can be achieved. Large cross sections do on the other hand limit the sample size as the beam intensity is attenuated more rapidly. The aperture (D) of the beamline is an important part that together with the distance (L) to the samples position defines the the geometric unsharpness (d) of the imaging system, which can be described by pin-hole optics.

$$d = \frac{l}{L/D} \qquad [2]$$

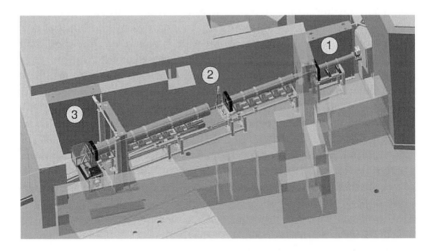

Figure 2: The ICON neutron imaging beamline. The components are; shutter system and an area for beam preconditioning (1), evacuated flight tubes leading to the first experimental position for high resolution imaging (2), followed by flight tubes leading to the second experimental position for voluminous and heavy samples and high L/D (3).

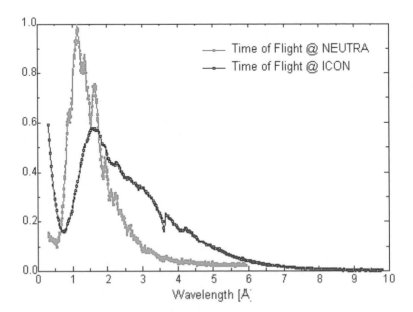

Figure 3: Neutron spectra of ICON and NEUTRA as measured with a time of flight method.

where l is the sample detector distance. The NEUTRA beamline has a fixed neutron aperture of 20 mm. ICON is constructed with an aperture drum providing five apertures from 1 mm to 80 mm. This gives L/D up to 10000. Vacuum flight tubes are installed to avoid flux losses from the source to the experimental position.

The experimental positions are equipped with linear positioning units and turntables. The linear unit is used to translate the sample in the plane perpendicular to the beam. The turntable allows acquisition of tomographic data sets. A palette of camera and lens systems makes both beamlines versatile user instruments. The cameras are mounted in light impermeable boxes that include tilted mirrors. With this arrangement the camera is placed outside the neutron beam to avoid direct neutron and gamma irradiation of the hardware. The cameras are focused on the scintillator which is placed directly behind the sample. The field of view can be chosen from 27-300 mm and depending on used scintillator-lens-camera combination resolutions up to 13.5 μm/pixel can be achieved. With a special device, an anisotropic resolution up to 10 μm in one direction can be reached.

A frequent imaging task of the beamlines is computed tomography. Depending on sample dimensions and requested resolution, scan times between a few hours and twenty hours can be expected. Tomography requires a processing step that transforms the projection data into a 3D volume. Normally, the filtered back-projection algorithm [KAK 01] is used. Depending on sample size and resolution the projection data sets contain 375 or 625 projections, in rare cases of very high resolution also 1125 projections are used. The exposure time per projection varies from a few seconds to a few minutes.

Recently an energy selector turbine was installed at the ICON beamline as a permanent part of the instrumentation. This turbine narrows the neutron spectrum to $\Delta\lambda/\lambda$=15%. The installation allows instant operation of the selector without a time consuming calibration step. Using energy selective imaging opens investigations of scattering in polycrystalline materials. The energy selector is also a fundamental component of the differential phase contrast setup. An upcoming addition to ICON's instrumentation is a setup for grating based differential phase contrast (DPC) imaging [PFE 06]. This setup will allow investigations of low-contrast samples if their phase shifting properties are strongly differing from each other. In this case the DPC method will provide better contrast and an alternative data set containing the phase information of the sample.

2.4. Quantitative neutron imaging

Water is often the interesting component in neutron imaging experiments. One of the objectives is then to quantify the spatial distribution and amount of water in the sample. The problem is that water is a strong incoherent neutron scatterer. Most of the neutrons hitting the water molecules are scattered and only a small fraction is absorbed. Scatter has the effect that more neutrons are registered behind the sample than would be motivated by the sample thickness and composition if all neutrons were absorbed by the sample. In the reality the neutrons are scattered or even multiply scattered in the sample. This in turn results in misreading of the water content of the sample. The error can be up to 50% of the true water content. In tomograms scatter is identified by a cupped intensity profile, *i.e.* the central regions of the sample have lower intensity than the periphery.

The solution is to estimate the scatter component in the images and remove it as a correction procedure. The procedure developed by R. Hassanein [HAS 06] provides tomograms where the water content can be estimated with an accuracy of about 5%. The price of such correction methods is increased processing time and a decreased signal to noise ratio.

3. Comparing neutron and X-ray imaging

As pointed out in the previous section the difference between neutron and X-ray imaging is the attenuation mechanism that gives a completely different set of attenuation coefficients. For low energy X-rays the photo-electric effect dominates which gives pure attenuation. For neutrons, scattering contributes to the attenuation coefficient and is sometimes responsible for the main contribution of the attenuation. In figure (4) attenuation coefficients for some minerals are plotted to show which minerals are more suited for the two modalities [VON 05]. A first observation is that the attenuation is mostly lower for neutrons than for X-rays. This is relevant for large samples. Secondly there are some minerals for which the attenuation coefficients for both modalities increase in a similar manner. Finally there are some minerals for which the X-ray contrast is rather low but neutrons provide high contrast. In cases of mixed minerals some components may have high contrast with neutrons and others with X-rays. This can be used to increase the contrast of the images. An excellent example of the complementarity of the two image types is the ammonite in figure (5), [CAR 06]. Here, the thin septa between the chambers are altered due to iron hydroxides. The high hydrogen content provides high contrast in the neutron images. For X-rays the small mass density differences give relatively low contrast in the sample, only the walls have a good contrast.

With the XTRA option of the NEUTRA beamline a sample can be imaged with both neutrons and X-rays using the same setup and without moving the sample. This allows convenient pixel-wise comparison between images from the two modalities. The X-ray source works up to 320 kV and allows imaging of relevant samples.

A final difference to mention are the scan times required for the two methods. Using the XTRA option at NEUTRA the scan times are in the same order of magnitude, *i.e.* some hours. To achieve the highest resolution with neutron micro-CT often 10-20 hours are needed to complete a scan, this is an order of magnitude more than the time required to complete a scan at a synchrotron beamline like TOMCAT at the Swiss Light Source [STA 06].

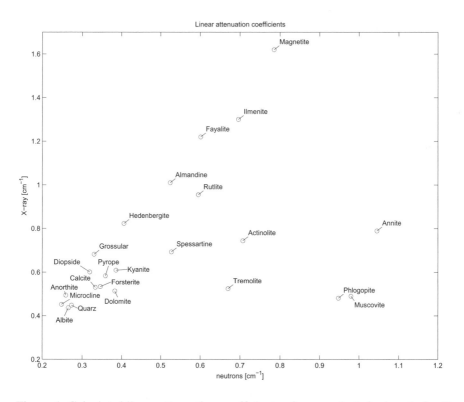

Figure 4: Calculated linear attenuation coefficients of some selected minerals for X-rays and neutrons. The coefficients are based on information from [SEA 92]

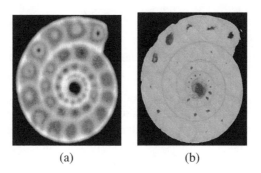

Figure 5: Tomography of a pyritized ammonite using thermal neutrons (a) and X-rays at 150 kV(b).

4. Applications

4.1. *Water movements and real-time imaging*

Dynamic processes with fluids in porous media are often studied using neutron imaging. Especially interesting is water since it has a high contrast relative to the porous medium. Figure (6) shows an example of how a small amount of water is distributed in sand. This sequence was captured at eight frames per second.

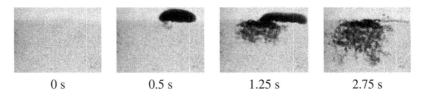

| 0 s | 0.5 s | 1.25 s | 2.75 s |

Figure 6: A time series showing a droplet of water released on a sand sample. The lighter upper region is air and the darker region is a partly hydrophobic sand material.

To quantify the amount of water in a sample at time t, the following relation is used

$$I_t = I_0\, e^{-(L_{Fluid}\,\Sigma_{Fluid} + L_{Sample}\,\Sigma_{Sample})} \qquad [3]$$

$$I_{t_0} = I_0\, e^{-(L_{Sample}\,\Sigma_{Sample})} \qquad [4]$$

where L_{index} are the lengths of the medium and fluid respectively and Σ_{index} are the macroscopic cross sections of the medium and fluid. These equations can be used together to compute the amount of water at each pixel in the image.

$$L_{Fluid}\,\Sigma_{Fluid} = -\log\left(\frac{I_t}{I_0}\right) - \left(-\log\left(\frac{I_{t_0}}{I_0}\right)\right) = -\log\left(\frac{I_t}{I_{t_0}}\right) \qquad [5]$$

If Σ_{Fluid} is known it is possible to compute the amount of water for each pixel in the image.

For slow processes it is even possible to acquire low resolving tomographies as time series[KAE 05][SCH 08]. This allowed the study of the water movement during drainage and wetting of a constructed heterogeneous sand column. The column was constructed as an arrangement of cube shaped region. The sample was drained and wetted by elevating the water level. The cubes were identified in the images and the water content in each cube could be estimated. Figure (7) shows the water distribution in the sample column for different levels of the water table.

The feasibility of real-time imaging is a question of relating the time constant of the imaged process the possible acquisition time. If the process is too fast the camera may only be able to capture the dry and final state. Furthermore, tomographies of very fast processes would produce motion artifacts that makes any quantification impossible.

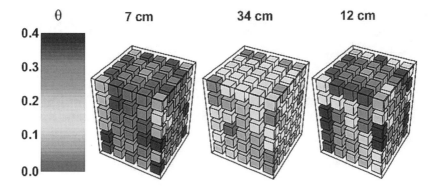

Figure 7: Wetting and draining of a heterogeneous sand column. The colors represent the water content in each cube when the water table is located 7, 34, and 12 cm below the sample.

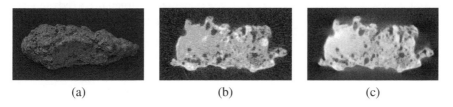

Figure 8: Tomographic cross sections of a bog ore sample (a) using thermal neutrons (b) and using X-rays at 250 keV, and with a 2 mm Cu filter (c).

4.2. *Minerals*

Figure 4 shows the linear attenuation coefficients for some minerals. This plot shows that many of the minerals have a higher attenuation with X-rays. The direct effect is that larger samples can be investigated with neutrons. Most interesting investigations can be made when the contrast between two materials is high then is possible to easily separate structures in the sample. This applies to both X-rays and neutrons. A second effect that can be investigated is when a beam-line is equipped with an X-ray source as previously described in section 3.

The first example in figure 8 shows the tomography of a piece of bog ore. The sample dimensions are about 30 mm × 50 mm × 100 mm. The main components of bog ore are iron oxides. The tomograms show first that neutrons are more suited due to the better penetration and second, the structures near the sample surface are displayed with a higher contrast.

The second example is also iron based. In this example, a piece of pyrite was investigated, see figure 9. The sample is approximately 40 mm thick. This sample was even harder for the X-rays to penetrate. This resulted in the streaks near the edges and the fuzzy edges. The acquisition with neutrons, on the contrary, did not suffer

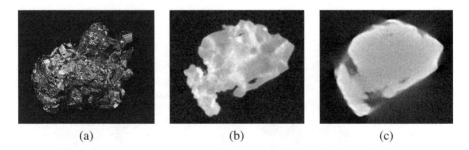

(a)   (b)   (c)

Figure 9: Tomographic cross sections of a pyrite sample (a) using thermal neutrons (b) and using X-rays at 250 keV, and with a 2 mm Cu filter (c).

from intensity starvation and hence good tomograms resulted. Further on, the sample composition appeared to be ideal for neutron imaging. The larger crystal fragments appear with good contrast to the connecting regions. These structures do not appear in the X-ray tomogram which only reveals the shape of the sample.

4.3. *Clays*

Toxic waste deposal requires stable containment to avoid contamination of the ground water. Currently, in Switzerland the Opalinus clay is in focus for this task. The investigations are both considering the layered structure of this clays and the wetting of the clays. This is a novel task for neutron imaging that already has shown promising results. In the future, experiments showing the wetting of the sample will be used to verify models of reactive transport phenomena,[GRO 09]. The tomogram in figure (10) shows that the micro setup at ICON is capable of resolving the Illite and Calcite regions of the sample. These components both have lower attenuation coefficients than water, which makes neutron imaging the optimal method for future wetting and transport experiments. Complementary measurements will also be made using synchrotron based micro-tomography, micro X-ray absorption spectroscopy, and micro X-ray diffraction.

4.4. *Concrete*

A technical application of neutron imaging is the study of concrete. Again water is central for the investigation. In an experiment reported by [ZHA 09], the water movements initiated by cracks in reinforced concrete were studied using neutron real-time imaging. Figure (11) shows a time-series of this experiment that clearly shows the water first entered the sample through the crack and then found preferential paths at the reinforcement. These kinds of experiments will provide models to predict the service life of reinforced concrete.

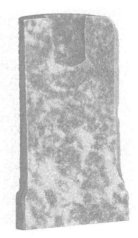

Figure 10: A tomogram of an Opalinus clay sample, courtesy of D. Grolimund, PSI. The sample diameter is 3 mm. The dark regions represents the Calcite domain and the lighter regions represent the Illite clay matrix.

Figure 11: Water penetration in reinforced concrete as function of time. Initially the water enters the crack, later the water finds preferential paths along the reinforcements. The yellow bars in the first frame indicate the location of the reinforcement in the dry sample.

5. Neutron imaging world-wide

Neutron imaging is available at about 15 different locations world-wide. The instrumentation vary in technical standard from beamlines with basic radiography capability and manual sample positioning to beamlines with advanced camera systems and automated sample manipulation that allows tomography. These beamlines are distributed on all continents giving the user community access to imaging beam-time for their experiments. The current status and progress of imaging beamlines world wide is reported by IAEA[IAE 08]. A list of neutron sources which are equipped with advanced neutron imaging beamlines is given in [LEH 09].

6. Summary

An overview of different applications of neutron imaging in geoscience has been given. Important for many applications is the presence of water which appears with high contrast compared to the medium it wets. The objective is often to study the distribution of water in the sample both at steady state and in dynamic processes. The important aspect is the ability to do this without adding any tracer element to water which would change behavior of the studied system.

In other applications the different set attenuation coefficients for neutrons may provide additional information to previous X-ray investigations. For many minerals the attenuation coefficient is lower for the neutrons; this allows the user to investigate inner structure of larger samples than with X-rays. These are some strengths of neutron imaging, but it must not be forgotten that neutron and X-ray imaging often are complementary and the use of one method does not exclude the other.

7. Acknowledgements

The the authors want to acknowledge the contributions from P. Lehmann, ETH Zürich, D. Grolimund, Paul Scherrer Institut, W.D. Carlson, The University of Texas at Austin, and F.H. Wittmann, Aedificat Institute Freiburg. Finally, the authors also want to thank L. Butler, Louisiana State University, for interesting discussions and suggestions.

8. References

[CAR 06] CARLSON W., "Three-dimensional imaging of planetary materials", *Earth and Planetary Science Letters*, vol. 249, num. 3–4, 2006, p. 133–147.

[CAR 07] CARMINATI A., KAESTNER A., HASSANEIN R., VONTOBEL P., FLÜHLER H., "Infiltration through series of soil aggregates", *Advances in Water Resources*, vol. 30, num. 5, 2007, p. 1168–1178.

[EIJ 04] VAN EIJK C., "Inorganic scintillators for thermal neutron detection", *Radiation Measurements*, vol. 38, num. 4–6, 2004, p. 337–342, Proceedings of the 5th European Conference on Luminescent Detectors and Transformers of Ionizing Radiation (LUMDETR 2003).

[FIS 97] FISCHER W., "SINQ – The spallation neutron source, a new research facility at PSI", *Physica B*, vol. 234–236, 1997, p. 1202–1208.

[GRO 09] GROLIMUND D., "Personal communication: Neutron tomography of Opalinus clay", Personal communication, August 2009.

[HAS 06] HASSANEIN R., "Correction methods for the quantitative evaluation of thermal neutron tomography", Diss. ETH No. 16809, Swiss Federal Institute of Technology, 2006.

[IAE 08] IAEA, "Neutron Imaging: A non-destructive tool for materials testing", Research report num. IAEA-TECDOC-1604, September 2008, International atomic energy agency, Wagramer Strasse 5, PO Box 100, 1400 Wien, Austria.

[KAE 05] KAESTNER A., VONTOBEL P., HASSANEIN R., LEHMANN P., SCHAAP J., LEHMANN E., FREI G., LAESER H., FLUEHLER H., "Mapping the three dimensional water dynamics in heterogeneuos sands using thermal neutrons", *Proc. 4th World Congress on Industrial Process Tomography*, 2005.

[KAK 01] KAK A., SLANEY M., *Principles of computerized tomographic imaging*, SIAM, 2001.

[KRA 88] KRANE K., *Introductory nuclear physics*, John Wiley & sons, 1988.

[LEH 09] LEHMANN E., KAESTNER A., "3D neutron imaging", *Encyclopedia of Analytical Chemistry*, vol. a9123, 2009, John Wiley & Sons Ltd.

[PFE 06] PFEIFFER F., GRUNZWEIG C., BUNK O., FREI G., LEHMANN E., DAVID C., "Neutron phase imaging and tomography", *Physical Review Letters*, vol. 96, num. 6, 2006, Page 215505.

[PLE 95] PLEINERT H., DEGUELDRE C., "Neutron radiographic measurement of porosity of crystalline rock samples: a feasibility study", *Journal of Contaminant Hydrology*, vol. 19, num. 1, 1995, p. 29–46.

[SCH 08] SCHAAP J., LEHMANNN P., KAESTNER A., VONTOBEL P., HASSANEIN R., FREI G., DE ROOIJ G., LEHMANN E., FLÜHLER H., "Measuring the effect of structural connectivity on the water dynamics in heterogeneous porous media using speedy neutron tomography", *Advances in Water Resurces*, vol. 31, num. 9, 2008, p. 1233–1241.

[SEA 92] SEARS V., "Neutron scattering lengths and cross sections", *Neutron News*, vol. 3, num. 3, 1992, p. 29–37.

[STA 06] STAMPANONI M., GROSO A., ISENEGGER A., MIKULJAN G., CHEN Q., BERTRAND A., HENEIN S., BETEMPS R., FROMMHERZ U., BÄHLER P., MEISTER D., LANGE M., ABELA R., "Trends in synchrotron-based tomographic imaging: the SLS experience", *Proc. SPIE, Developments in X-Ray Tomography V*, vol. 6318, SPIE, 2006.

[VON 05] VONTOBEL P., LEHMANN E., CARLSON W., "Comparison of X-ray and Neutron Tomography Investigations of geological materials", *IEEE trans. on Nuclear Science*, vol. 52, num. 1, 2005, p. 338–341.

[ZHA 09] ZHANG P., WITTMANN F., ZHAO T., LEHMANN E., "Penetration of water into uncracked and cracked steel reinforced concrete elements; visualization by means of neutron radiography", *Restoration of Buildings and Monuments*, vol. 15, num. 1, 2009, p. 67–76.

Progress Towards Neutron Tomography at the US Spallation Neutron Source

L. G. Butler

Department of Chemistry
Louisiana State University
Baton Rouge, LA, 70803
USA
lbutler@lsu.edu

ABSTRACT. *The new US Spallation Neutron Source provides a world-class source of pulsed neutrons which may be useful for neutron tomography experiments in the geosciences. An instrument development team is proposing the construction of a tomography beamline at SNS, a project estimated at about $15M US. Progress on the technical, scientific, user community, and business cases is presented.*

KEYWORDS: *spallation neutron source, tomography, VENUS, geoscience*

1. Introduction

The geosciences have been quick to use new imaging methods, both in the field and in the laboratory. As an example of the latter, we point to the success of the several tomography beamlines at the US Advanced Photon Source X-ray synchrotron (http://www.aps.anl.gov). While x-ray imaging is quite powerful, neutron imaging offers new options for geomaterials [Vontobel 2005]. At the time of writing, access to neutron imaging in the US is quite limited: the National Institute of Standards and Technology (http://physics.nist.gov/MajResFac/Nif/) and the University of California at Davis (http://mnrc.ucdavis.edu/imaging.html) have established user access programs and North Carolina State University and Indiana University are developing imaging facilities. The recent opening of the US Spallation Neutron Source (SNS, http://neutrons.ornl.gov/) at Oak Ridge National Laboratory in Tennessee may offer to the geosciences additional access with greatly improved imaging methods. In brief, the SNS experimental hall has room as of Fall, 2009, for a neutron tomography beamline and its neutron science advisory committee is considering the application for space on the floor. If laboratory space is granted, the next milestones are securing funds for an engineering design, then funds for construction. Altogether, the SNS neutron tomography beamline project may cost more than $15 M US and require several years to design and construct. At this writing, the project has a logo and a name, VENUS (VErsatile Neutron Imaging InstrUment at SNS).

The VENUS neutron tomography beamline design takes advantage of the unique pulse neutron flux produced at SNS. Members of the geoscience community are already familiar with time-of-flight methods as used in seismic imaging; many of the same issues – finite time duration of the initial pulse, propagation distances and media, and detector time resolution – are found in imaging with pulsed neutrons sources.

2. Spallation Neutron Source and proposed VENUS tomography beamline

In context, the US SNS is one of several pulsed neutron sources world-wide. In Oxfordshire, Great Britain, the ISIS-2 neutron source (http://ts-2.isis.rl.ac.uk/) will soon have the IMAT imaging beamline. The Japan Proton Accelerator Research Center (http://j-parc.jp/) is using the NOBORU beamline to evaluate tomography methods. The European Spallation Neutron Source (http://www.ess-neutrons.eu/) has recently completed the site selection process, opting for construction in Sweden. The China Spallation Neutron Source is proposed for construction in Dongguan, Gaungdong province [Wei 2009]. The above sources operate, or will operate, in a pulsed mode, typically 25 to 60 Hz. The SINQ neutron source at the Paul Sherrer

Institute in Switzerland (http://sinq.web.psi.ch/) is a continuous neutron source creating neutrons via a spallation process.

In brief, a spallation neutron source starts with a near-GeV proton beam striking a high atomic number target. Each proton-nucleus collision generates 20-30 neutrons in the "spallation" process [Bauer 2001]. The high-energy neutrons are slowed with a moderator, often liquid hydrogen, to yield a flux with maximum intensity at 1-5 Å wavelength; see, for example Figure 15 in [Lu 2008] and note that 20 meV = 2.022 Å. The De Broglie equation allows one to convert neutron wavelengths into velocity as needed for time-of-flight measurements.

2.1. Properties of neutrons and time-of-flight imaging

Materials science is replete with imaging methods, ranging from optical, MRI, x-ray, positron, ultrasound, scanning probe microscopes to atom-scale aberration-corrected electron microscopy. In this context, neutrons are fascinating probes for several unique reasons:

– As neutral particles, their transmission through dense matter is exceptional. Neutrons can image structure inside solid lead objects.

– Neutrons have exceptional sensitivity to hydrogen, deuterium and other light atoms. One centimeter of water greatly attenuates 2 Å (20 meV, 2,000 m/s) neutrons.

– Neutrons in near thermal equilibrium with liquid hydrogen have wavelengths on the order of common crystallographic dimensions, so transmission through an object is sensitive to the various crystalline phases.

– Neutrons have a large magnetic spin moment, and are sensitive to magnetic fields and gradients.

Existing neutron tomography beamlines at reactor and continuous spallation sources have imaged samples spanning a huge range of science and culture. Excised rat lungs show changes as a function of inflation, key data needed for fundamental knowledge of human artificial ventilation systems. The images of trapped magnetic fields in superconducting lead provide tantalizing views of more applications to come in the field of high T_c materials [Kardjilov 2008; Dawson 2009; Strobl 2009]. Cultural heritage artefacts, as recovered from the depths of the Mediterranean Sea, have been examined to find what lies underneath the centuries-old encrustations [Kockelmann 2006].

Time-of-flight neutron imaging has been tested at the ISIS pulsed neutron source for metals [Kockelmann 2007]. Iron and copper were readily distinguished based upon their differing crystallographic d-spacings, when imaged with neutrons in the range of 1 to 5 Å. They found that in transmission-mode imaging, the crystallographic structure strongly affects the neutron scattering cross sections. Neutron scattering cross sections are usually tabulated for set velocity, typically 2,200 m/s which corresponds to 1.7982 Å or 25.299 meV. A few materials have been studied as a function of neutron energy, as shown in Figure 1. We see that crystalline materials show distinct "Bragg edges" where scattering cross sections change several-fold. Kockelmann *et al.* found that metal stress significantly affects the amplitude and sharpness of the Bragg edges. Figure 1 also shows the extremely large scattering cross section for liquid water, a fact that has led to experiments in wetting front imaging in geomaterials [Carminati 2007; Tullis 2007]. The data in Figure 1 are from "Experimental Nuclear Reaction Data" (EXFOR/CISISRS) web site (http://www.nndc.bnl.gov/exfor/exfor00.htm) and are identified in the database as: Al, 1958 R.J.Brown+; C, 1957 P.A.Egelstaff; Fe, 1982 J.A.Harvey+; P, 1952 W.W.Havens Jr; Ti, 1960 R.E.Schmunk+; and water, 1966 J.L.Russell Jr+.

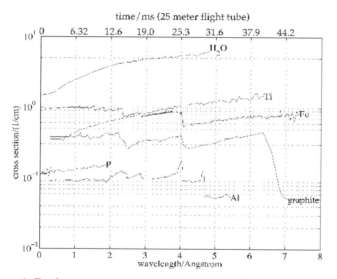

Figure 1. *Total neutron scattering cross-sections as a function of wavelength and as a function of flight time for a 25 m long beamline*

The upper axis in Figure 1 indicates neutron travel time for a hypothetical 25 m flight path between the pulsed neutron source and the sample. The initial 2 Å neutron pulse as it enters the flight tube has a time-width on the order of 20 μs. Neutron detectors with equivalent time resolution have recently become available [Dangendorf 2009; Tremsin 2009]. Flight tubes shorter than 25 m will, in principle, benefit from higher neutron flux (less $1/r^2$ losses) while longer tubes will yield better wavelength accuracy and allow for larger samples (more time evolution relative to the initial pulse; more space in the experimental hall).

2.2. *Workshops and user community input*

Based on attendance at recent workshops held at SNS, VENUS is already appreciated by many. The October 2006 "Imaging and Neutrons" workshop (http://neutrons.ornl.gov/workshops/ian2006/) drew 200 participants, more than for any other kick-off workshop for an SNS beamline. The VENUS Instrument Development Team (IDT) was created at the 2006 workshop and then drafted a Letter-of-Intent for the ORNL neutron science advisory committee. The VENUS LoI was approved with its first reading at the November 2008 meeting. To refine the LoI, another workshop was held in November 2008 (http://neutrons.ornl.gov/conf/nisns2008/). Comments from the workshops, the IDT, and the science advisory committee have lead to the following conceptional design for VENUS.

2.3. *Conceptional design of the VENUS beamline*

Relative to existing beamlines at reactor and continuous spallation sources, the proposed VENUS neutron tomography beamline will be a next-generation instrument. Key features include:

– Neutrons from the pulsed source will be observed with time-of-flight (TOF) strategies. This gives us access to a wide neutron energy range in a very short time. TOF may also offer new strategies for imaging scattered neutrons, especially in thicker samples.

– The neutron energy range is particularly broad, with ~23 MeV spallation neutrons in a 695 ns time window as well as the moderated 0.5 Å (7912 m/s, 327 meV) to 10 Å (396 m/s, 0.818 meV) originating from an approximately 20 to 40 μs time window.

– The pulsed beam allows synchronization with repetitive processes for stroboscopic imaging.

– The wave-particle dualism allows one to treat the neutron as a particle having a defined mass or to regard it as a propagating wave with corresponding amplitude and wavelength. In conventional radiography the image formation is given in terms of attenuation of the propagated neutron beam. In the case of phase-contrast imaging, the phase variations obtained by the propagation of coherent radiation are transformed to intensity variations. Phase-contrast imaging with neutrons was reported by implementing a free-path propagation technique where a beam with high order of spatial coherence was used [Allman 2000]. A new technique using spatial gratings was presented recently [Pfeiffer 2006] where the low beam intensity problems of the former methods are partially solved.

–   The spin rotation of the component perpendicular to the magnetic field B can be described by the spin phase

$$\varphi_s = \frac{\gamma\,\lambda\,m}{h} \int\limits_{path} B\,ds$$

[1]

where γ is the gyromagnetic ratio of the neutron (-1.8324×10^8 rad s^{-1} T^{-1}) and m is the mass and λ the wavelength of the neutron. The total spin rotation in simple monochromatic transmission imaging can be deduced with an uncertainty with respect to the periodicity of the cosine function, if no other *a piori* knowledge about the sample can be used. In contrast, a time-of-flight approach, could overcome this drawback efficiently, due to the additional wavelength-dependent information [Strobl 2009].

–   At this point, it is important to discuss the relative merits of monochromatic reactor sources versus time-of-flight spallation sources. Time-of-flight allows the experimentalist to adjust data collection for either very fine or very broad wavelength ranges; a monochromator generally has a fixed bandwidth. Data from FRM II [Schulz 2009] and SNS flux simulations show that FRM and SNS have equivalent data acquisition rates when VENUS is operated in a 10-point multiplex mode for each neutron pulse (assumes $\Delta\lambda=0.1$ Å and a bandwidth of 1 Å). The VENUS advantage increases with bandwidth and/or larger $\Delta\lambda$.

Combining all of these thoughts has lead to a preliminary design which was presented to the SNS neutron science advisory committee as part of a beamline letter-of-intent (approved). A sketch of the proposed VENUS beamline is shown in Figure 2. The sketch is based on a flight tube of 15 to 20 m with an upstream optics cave that may contain optional neutron focusing optics, gratings for phase contrast imaging, or spin-polarizers for magnetic field imaging. Not visible in the sketch are upstream collimators to establish beam quality, gamma ray filter, neutron shutter, and bandwidth choppers to prevent overlap from multiple neutron pulses at the 60 Hz SNS source. The experimental hutch is designed to accommodate large samples and ancillary sample environmental equipment.

The design, funding and construction of such a large instrument is challenging. The VENUS Instrument Development Team consists of 30+ members from research facilities, industry, and academia across the world (see SNS workshop web sites). The science and business cases continue to evolve and user input is welcomed.

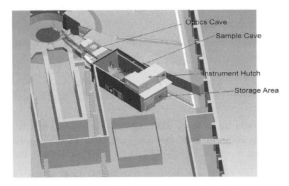

Figure 2. *The VENUS beamline as proposed in the letter of intent*

3. Acknowledgments

The author would like to thank Drs. Eberhard Lehmann and Burkhard Schillinger and their groups at the Paul Scherrer Institute and FRM II, respectively, for introducing the author to the technique of neutron tomography.

4. References

Allman, B. E., McMahon, P. J., Nugent, K. A., Paganin, D., Jacobson, D. L., Arif, M. and Werner, S. A., "Imaging - Phase radiography with neutrons", *Nature*, vol. 408, 2000, p.158-159.

Bauer, G. S., "Physics and technology of spallation neutron sources", *Nuclear Instruments & Methods in Physics Research Section a-Accelerators Spectrometers Detectors and Associated Equipment*, vol. 463, 2001, p.505-543.

Carminati, A., Kaestner, A., Ippisch, O., Koliji, A., Lehmann, P., Hassanein, R., Vontobel, P., Lehmann, E., Laloui, L., Vulliet, L. and Fluhler, H., "Water flow between soil aggregates", *Transport in Porous Media*, vol. 68, 2007, p.219-236.

Dangendorf, V., Bar, D., Bromberger, B., Feldman, G., Goldberg, M. B., Lauck, R., Mor, I., Tittelmeier, K., Vartsky, D. and Weierganz, M., "Multi-Frame Energy-Selective Imaging System for Fast-Neutron Radiography", *IEEE Transactions on Nuclear Science*, vol. 56, 2009, p.1135-1140.

Dawson, M., Manke, I., Kardjilov, N., Hilger, A., Strobl, M. and Banhart, J., "Imaging with polarized neutrons", *New Journal of Physics*, vol. 11, 2009, p.art. no. 043013.

Kardjilov, N., Manke, I., Strobl, M., Hilger, A., Treimer, W., Meissner, M., Krist, T. and Banhart, J., "Three-dimensional imaging of magnetic fields with polarized neutrons", *Nature Physics*, vol. 4, 2008, p.399-403.

Kockelmann, W., Frei, G., Lehmann, E. H., Vontobel, P. and Santisteban, J. R., "Energy-selective neutron transmission imaging at a pulsed source", *Nuclear Instruments & Methods in Physics Research Section a-Accelerators Spectrometers Detectors and Associated Equipment*, vol. 578, 2007, p.421-434.

Kockelmann, W., Siano, S., Bartoli, L., Visser, D., Hallebeek, P., Traum, R., Linke, R., Schreiner, M. and Kirfel, A., "Applications of TOF neutron diffraction in archaeometry", *Applied Physics a-Materials Science & Processing*, vol. 83, 2006, p.175-182.

Lu, W., Ferguson, P. D., Iverson, E. B., Gallmeier, F. X. and Popova, I., "Moderator poison design and burn-up calculations at the SNS", *Journal of Nuclear Materials*, vol. 377, 2008, p.268-274.

Pfeiffer, F., Grunzweig, C., Bunk, O., Frei, G., Lehmann, E. and David, C., "Neutron phase imaging and tomography", *Physical Review Letters*, vol. 96, 2006, p.art. no. 215505.

Schulz, M., Böni, P., Calzada, E., Mühlbauer, M. and Schillinger, B., "Energy-dependent neutron imaging with a double-crystal monochromator at the ANTARES facility at FRM II", *Nuclear Instruments & Methods in Physics Research Section a-Accelerators Spectrometers Detectors and Associated Equipment*, vol. 605, 2009, p.33-35.

Strobl, M., "Future prospects of imaging at spallation neutron sources", *Nuclear Instruments & Methods in Physics Research Section a-Accelerators Spectrometers Detectors and Associated Equipment*, vol. 604, 2009, p.646-652.

Strobl, M., Kardjilov, N., Hilger, A., Jericha, E., Badurek, G. and Manke, I., "Imaging with polarized neutrons", *Physica B*, vol. 404, 2009, p.2611-2614.

Tremsin, A. S., McPhate, J. B., Vallerga, J. V., Siegmund, O. H. W., Hull, J. S., Feller, W. B. and Lehmann, E., "Detection efficiency, spatial and timing resolution of thermal and cold neutron counting MCP detectors", *Nuclear Instruments & Methods in Physics Research Section a-Accelerators Spectrometers Detectors and Associated Equipment*, vol. 604, 2009, p.140-143.

Tullis, B. P. and Wright, S. J., "Wetting front instabilities: a three-dimensional experimental investigation", *Transport in Porous Media*, vol. 70, 2007, p.335-353.

Vontobel, P., Lehmann, E. and Carlson, W. D., "Comparison of X-ray and neutron tomography investigations of geological materials", *IEEE Transactions on Nuclear Science*, vol. 52, 2005, p.338-341.

Wei, J., Chen, H. S., Chen, Y. W., Chen, Y. B., Chi, Y. L., Deng, C. D., Dong, H. Y., Dong, L., Fang, S. X., Feng, J., Fu, S. N., He, L. H., He, W., Heng, Y. K., Huang, K. X., Jia, X. J., Kang, W., Kong, X. C., Li, J., Liang, T. J., Lin, G. P., Liu, Z. N., Ouyang, H. F., Qin, Q., Qua, H. M., Shi, C. T., Sun, H., Tang, J. Y., Tao, J. Z., Wang, C. H., Wang, F. W., Wang, D. S., Wang, Q. B., Wang, S., Wei, T., Xi, J. W., Xu, T. G., Xu, Z. X., Yin, W., Yin, X. J., Zhang, J., Zhang, Z., Zhang, Z. H., Zhou, M. and Zhu, T., "China Spallation Neutron Source: Design, R&D, and outlook", *Nuclear Instruments & Methods in Physics Research Section a-Accelerators Spectrometers Detectors and Associated Equipment,* vol. 600, 2009, p.10-13.

Synchrotron X-ray Micro-Tomography and Geological CO$_2$ Sequestration

P. S. Nico* — J. B. Ajo-Franklin* — S. M. Benson —
A. McDowell* — D. B. Silin* — L. Tomutsa* — Y. Wu***

*Lawrence Berkeley National Laboratory
One Cyclotron Road
Berkeley, CA 94720
USA
psnico@lbl.gov
jbajo-Franklin@lbl.gov
aamacdowell@lbl.gov
dsilin@lbl.gov
l_tomutsa@yahoo.com
YWu3@lbl.gov

**Global Climate & Energy Project
Stanford University
Stanford, CA 94305
USA
smbenson@stanford.edu

ABSTRACT. *We used beamline 8.3.2 of the Advanced Light Source at Lawrence Berkeley National Lab to gain insights into processes important to the geological sequestration of CO$_2$. Beamline 8.3.2 is a dedicated hard x-ray CT beamline with a superbend magnet source that provides a monochromatic beam in the 8 to 45 keV range. Synchrotron based x-ray CT has several unique capabilities including a tuneable coherent x-ray source which allows for multi-energy chemically sensitive imaging. Data from 8.3.2 were analyzed using a novel Maximal Inscribed Sphere (MIS) algorithm in order to predict the distribution of CO$_2$ within a multi-phase system. In addition, time lapse imaging was used to characterize microbially driven precipitation events.*

1. Introduction

Understanding and modeling processes in the subsurface requires an accurate understanding of the three dimensional structure of solid media and how that structure changes with time. Because x-ray CT can provide structural information without disrupting or destroying the matrix, it is an ideal tool for parameterizing and verifying models and for studying dynamic precipitation/dissolution processes. Unlike most lab-based tomographic instruments, in synchrotron tomography the x-ray source usually has very high flux and is monochromatic, tuneable, and largely coherent. This provides for several useful capabilities including phase contrast imaging and chemically sensitive imaging.

With the growing concern over the climate impacts of atmospheric CO_2 has come increasing interest in carbon capture and storage (CCS) technologies. Geologic carbon storage involves injection of CO_2 into an appropriate subsurface formation such that the injected CO_2 will be stable for time frames of thousands to tens of thousands of years. Effectively injecting large quantities of CO_2 necessitates a thorough understanding of and the ability to predict the movement of CO_2 within porous media. In addition, carbon stored in the form of solid carbonates (e.g. calcite) is impervious to borehole or seal failure hence techniques. Mineralization also significantly modifies rock permeability and mechanical properties by clogging pore throats and strengthening grain-to-grain contacts; these alterations make understanding the 3D distribution and morphology of carbonate precipitation crucial when attempting to predict or enhance the long-term behaviour of CO_2 storage units.

Below, we highlight two examples of how synchrotron x-ray tomography can help improve our understanding of these processes. First, we demonstrate the use of a novel algorithm for analyzing tomographic data in order to predict the distribution of CO_2 within a two phase CO_2/brine system. Second, we use time lapse imaging to dynamically characterize microbially-mediated precipitation in a porous medium.

2. Description of beamline 8.3.2

The x-ray micro-tomography facility is based at beamline 8.3.2 at the Advanced Light Source, Berkeley, California. The setup is similar to the standard setup for this technique developed in the 1990's (Kinney *et al.*, 1992). The x-rays are produced from a superbend magnet source and pass through a monochromator comprised of two multilayer mirrors, which can be altered in angle to select the required x-ray energy. X-rays of energy 8 KeV to 45 KeV are available. The sample is mounted on an air bearing stage that can be positioned in the x-ray beam. The x-rays transmitted through the sample then interact with a $CdWO_4$ single crystal scintillator that fluoresces the shadowgram x-ray image as visible light. This image is then magnified through a choice of microscope objectives and relayed onto a 4008x2672 pixel CCD camera (Cooke PCO 4000). The ccd pixel size is 9 μm, thus with a 10x objective, pixel sizes of 0.9 μm at the sample image can be mapped onto the ccd. The samples are rotated in the x-ray beam from angles 0 to 180 degrees in the desired increment. A typical scan takes about 20-40 minutes. Samples are generally scanned in absorption mode and the reconstructed images obtained following normalization and the application of a filtered back projection algorithm.

2.1. *Multi-energy imaging*

The ability to image using monochromatic x-ray beans of different energies can be exploited to introduce a degree of chemical sensitivity into the imaging process. This process is presented here as a general beamline capability but is applicable to the issues of precipitate identification has discussed in the $CaCO_3$ precipitation section below. In this example MoS_2 coated SiO_2 sand was chosen as a test material because MoS_2 is known to form thin coatings similar to those expected from surface precipitation.

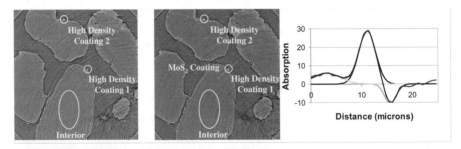

Figure 1. *Images taken at 19,500 eV (left) and 20,500 eV (right). Graph shows trace across Mo containing grain showing FHWM of ~4 microns. (The observable ring artefacts result either from non-linearities in beamline optics or imperfections in the reconstruction algorithm. No further attempt was made to remove them because they did not interfere with the desired analysis).*

Figure 1 shows two slices from the same region of the MoS_2 coated SiO_2 sample. One image was taken at 19,500 eV, ~500 eV below the Mo K edge. The other was taken at 20,500 eV, ~500 eV above the Mo K edge. There regions within the image were analyzed for changes as a function of energy: the interior of a grain and two "high density" regions on grain surfaces. The change in calculated linear attenuation coefficient between the two energies was undetectable for the grain interior as expected. The first "high density" coating showed a 75% increase in attenuation coefficient at higher energy indicating that this feature contains Mo. The second high density coating feature showed a 22% decrease in attenuation coefficient over the same energy shift indicating that while in any single image these two features appear similar, using multi-energy imaging they can be distinguished as they have distinctly different compositions.

3. Imaging of Frio sandstone

Figure 2 shows a single slice from a set of tomographic images collected from a sample of Frio sandstone. The imaged was processed via a simple threshold segmentation in order to separate the grains from the pore space.

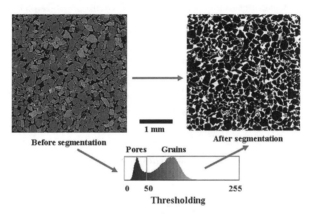

Figure 2. *CT image before and after threshold segmentation with histogram of voxel values showing selection of threshold value*

3.1. *Maximal inscribed spheres calculation*

The maximal inscribed sphere calculation is a method for determining the morphological properties of the rock directly from a 3D image data set (Silin, *et al*, 2006). The method is also suitable for characterization of the geometry and connectivity of the pore structure in a solid material in terms of "pore bodies" and "pore throats". The algorithm works by assigning each voxel within the data set the radius associated with the "maximal inscribed sphere" as shown schematically in Figure 3.

3.2. *Simulation of CO_2 distribution in Frio sandstone*

Once each voxel in the data set has been assigned an appropriate radius according to the MIS calculation, this information can be used to calculate physical properties such as capillary pressure curves and fluid distribution. The distribution of two immiscible fluids is calculated by comparing the MIS assigned radius of a given voxel to the theoretical radius of curvature at the interface of the two fluids at a give pressure. Those voxels with radii larger than the theoretical value are assigned to the non-wetting phase, with the balance of the pore space being assigned to the wetting phase. Figure 4 compares the calculated distribution of CO_2 and brine within the assumed water wet sandstone sample compared with the distribution as imaged by microtomography. The similarity of the two images can be quantified with the Hausdorff distance (Huttenlocher *et al.*, 1993). Analysis shows that 80% of gas-occupied voxels of the microtomography image are within a 50-voxel Hausdorff neighborhood of the computed gas voxel clusters. (Silin *et al.*, 2009).

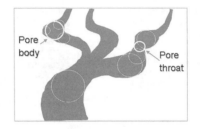

Figure 3. *Conceptual drawing of sphere assignment in maximal inscribed sphere (MIS) calculation*

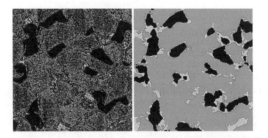

Figure 4. *Comparison of CT image (left) and MIS calculated (right) distribution of rock (red), brine (green), and CO2 (black) distribution within a slice of Frio sandstone (adapted from Silin et al., 2008)*

4. Imaging of CaCO₃ precipitation

Microbial enhancement is one possible approach to increasing carbonate precipitation rates and has additional applications in both environmental remediation and geotechnical contexts (Cunningham *et al.*, 2009). To facilitate these studies, we have developed a family of flow-through bioreactors which can be scanned continuously during precipitation experiments or accurately remounted if longer incubation periods are required for slowly growing cultures.

Bacterial hydrolysis of urea results in the production of ammonium and carbonate ions and an increase in pH, which favors carbonate precipitation if appropriate metal cations (e.g. Ca^{2+}) are available. A first series of tests (Ajo-Franklin et al., 2008) were performed on natural aquifer samples from a site at Idaho National Laboratory (INL). During these studies we scanned wet-packed samples in the bioreactor after equilibration with site groundwater. This step was followed by removal of the reactor from the beamline, injection of 5 mg/L molasses, and then 2 weeks of injection with 10 mM urea dissolved in groundwater to promote calcite precipitation.

After the urea injection was completed, the bioreactor was remounted in the beamline and scanned a second time. Figure 5 shows x-ray attenuation images of the same vertical slice at baseline and after 2 weeks of stimulation; high density precipitate is visible at grain boundaries in a localized patch. Due to lack of chemical sensitivity the identity of the newly formed precipitate can not be uniquely assigned. However, based on a drop in effluent Ca^{2+} concentration from 1.1 to 0.8 mM during urea injection in concurrent macro-column experiments, it is likely that the precipitate is $CaCO_3$. Interestingly, precipitation in this case took the form of a small number of dense patches rather than throughout the sample, a geometry which has less impact on flow and mechanical properties than uniform dissemination across narrow pore-throats.

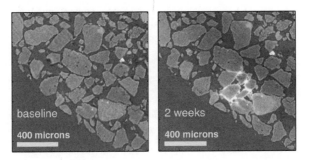

Figure 5. *Images show an example of microbial mineralization within a loose granular sample from the INL site. (Flow rate is 2ml/day perpendicular to the page toward the reader)*

5. Conclusions

Our initial studies indicate that synchrotron micro-tomography will be a powerful tool for monitoring mineralization processes relevant to CO_2 sequestration. Future investigations will integrate bioreactors capable of sustaining higher pressures with a wide range of secondary sensors for measuring electrical and mechanical properties in addition to 3D structure on the micron scale. In addition, multi-energy imaging focused on identifying Sr containing precipitates as a proxy for $CaCO_3$ will be used to support chemical identification of newly formed precipitates. Similarly, micro-tomography data in combination with MIS calculation analysis has been shown to be a reliable approach for understanding the behavior of liquid CO_2 within a multi-phase system. In combination, these two approaches will help to address the knowledge gaps that must be bridged in order for geological sequestration of CO_2 to make an important contribution to the world's energy future.

6. Acknowledgements

The authors would like to thank James Nasaitka for his tremendously valuable assistance with the *in situ* reaction cell. The use of the ALS is supported by the Director, Office of Science, Office of Basic Energy Sciences, of the U.S. Department of Energy under Contract No. DE-AC02-05CH11231. We also thank Office of Science, Office of Biological and Environmental Research for support of the Sustainable Systems: Science Focus Area at LBL. Lastly, we appreciate the support of the Research Partnership to Secure Energy for America, RPSEA.

7. References

Ajo-Franklin J., Hubbard S.S., Wu Y., Nico P., "Using Synchrotron Micro-CT To Monitor Microbially-Induced Calcite Precipitation on the Pore Scale", *AGU Biogeophysics Chapman Conference,* Portland, ME, Oct 13-16, 2008

Cunningham A.B., Gerlach R., Spangler L., and Mitchell A.C., "Microbially enhanced geologic containment of sequestered supercritical CO_2", *Energy Procedia*, vol. 1 no. 1, 2009, p. 3245-3252

Huttenlocher, D.P., Klanderman, G.A., Rucklidge, W.J., "Compariong images using the Hausdorff Distance", *IEEE Transactions on Pattern Analysis and Machine Intelligence*, vol. 15, 1993, p. 85-863.

Kinney J.H., Nichols M.C., "X-Ray Tomographic Microscopy (Xtm) Using Synchrotron Radiation", *Annual Review of Materials Science*, vol. 22, 1992, p. 121-152.

Silin D., Patzek T., "Pore space morphology analysis using maximal inscribed spheres", *Physica A,* vol. 371, 2006, p. 336-360.

Silin D., Tomutsa L., Benson S M., Patzek T W., "Pore-Scale Analysis of Microtomography Images of the Rock in Geosequestration Research", *Eos Trans. AGU,* vol. 89 no. 53, Fall Meet. Suppl., Abstract H12C-07

Silin D., Tomutsa L., Benson S M., Patzek T W., "Microscale imaging and Pore-Scale Modeling of Two-Phase Fluid Distribution", *In Review,* 2009.

Residual CO$_2$ Saturation Distributions in Rock Samples Measured by X-ray CT

H. Okabe* — Y. Tsuchiya* — C. H. Pentland** — S. Iglauer** — M. J. Blunt**

*Japan Oil, Gas and Metals National Corporation
1-2-2 Hamada, Mihama-ku, Chiba, 261-0025, Japan
okabe-hiroshi@jogmec.go.jp
tsuchiya-yoshihiro@jogmec.go.jp

** Department of Earth Science and Engineering, Imperial College London
Prince Consort Road, SW7 2BP, United Kingdom
c.pentland07@imperial.ac.uk
s.iglauer@imperial.ac.uk
m.blunt@imperial.ac.uk

ABSTRACT. *Understanding multiphase flow and trapping mechanisms in porous rocks at the laboratory-scale is required to design carbon dioxide (CO$_2$) geological sequestration at the field-scale. Coreflood laboratory experiments on sandstones and carbonate rocks using supercritical CO$_2$ and brine were conducted with in-situ saturation monitoring by X-ray Computed Tomography (CT). The residual CO$_2$ saturation obtained by capillary trapping is evaluated experimentally. Supercritical CO$_2$ is injected into a brine saturated core followed by injection of CO$_2$-saturated brine. This prevents the dissolution of carbon dioxide into the brine and the amount of CO$_2$ residual trapping in the rock can be measured. The experimental results show the trapped CO$_2$ saturation ranging from 0.2 to 0.4 for the rocks with permeability ranging from 6 to 220 md. The proposed method with in-situ saturation monitoring reveals piston-like displacement inside homogeneous sandstone, while heterogeneous flow behaviour is observed on a carbonate rock. The relationship between porosity distribution and residual CO$_2$ saturation on the carbonate rock is inferred. Injected CO$_2$ flows through more porous regions but is preferentially trapped in lower porosity zones. In-situ saturation monitoring, therefore, is the key to understand fluid movement and trapping in the rock samples. This work can then be applied to CO$_2$ storage in saline aquifers, where the CO$_2$ may be rapidly and securely trapped as a residual phase.*

KEYWORDS: *geological sequestration, coreflood, CO$_2$ injection, residual trapping, x-ray CT scanner, in-situ saturation monitoring.*

1. Introduction

Intergovernmental Panel on Climate Change (IPCC) has published a special report on carbon dioxide capture and storage (IPCC, 2005). It refers to CO_2 capture and storage (CCS), which is one of the technical options to reduce greenhouse gas emissions to the atmosphere. CO_2 geological sequestration is considered to be a long-term storage option for the gas, and the petroleum industry, which has CO_2 Enhanced Oil Recovery (EOR) experiences, can contribute to the advancement of this technology. In geological structures, four trapping mechanisms are present: (1) structural and stratigraphic trapping: the buoyant CO_2 remains as a mobile fluid but is stored under impermeable cap rocks; (2) residual phase trapping: disconnection of the CO_2 phase into an immobile fraction; (3) solubility trapping: dissolution of the CO_2 into the brine, possibly enhanced by gravitational instabilities due to the larger density of the CO_2-saturated brine; and (4) mineral trapping, geochemical binding to the rock due to mineral precipitation. In this paper, residual CO_2 saturation distributions in rocks are evaluated experimentally with application to CO_2 injection into saline aquifers. CO_2 is the non-wetting phase case, and capillary trapping of the CO_2 is an important mechanism to avoid the leakage.

2. Experiments

2.1. *Rock samples*

Three sandstones and a carbonate rock shown in Table 1 are prepared as standard plug cores with 38 mm in diameter and between 70 and 140 mm long. Berea sandstone samples with different porosities and permeabilities are firstly compared to identify the effects of porosity and permeability on the saturation. The sandstone samples are composed primarily of quartz but also contain minor amounts of calcite, dolomite, clay minerals and feldspar. In addition, a heterogenous sandstone from Japan and a Middle Eastern carbonate sample are used. The carbonate rock is composed mainly of bioclastic grainstone and packstone with some algal fragments which may form vuggy pore spaces. It exhibits a substantial presence of sub-micron porosity confirmed by the micro-CT image and Mercury Injection Capillary Pressure (MICP) curves (Sok *et al.*, 2007). Prior to the injection, effective porosities of the plug cores are measured with an x-ray CT scanner system (resolution 0.35mm/pixel), which gives the average porosity ranging from 0.14 to 0.27 as shown in Table 1. Typical three-dimensional (3D) porosity distributions are illustrated in Figure 1.

sample	diameter, mm	length, mm	CT derived porosity	Air permeability, md
Berea 1	37.83	75.08	0.218	220.3
Berea 2	37.89	114.17	0.197	89.1
sandstone	36.80	145.00	0.269	6.8
carbonate	38.00	70.50	0.145	14.4

Table 1. *Properties of the rock samples under confining conditions*

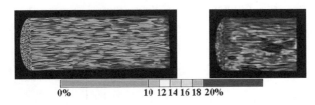

0%   10 12 14 16 18 20%

Figure 1. *Vertical cuts of 3D porosity distributions of relatively homogenous Berea sandstone 2 (left) and heterogenous Middle Eastern carbonate rock (right), which are imaged by the x-ray CT scanner. Color scale indicates porosity*

2.2. *Experimental setup/procedure*

The experimental apparatus with the x-ray CT scanner is shown in Figure 2. The system was previously used for a similar study (RITE, 2007), but it is improved in terms of preparing CO_2-saturated brine. Cleaned cores are used as we assume CO_2 storage in saline aquifers. The experimental temperature is kept at 40°C and the overburden pressure is 3,000 psig (approximately 20 MPa), excepting the carbonate rock experiment under 2,000 psig (14 MPa). The key steps of the coreflood experiment are summarized below.

1) Prepare injectant (Wet condition supercritical CO_2 and CO_2 saturated brine).

2) Take a CT image of the plug core under dry conditions.

3) Inject supercritical CO_2, and take a CT image under 100% CO_2 saturation.

4) Inject brine, and take a CT image under 100% brine saturation to calculate the porosity distribution.

5) Inject CO_2 saturated brine, and take a CT image under 100% CO_2 saturated brine saturation.

6) Inject brine again to saturate the core.

7) Inject supercritical CO_2 at the designed flow rate (capillary number~10^{-7}) until no more brine is produced and the pressure difference becomes stable (more than 5PV injected). *In-situ* saturation monitoring is conducted by x-ray CT during the CO_2 flooding.

8) Inject CO_2 saturated brine at the designed flow rate (capillary number~10^{-7}) to confirm the CO_2 trapping with in-situ saturation monitoring (more than 5PV injected).

Further details of the coreflood procedure with *in-situ* saturation monitoring by x-ray CT are provided elsewhere (Oshita *et al.*, 2000, Okabe *et al.*, 2006). The accuracy of the CT derived porosity/saturation is dependent on an attenuation contrast of x-ray between fluid phases. In the study, 15wt% NaI is used as a dopant to the brine phase, and the accuracy of the measurements of porosity and saturations is evaluated at ±1 Bulk Volume% and ±2 Pore Volume%, respectively.

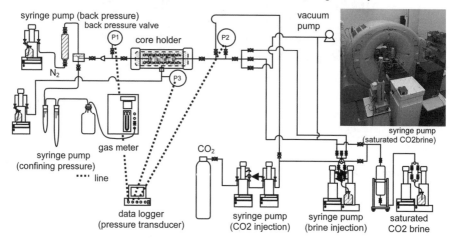

Figure 2. *X-ray CT coreflood system with developed temperature control units. The rubber heater covers the coreholder, and the line heaters are attached to the fluid injecting lines*

3. Results

Table 2 summarizes the results of average CO_2 saturation measured by x-ray CT after CO_2 injection for the initial gas saturation (S_{gi}) and after post-injection with CO_2 saturated brine for the residual gas saturation (S_{gr}). Figure 3 shows some trends between porosity, S_{gi} and S_{gr} although it is difficult to relate porosity and saturations from the figure since a limited number of different rock samples are used so far. The residual CO_2 saturations of Berea sandstone are in good agreement of the results of trapped gas saturations estimated from the relationship between pressure and volume under an isothermal expansion (Suekane *et al.*, 2008), which results in the trapped gas saturation in the range from 0.248 to 0.282.

sample	porosity	permeability, md	S_{gi}	S_{gr}
Berea 1	0.218	220.3	0.355	0.259
Berea 2	0.197	89.1	0.398	0.274
sandstone	0.269	6.8	0.483	0.356
carbonate	0.145	14.4	0.395	0.226

Table 2. *Results of CO_2 saturations on the samples. S_{gi} and S_{gr} are CO_2 saturations before and after post-injection with CO_2 saturated brine, respectively. Note: more than 5PV CO_2 and brine are injected except the Berea 1, which has 2-3PV injection*

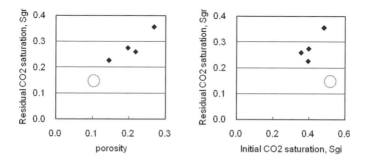

Figure 3. *Experimental results of relationships between porosity, initial CO_2 saturation and residual CO_2 saturation for the rock samples. Red circle indicates results on a carbonate sample and others are measured on sandstones*

In-situ saturation monitoring can reveal 3D saturation distributions within Berea sandstone sample as shown in Figure 4. The Berea 2 sample has relatively homogeneous porosity distribution as shown in Figure 1 (left). Therefore, homogeneous piston-like displacements are observed both for supercritical CO_2 injection (Figure 4 upper) and post-injection with CO_2 saturated brine (Figure 4 lower). Figure 5 shows the water saturation profile along the Berea 2 sample. To achieve the designed capillary number ($\sim 10^{-7}$) for the sample, CO_2 injection and post-injection are conducted at the flow rates at 0.93 cc/min and 0.03 cc/min, respectively. In this figure, the fluid front movement is clearly captured. Saturation change becomes small after 0.5 PV CO_2 injection, but continues to decrease around 3 PV. While the post-injection with CO_2 saturated brine decreases CO_2 saturation, but large saturation change cannot be observed after 0.4 PV brine injection. In order to confirm additional saturation change, brine injection is continued to reach 10 PV injection. Since the rock has relatively homogenous porosity profile as shown in Figure 5 right, the saturation distributions before/after the post-injection are also homogenous. The trapped CO_2 saturation in the rock is 0.27 on average.

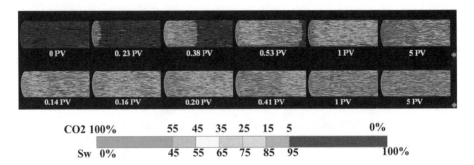

CO2 100%   55   45   35   25   15   5   0%

Sw   0%   45   55   65   75   85   95   100%

Figure 4. *X-ray CT derived water saturation distributions during CO_2 injection (upper) and post-injection with CO_2 saturated brine (lower) on the Berea 2. Homogenous piston-like displacements are observed for the relatively homogenous sandstone. After 5PV injection of CO_2 at the upper right, injectant is switched from CO_2 to brine and 0.14PV brine injection is firstly shown at the lower left.*

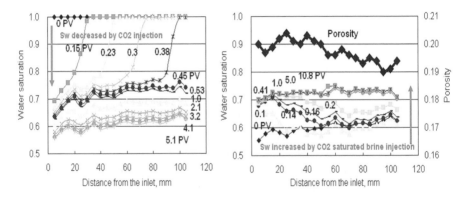

Figure 5. *X-ray CT derived water saturation profiles during CO_2 injection (left) and post-injection with CO_2 saturated brine (right) on the relatively homogenous Berea 2 sample*

To compare with homogenous rock samples, a carbonate core plug is used and it shows strong heterogeneity in terms of the porosity and the saturation profiles illustrated in Figure 6 (Okabe and Tsuchiya, 2008). X-ray CT shows CO_2 flow through more porous areas, which is similar to the water injection to the mixed-wet carbonate rock. The average water saturation is decreased to 0.6 although the flow rate is increased to 4.5 cc/min at 10PV in order to see the additional saturation change due to the rate effect. The water saturation cannot be restored to the initial condition even after 7.5PV post-injection with CO_2 saturated brine, and the heterogeneous saturation profile is observed in Figure 6(b). The residual CO_2 saturation profile correlates the porosity profile as shown in Figure 6(b). The trapped CO_2 saturation in the carbonate rock is 0.23 on average.

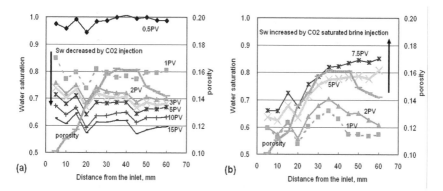

Figure 6. *X-ray CT derived water saturation profiles during the CO_2 injection (a) and post-injection with CO_2 saturated brine (b) on the carbonate rock*

4. Conclusions

Residual CO_2 saturation by capillary trapping is evaluated experimentally. Supercritical CO_2 is injected into the brine saturated core followed by the post-injection with CO_2 saturated brine. This prevents the dissolution of carbon dioxide into the brine to correctly measure the amount of CO_2 residual trapping. The experimental results show the trapped CO_2 saturation ranging from 0.2 to 0.4 for the rocks with the permeability range from 6 to 220 md. *In-situ* saturation monitoring reveals piston-like displacement inside homogeneous sandstone, while heterogeneous flow behavior is observed on a carbonate rock. A relationship between porosity distribution and residual CO_2 saturation on the carbonate rock is inferred. Injected CO_2 flows through more porous regions, while it is trapped in lower porosity zones. Further studies are required to relate the petrophysical properties of rocks and the residual saturation.

5. Acknowledgements

We would like to acknowledge JOGMEC for granting permission to publish this paper and our colleagues in EOR Research Division for their advices. Special thanks go to Isao Iwasaki and Yasuyuki Akita for their experimental supports and Shigeru Kato for his dedicated visualization works.

6. References

IPCC. *Special Report on Carbon Dioxide Capture and Storage*, Cambridge University Press, Cambridge, UK. 2005.

Okabe, H. and Tsuchiya, Y. "Experimental investigation of residual CO2 saturation distribution in carbonate rock", *International Symposium of the Society of Core Analysts, SCA2008-54*. Abu Dhabi, UAE. 29 October - 2 November, 2008.

Okabe, H., Tsuchiya, Y., Oseto, K. and Okatsu, K. "Development of X-ray CT coreflood system for high temperature condition", *Advances in X-Ray Tomography for Geomaterials*, ISTE Ltd., 309-314. 2006.

Oshita, T., Okabe, H. and Namba, T. "Early water breakthrough - X-ray CT visualizes how it happens in oil-wet cores", *SPE Asia Pacific Conference on Integrated Modelling for Asset Management, SPE 59426*. Yokohama, Japan. 25-26 April, 2000.

RITE. Report on research and development of carbon dioxide geological storage technologies, Fiscal Year 2006 Report of Research Institute of Innovative Technology for the Earth (RITE), http://www.rite.or.jp/ (in Japanese). 461-509. 2007.

Sok, R. M., Arns, C. H., Knackstedt, M. A., Senden, T. J., Sheppard, A. P., Averdunk, H., Pinczewski, W. V. and Okabe, H. "Estimation of Petrophysical Parameters from 3D images of Carbonate Core", *SPWLA Middle East Regional Symposium*. Abu Dhabi, UAE. April 15-19, 2007.

Suekane, T., Nobuso, T., Hirai, S. and Kiyota, M. "Geological storage of carbon dioxide by residual gas and solubility trapping", *International Journal of Greenhouse Gas Control*, 2, 1, 58-64. 2008.

X-ray CT Imaging of Coal for Geologic Sequestration of Carbon Dioxide

D. H. Smith — S. A. Jikich

National Energy Technology Laboratory
U.S. Department of Energy
Morgantown, WV 26507-0880
duane.smith@netl.doe.gov
sinisha.jikich@pp.netl.doe.gov

ABSTRACT. Concerns about global warming have motivated research and field projects for geologic sequestration of carbon dioxide by its injection into unmineable coal seams. We report x-ray CT measurements of coal heterogeneities and high-permeability regions in coal, bulk compressibilities and carbon dioxide concentration gradients produced by its diffusion through the coal matrix. Gravimetric measurements performed to examine the accuracy of the CT measurements of carbon dioxide concentrations also are discussed.

KEYWORDS: coal, heterogeneities, compressibilities, carbon dioxide, sorption

1. Introduction

The study of coal by X-ray computerized tomography almost certainly began in 1982 (Maylotte *et al.*, 1982). Since then, about twenty other studies have appeared.

In recent years, X-ray CT studies of coal frequently have been motivated by the potential for sequestration of carbon dioxide to reduce atmospheric concentrations of this greenhouse gas (White *et al.*, 2005). Injection of carbon dioxide into coal seams may also lead to enhanced production of coalbed methane. For sequestration, the two technical parameters of most interest are the sequestration capacity, that is, how much carbon dioxide can be "stored" in a seam of coal, and the injectivity, which is a quantitative measure of how rapidly carbon dioxide can be injected into a seam. In this paper we present and consider X-ray CT imaging data for the heterogeneities, carbon dioxide diffusivities, sorption isotherms, and elastic behavior of coal. These kinds of measurements are of considerable importance for sequestration of carbon dioxide in coal.

2. Experimental

The samples of bituminous coals, methods of sample preparation, sample holders, CT imaging equipment and methods, and equipment and methods for sample compression or introduction of carbon dioxide have all been described elsewhere (Jikch *et al.*, 2009a, 2009b; Smith *et al.*, 2007, 2009).

3. Results

3.1. *Coal heterogeneities*

Figures 1(a) and (b) illustrate computer tomography images of a Pittsburgh coal core, containing two different amounts of carbon dioxide, measured at atmospheric pressure. These images illustrate the very great heterogeneities exhibited by coal, and the ability of x-ray CT to measure different carbon dioxide concentrations.

3.2. *Coal compressibilities*

Figure 2 illustrates the compressibilities at different confining pressures as obtained from CT data of Upper Freeport sample #3. The permeability and thus the injectivity depend, in part, on the coal compressibility (McKee *et al.*, 1988).

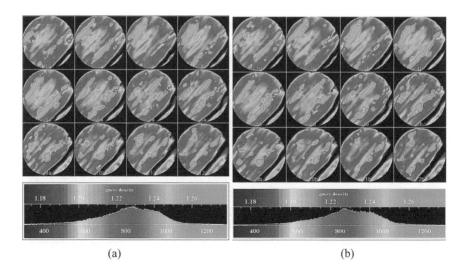

Figure 1. *Computer tomography images of a 2.5 cm diameter coal core, measured at 130 kV and atmospheric pressure: (a) before degassing, "immediately" after depressurization from exposure to CO_2 at 1.38 Ma (200psi); (b) after gravimetric degassing for about one hour. Each image represents a trans-axial disc 0.20 mm thick.*

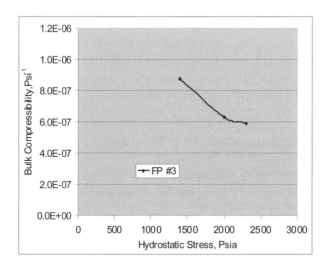

Figure 2. *Bulk compressibilities as obtained from CT data at different confining pressures for Upper Freeport sample #3*

3.3. *Calibration of CT measurements of carbon dioxide content*

It is very desirable that the accuracy of carbon dioxide concentrations obtained by CT measurements first be confirmed by comparison with some other measurement technique. For this purpose, the technique of "ambient pressure gravimetry" was developed (Smith *et al.*, 2007). In this technique, carbon dioxide is allowed to escape from a sample previously exposed to CO_2 at elevated pressure, and the resultant changes of sample weight and changes of carbon dioxide content as measured by x-ray CT are compared.

Figure 1 illustrates computer tomography images of a Pittsburgh coal core, measured at 130 kV and atmospheric pressure: (a) "immediately" after depressurization to ambient, atmospheric pressure following prolonged exposure to CO_2 at 1.38 Ma (200 psia) and (b) after it had (partially) degassed into the atmosphere for about one hour while resting on an automatic analytical balance to record its changes of weight as it lost carbon dioxide. The differences between the images can provide an accurate measurement of the difference in the amounts of carbon dioxide in the sample before and the degassing period.

Figure 3 is a plot vs. the square root of elapsed time of the mass of a coal sample as it degassed previously sorbed carbon dioxide into the atmosphere while sitting on the pan of the recording balance. Also shown on the plot is the mass of the sample as measured by CT measurements made immediately before and after the weighing interval, plotted on the same time scale. (The choice of time$^{1/2}$ linearizes the gravimetric data, as expected for Fickian diffusion.) For this sample of Pittsburgh coal, the carbon dioxide content measured by x-ray CT was about 6% less than measured by ambient-pressure gravimetry, confirming with reasonable agreement the accuracy of the carbon dioxide content as obtained from the measurements.

3.4. *Carbon dioxide sorption and density histograms*

In the measurements of this paper, there were *circa* 9900 voxels in each slice, each voxel representing a parallelepiped of (measured) dimensions 0.21 mm x 0.21 mm x 2.0 mm. Figure 4 represents the histogram for the *circa* 9900 voxels density of "slice #1" of the evacuated Pittsburgh coal sample (before gas injection), that is the 2 mm thick disc of coal nearest the eventual carbon dioxide inlet. The average density for the slice is 1197.9 mg/cm^3; the skewness factor is 0.68. (For a perfectly symmetrical distribution the skewness factor is 0.) There is a distinct indication of an overlapping, bimodal distribution, with apparent maxima at about 1,193 mg/cm^3 and 1209 mg/cm^3. Hence, the average skewness factor (measure of the asymmetry) of the total distribution, 0.68, appears to represent the influence more than one, perhaps individually symmetric, distributions.

As part of the attempts to better understand and predict the sorption of carbon dioxide by coal and the effects of sorption on coal properties, an outstanding question is whether and how sorbed carbon dioxide changes the coal structure (Larsen 2004). Figure 5 is a histogram for the voxel densities of Figure 4 after 1065 total hours of exposure to CO_2, at a succession of increasing pressures, culminating in sorption at 903 MPa (614 psia), followed by degassing of the sample.

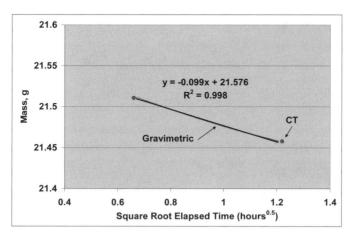

Figure 3. *For the sample of Pittsburgh coal, carbon dioxide content measured by CT was about 6% less than measured by ambient-pressure gravimetry*

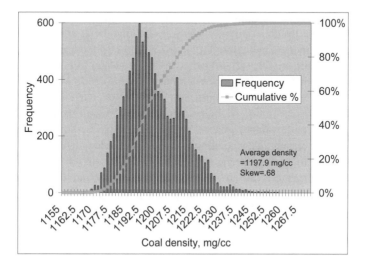

Figure 4. *Histogram for the voxel densities of slice #1 of the Pittsburgh coal sample before injection of CO_2*

The average density in slice #1 was 1198.9 mg/cm^3 after degassing, vs. 1197.9 mg/cm^3 initially. This very small difference (0.08%) seems to indicate that, at most, a very small amount of CO_2 remained in the core; and it could even reflect irreversible mechanical compaction by the confining pressure applied to the sample. However, while the average density of the slice after degasification returned to almost its pre-injection value, the frequency distribution of the voxel densities in the slice did not. The most-frequent density before sorption and after degasification remained virtually the same. However, the calculated average skewness of the distribution was 0.82 after degasification, 0.67 before sorption of CO_2. Comparison of Figures 4 and 5 reveals obvious changes in the frequency distribution; in particular, the notable "second peak" at higher densities and bimodal distribution before sorption and compression appears to have disappeared in Figure 5. The latter change suggests a structural re-arrangement in the slice may have occurred due to CO_2.

3.5. Carbon dioxide diffusion

The rate of diffusion of carbon dioxide through the coal matrix is very important, since it may affect the amount of methane produced and carbon dioxide sequestered (Smith et al., 2005). Figure 6 is a plot of slice-average CO_2 concentrations vs. nominal diffusion distance for diffusion, at different pressures of carbon dioxide, in a sample of Pittsburgh #8 coal from the Emerald Mine. For each gas pressure, the confinement pressure was 200 psia greater than the gas pressure, so that the effective pressure was nominally constant. Carbon dioxide first diffused for 165 hours at gas pressure 210 psia into the initially gas-free sample; next diffusion occurred at 340 psia for 392 hours (total elapsed time 557 hours) into the non-degassed sample; etc. Hence, only for the data at 210 psia can the convenient boundary condition of zero carbon dioxide concentration at infinity be assumed. Moreover, the local minima and maxima in the data illustrate that the convenient assumption of one-dimensional diffusion into a homogenous solid also would be only an approximation. For example, places where the carbon dioxide concentration increased with increasing distance imply that carbon dioxide diffusion occurred from less-concentrated to more-concentrated regions. While this is thermodynamically possible (since diffusion is driven by the Gibbs free energy, not the concentration), it seems unlikely. The more probable reason is that Figure 6 represents average concentrations for cross-sections orthogonal to the nominal direction of flow. Fits of simple analytical equations to the smoothed data of Figure 6 would give only approximate measures of the effective diffusivity.

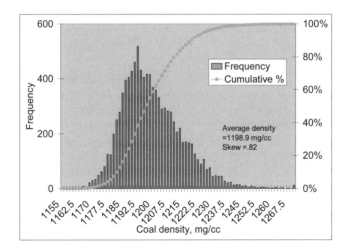

Figure 5. *Histogram for the voxel densities after the core was degassed*

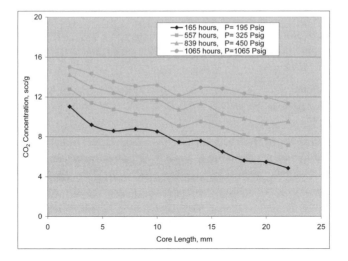

Figure 6. *CO_2 concentrations vs. diffusion distance in Pittsburgh #8 coal from the Emerald Mine*

4. Conclusion

In support of the development of geologic sequestration of carbon dioxide in coal, density heterogeneities of (bituminous) Pittsburgh coal have been measured by x-ray CT imaging. Spatial distributions of carbon dioxide sorbed within the coal also are reported. The effects of sorption followed by desorption on changes of coal density are examined. Measurements that illustrate the slow diffusion of carbon dioxide have been made. To ensure that the measured CO_2 concentrations were accurate, concentrations of sorbed carbon dioxide as measured by x-ray CT were compared with concentrations measured by a novel gravimetric technique.

5. References

Jikich, S. S., McLendon, R., Seshadri, K., Irdi, G., Smith, D. H., "Carbon dioxide transport and sorption behavior in confined coal cores for carbon sequestration", *SPE Reservoir Engineering & Evaluation*, vol. 12 no.1, p.124-36, 2009a.

Jikich, S. A., McLendon, T. R., Smith, D. H. "Permeability variations in Upper Freeport coal cores due to changes in effective stress and sorption", *Proc. SPE Annual Technical Conference and Exhibition*, New Orleans, LA, 4-7 October, 2009b.

Larsen, J. W., "The effects of dissolved CO_2 on coal structure and properties", *International Journal of Coal Geology*, vol. 57 no. 1, p. 63-70, 2004.

Maylotte, D. H., Kosky, P. G., Spiro, C. L., Lamby, E. J., Computed tomography of coals, quarterly technical progress report no. 1, August 16, 1982 - November 28, 1983, General Electric Company: Schenectady, New York (1982) 13 pages, DOE/MC/19210-1466, DOE contract number: AC21-82MC19210. [Available through NETL library, Morgantown, WV]

McKee, C. R., Bumb, A. C., Koenig, R. A., "Stress-dependent permeability and porosity of coal and other geologic formations", SPE 12858, *SPE Formation Evaluation*, vol. March 1988, p. 81-91.

Smith, D. H., Sams, W. N., Bromhal, G., Jikich, S., Ertekin, T., "Simulating carbon dioxide sequestration/ECBM production in coal seams: effects of permeability anisotropies and diffusion time constant", *Soc. Petrol. Engrs. Reserv. Eng. Eval.*, vol. 8, p. 153-163, 2005.

Smith, D. H., Jikich, S. A., Seshadri, K., "Carbon dioxide sorption isotherms and matrix transport rates for non-powdered coal", *2007 Proc. International Coalbed Methane Symposium*, Tuscaloosa, AL, 21-25 May 2007.

Smith, D. H., Jikich, S. A., "Permeability, elastic, and carbon dioxide-sorption properties of Upper Freeport cores from a carbon dioxide sequestration/enhanced coalbed methane field project", *Proc. International Coalbed & Shale Gas Symposium*, Tuscaloosa, AL, 18-22 May 2009.

White, C.M., Smith, D. H., Jones, K. L., Goodman, A. L., Jikich, S. A., LaCount, R. B., DuBose, S. B., Ozdemir, E., Morsi, B. I., Schroeder, K. T., "Sequestration of carbon dioxide with concomitant enhanced coalbed methane recovery--A review", *Energy & Fuels*, vol. 19, p. 659-724, 2005.

Comparison of X-ray CT and Discrete Element Method in the Evaluation of Tunnel Face Failure

B. Chevalier* — D. Takano** — J. Otani*

*X-Earth Center
Kumamoto University
2-39-1, Kurokami, Kumamoto,
Kumamoto 860-8555,
Japan
junotani@kumamoto-u.ac.jp

**Grenoble Universités
Laboratoire 3S-R
BP 53
38041 Grenoble Cedex 09,
France
Daiki.Takano@hmg.inpg.fr

ABSTRACT. *A comparison of x-ray computed tomography and three-dimensional Discrete Element Method (DEM) analysis used in a boundary problem is presented. A tunnel face failure was simulated in a model test by pulling a cylinder out of a granular layer. The granular layer was investigated by X-ray computed tomography after pulling the cylinder and revealed a three dimensional face failure. The difference value of displacement of the cylinder made it possible to observe the mechanisms of development of the failure mechanisms such as arching effect. This experiment was reproduced with DEM. The mechanisms observed in the experimental study were well performed in one case. However, a second numerical analysis involving different particle shape (clump particles) was performed and showed a self-stable arch in the granular material, which was not observed in the experiment.*

KEYWORDS: *computed tomography, discrete element method, tunnel*

1. Introduction

The study of boundary problems in geotechnical engineering is very important for a better understanding of the mechanisms involved in the interaction between soil and structures. The use of non destructive analysis techniques like X-ray computed tomography (CT) is a big issue for the study of boundary problems: experimental conditions very close to the *in-situ* conditions can be reproduced and are not disturbed by measurements apparatus located in the mass. In the other hand, Discrete Element Method (DEM) is a well-tried numerical tool for the numerical modeling of problem involving granular materials. Boundary problem can then be studied with CT from the point of view of the strain fields and DEM gives access to data related to both strain and stress fields. The aim of the study presented here was to compare both analysis techniques on an applied boundary problem: the evaluation of tunnel face failure. A tunnel face failure was simulated by pulling a cylinder out of a mass of sand. The failure zone induced by the removed cylinder was then investigated with x-ray CT. A three-dimensional slip surface was identified marking the boundary between the failure volume and the stable part of the granular material. In a second step, a DEM model was developed in order to analyze the same failure mechanism from a numerical point of view. Two different particle assemblies were considered in the numerical analysis, mainly differing by the shape of the particles involved. Calculation of the strain and stress fields in the samples was performed, so that the failure mechanisms were observed precisely in the numerical analysis and compared to the experimental results.

2. Methods

2.1. *Experimental process*

The test method consists of reproducing a tunnel face failure by pulling a cylinder out of a sand layer. First, a 125 mm diameter tank was filled (free-fall method) with sand with a relative density equal to 80%. Then, the soil tank is set in the pull-out system (Figure 1) and the cylinder was pulled with a constant velocity 0.1 mm/s. Once the desired displacement of cylinder was reached; the tank was placed in an industrial x-ray CT scanner booth (Otani *et al.* 2000; Otani *et al.*, 2002) and scanned from the tunnel bottom to soil surface with a 1 mm pitch.

Two different sands were used differing by their grain size. The first one is a silica sand No.3 characterized by a mean grain size of 2.0 mm; its peak friction angle is 40° for a 80% relative density. The second one is a silica sand No.8 characterized by a mean grain size of 0.12mm; its peak friction angle is 42° for a 80% relative density.

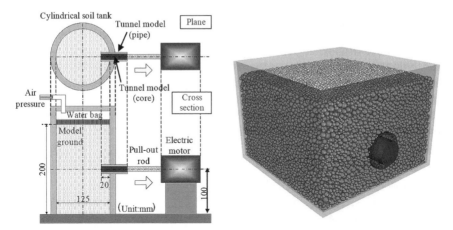

Figure 1. *View of the pull-out system*   **Figure 2.** *View of the test box for DEM analysis*

2.2. *Numerical model and process*

Discrete Element Method was initially developed for the numerical study of rock mechanics (Cundall, 1979) and its fields of application were constantly widened. The numerical analyses presented here were computed with a C++ code called SDEC (Magnier, 1997). The model used here is based on molecular dynamics: particles are not deformable but an artificial deformability is simulated by the possible overlap between particles at their contact points. The successive applications of the Newton's second law of motion and of contact laws make possible to reproduce the behavior of granular materials under various conditions. In our case, the contact force consists in two components. The normal component is perpendicular to the contact plane and is deduced from the overlap between contacting particles with a linear contact law of stiffness k_n. The tangential component of the contact force is defined as Cundall proposed (Cundall, 1982) and is calculated with a linear contact law from the incremental tangential displacement of the contact point and a tangential stiffness k_t. Normal and tangential components are linked by a Coulomb friction criterion of coefficient μ. A non viscous local damping (introduced by Cundall (Cundall, 1987)) is employed on the particles accelerations.

The three parameters of contact laws are $k_n=9.6\times10^4$N.m-1, $k_t/k_n=0.75$ and $\mu=30°$. The global friction angle of the particles assemblies were deduced from numerical modeling of triaxial tests under 50 kPa confining pressure. The first assembly tested was composed on 80,000 spheres with diameter randomly drawn between 1.37 mm and 5.55 mm. The assembly had a void ratio equal to e=0.55 and a density of 17.1 kN.m^{-3}. Its peak friction angle is 32°.

This value of friction angle is quite low compared to the friction angle of the real sand. However, friction angles greater than 35° can not be reproduced with simple spheres without introducing a resistance to rolling. Consequently, a second assembly was used made of clumps in order to obtain a particle assembly with mechanical properties closer to the real material. Due to their complex shapes (particularly local concavity), clumps assemblies can reach higher frictional strength.

The second assembly was composed on 80,000 clumps, made of two identical spheres which centers are separated by a distance of 50% of the sphere diameter. The sphere diameters were randomly drawn between 1.17 mm and 4.61 mm. The void ratio of the second assembly is equal to e=0.43 for a density of 18.5 kN.m^{-3} and a peak friction angle 46°. A test-box (Figure 2) with a 125x125 mm square section was used for the reproduction of the tunnel face failure. A cylinder with a 20 mm diameter was pulled out of the granular mass to induce a tunnel face failure.

3. Comparison and discussion

3.1. *Face failure visualization*

The DEM results were first compared with experimental pulling test on silica sand No. 3 (mean diameter 2.0 mm). This comparison was made because the grain size of this material is relatively close to the size of particles used in the DEM analysis. Figure 3 shows the distribution of the relative densities in a vertical cross section for the experimental test.

We can notice on this figure that pulling the cylinder out creates a face failure in front of the tunnel identified here with the progressive loosening of the soil mass. This loosening zone starts from the bottom of the tunnel and follows a slip line towards the surface of the soil layer (blue dotted lines). When the pull-out displacement is greater than 5 mm, the loosening zone expands vertically toward the surface, as observed in the trap-door tests.

Figures 4 and 5 represent the distribution of volumetric strains in cross sections respectively for the spheres and for the clumps particles. Strain tensors were calculated on tetrahedrons deduced from a Delaunay tessellation performed on the mass centers of spheres from the derivative of the displacement vectors of the vertex of the tetrahedrons (Cundall, 1989).

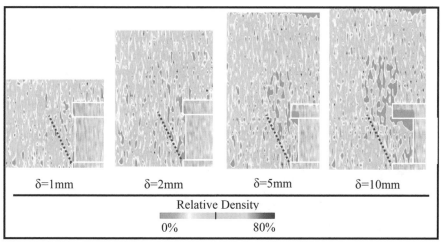

Figure 3. *Distribution of the relative densities deduced from CT in vertical cross-sections: case of silica sand No. 3 (mean diameter 2.0mm)*

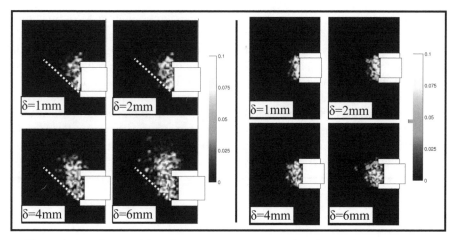

Figure 4. *Distribution of the volumetric strains in vertical (right) and horizontal (right) cross-sections: case of the assembly of spheres*

We can notice on Figure 4 (spheres) that the mechanisms obtained from numerical results are very close to the experimental results: the failure zone is also bound by a slip line starting at the bottom of the tunnel face. For greater displacement, the loosening zone expands vertically toward the surface. However, the angle between the slip line and the horizontal direction is less in the numerical results than in experimental tests. This could be related to the friction angles of both numerical and experimental samples: 32° for the numerical sample and between 40° and 42° for the sands.

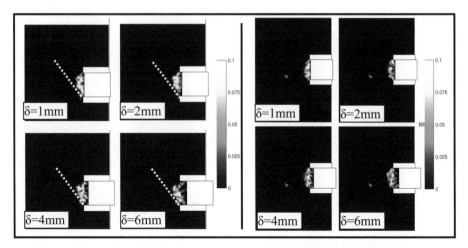

Figure 5. *Distribution of the volumetric strains in vertical (right) and horizontal (right) cross-sections: case of the assembly of clumps*

If we consider the other numerical sample (clumps, Figure 5), presenting a higher friction angle (45°), we observe that the angle of the slip zone with the horizontal directions is increased compared to the spheres samples. However, the loosening zone doesn't expand toward the surface of the layer: a real self stable arch is formed near the tunnel face and prevents the effect of pulling the cylinder to spread into the granular layer. In order to illustrate the difference of the arching effect obtained with the 2 numerical samples, the distributions in vertical cross sections of the principal directions of the stress tensor are shown on Figure 6 for δ=6mm. The local stress tensors were calculated with the contact forces with the Love formula.

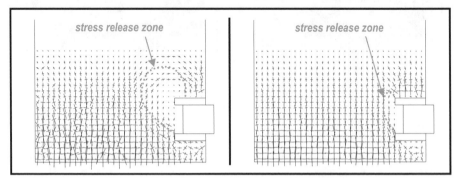

Figure 6. *Principal directions of stress in e vertical cross-section for δ=6mm: for the sphere sample (left) and for the clump sample (right)*

The stress release zone is much more reduced in the case of clumps, due to the formation of a self-stable arch. This perfect arching effect is due to the high level of imbrications between the clumps.

3.2. *Effect of particle size*

The comparison of the two different samples used in the numerical part of this study with each other on one side, and with the experimental results on the other side, suggests a strong effect of the particle size on the mechanisms observed. Indeed, each clump is made of 2 identical spheres, which centers are separated by a distance equal to 50% of the sphere diameter. Consequently, due to their concavity, clumps are very well imbricated and the loosening of the granular layer is then prevented. In addition, this effect of shape is coupled with an effect of the particle size, which also makes the gradual loosening of the layer harder. This effect of the particle size was also confirmed in the experiment by comparing the results obtained with the second sand (Keisa No. 8, mean particle size 0.12 mm). Figure 7 shows the distribution on vertical cross section of the relative densities in the case of Keisa sand No.8 to be compared with Figure 3.

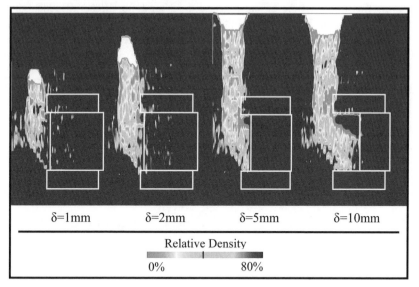

Figure 7. *Relative densities deduced from CT in vertical cross-sections: case of silica sand No. 8; white color corresponds to air voids areas*

Comparison between the two sands, shows that with a smaller particle size (Figure 8, Keisa No. 8), the loosening area reaches the surface of the granular layer for δ=5 mm while in the case of bigger particles (Figure 3, Keisa No. 3), the loosening did not reach the surface. This strong effect of particle size cannot be ignored when performing a numerical study of this problem, especially using Discrete Element Method.

4. Conclusion

The reproduction of the three-dimensional tunnel face failure mechanism was investigated from both experimental and numerical point of view, using X-ray CT on one hand and Discrete Element Method on the other hand. The numerical model showed its ability to reproduce the mechanisms observed experimentally but with some limitation, essentially related to the combined influence of size and shape of particles. Considering the particle size issue, the perturbation of the granular mass consists here in a horizontal displacement which induces afterwards a vertical perturbation. If this displacement is too small compared to particle size, its effects are greatly limited. In the same way, if the resulting displacement field is not sufficient to induce a reorganization of the particles in the sample (loosening), the mechanisms can not be well reproduced. Particle shape is one of the methods to model particles assemblies presenting frictional strength closer to real materials. However, the introduction of concave particles and its direct influence on the particle arrangement (particularly contact density) seems to have a great effect on the arching mechanisms involved in the present failure problem. Consequently, further investigations should be made in fields of particle size and on particle shapes.

5. Acknowledgements

This research was conducted under the fellowship of the Japanese Society for the Promotion of Science (JSPS 20.08818).

6. References

Cundall P. and Strack O. D. L. "A discrete numerical model for granular assemblies", vol. 29, pp. 47-65, *Géotechnique*, 1979.

Cundall P. A.; Dresher A. and Strack O. D. L. "Numerical experiments on granular assemblies: measurements and observations", *IUTAM Conference on Deformation and Failure of Granular Materials*, 1982.

Cundall P. *Analytical and Computational Methods in Engineering Rock Mechanics*, Chap. 4, pp.129-163, Allen & Unwin (Eds), 1987.

Cundall P. "Numerical experiments on localization in frictional materials" vol. 5, pp.148-159, *Ingenieur-Archiv*, 1989.

Magnier S.A., Donzé F.V. Discrete Element Project. Report of the University of Québec, Montréal, Canada, 1997.

Otani J., Mukunoki T. and Obara Y. "Application of X-ray CT method for the characterization of failure in soils", vol. 40(2), pp.111-118, *Soils and Foundations*, 2000.

Otani J., Mukunoki T. and Obara Y. "Characterization of failure in sand under triaxial compression using an industrial X-ray CT scanner", vol. 1, pp.15-22, *Journal of Physical Modeling in Geotechnics*, 2002.

Takano D., Otani J. and Nagatani H. "Evaluation of three dimensional tunnel face failure using X-ray CT", pp.1639-1642, *Proceedings of the 16th International Conference on Soil Mechanics and Geotechnical Engineering*, 2005.

Plugging Mechanism of Open-Ended Piles

Y. Kikuchi* — T. Sato** — T. Mizutani* — Y. Morikawa*

* Port & Airport Research Institute
3-1-1, Nagase, Yokosuka, Kanagawa 239-0826, Japan
kikuchi@pari.go.jp
mizutani-t@pari.go.jp
morikawa@pari.go.jp

**Kumamoto University
2-39-1 Kurokami, Kumamoto 860-8555, Japan
sato@tech.eng.kumamoto-u.ac.jp

ABSTRACT. *The mechanism of the plugging phenomenon at the toe of vertically loaded open-ended piles was observed in this study by using a micro-focus x-ray CT scanner. The behavior of the surrounding ground at the pile toe is discussed based on the observation of the movement of iron particles, which were mixed into the sand and made layers in the sand layer. The movement of the sand particles was extracted from visualized x-ray CT data. In addition, the movement of the sand particles was extracted using the PIV (Particle Image Velocimetry) method. The CT images of the experimental results showed that the condition of the wedge formation below the open-ended pile was clearly different from that below the closed-ended pile. Although the penetration resistance of the open-ended pile and the closed-ended pile was similar, the movement of soil inside the open-ended pile was not stopped but restricted, as shown by intermittent increases and decreases in penetration resistance during pile penetration.*

KEYWORDS: *open-ended pile, plugging effect, bearing capacity, PIV, x-ray CT scanner*

1. Introduction

For the past 50 years, the pile foundations of port facilities in Japan have been constructed by means of driving steel pipe piles into the ground. During these decades the diameter of piles and the embedded depth of piles have been enlarged. Although these changes have created considerable uncertainty about the toe-bearing capacity of piles, there have been insufficient studies on the bearing capacity of open-ended piles. This study focused on the mechanism of the plugging of open-ended piles to improve the accuracy of estimating the toe-bearing capacity of piles. The mechanism of the plugging at the toe of vertically loaded open-ended piles was observed in this study by using a micro-focus x-ray CT scanner (Kikuchi 2006). Three series of static penetration experiments with model piles were conducted. The model piles used in this study were open-ended piles and closed-ended piles. The behavior of the surrounding ground at the pile toe is discussed based on the observation of the movement of iron particles, which were mixed into sand and made layers in the sand layer. The movement of the sand particles was extracted from visualized x-ray CT data. In addition, the movement of the sand particles was extracted using the PIV (Particle Image Velocimetry) method.

2. Visualization of the plugging phenomenon

The first series of experiments was conducted to examine the plugging phenomenon. The piles used in this series were open-ended stainless steel piles with an outer diameter of 16 mm, a length of 80 mm, and a pile wall thickness of 0.3 mm. The container used was made of acrylic resin with an inner diameter of 85 mm and a height of 160 mm. The model ground was prepared with dry Toyoura sand (D_{50}=0.2 mm, U_C=1.6) by an air pluviation method. The thickness of the ground was 150 mm. Relative densities of the ground were set to 5, 70, and 98% by a vibration method. The pile penetrated the ground at a rate of 1 mm/min. The pile penetration experiment was conducted outside the x-ray CT scanner chamber. The penetration resistance and depth were measured at the pile top. When the pile had penetrated 30 and 60 mm, the load was released, the container was moved into the CT room, and x-ray CT scanning was performed.

The relationship between penetration resistance and depth is shown in Figure 1. Penetration resistance increases as the relative-density of the model ground increases.

Figure 2 shows a vertical section of CT images selected to obtain views through the central axis of the pile at each depth. The white lines are the pile, the gray area is the sandy ground, and the black area in the pile is air. The top of each CT image is the ground surface. It was observed that the ground surface, where the inner pipe pile was located, slid down with the pile in the test case of the low-density ground (Case 1, Dr=5%). Although it was speculated that the plug occurred at the pile toe, the penetration resistance did not increase. Although the ground surface did not slide down with the pile, the penetration resistance is relatively high in the test case of the high-density ground (Case 3, Dr=98%). These results showed that the occurrence of a ground invasion phenomenon into the pipe pile toe depends on the balance of ground reaction, the frictional resistance of the pile inside, and the weight of the soil. Therefore, the increment of resistance and the appearance of the plugging phenomenon do not have a one-to-one correspondence. In other words, there are some cases in which resistance occurred without the appearance of the plugging phenomenon.

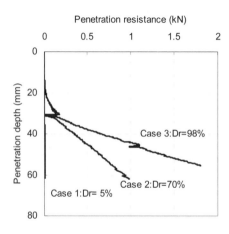

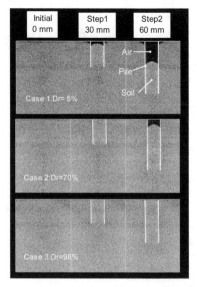

Figure 1. *Relationship between penetration resistance and depth*

Figure 2. *Vertical section of CT image.*

3. Ground behavior around the pile toe

In this series of penetration experiments, a difference in the movement of the sand around the pile between closed-end and open-ended piles was observed. A new penetration apparatus was made to improve test accuracy. The dimensions of the model piles were 15 mm in diameter, 40 mm in length, and 1 mm in thickness for the open-ended piles. The material for the pile was aluminum. The container was made of acrylic resin, 100 mm in inner diameter and 440 mm in height. The sand used for the ground was Toyoura sand (0.2 mm of D_{50} and 1.6 of U_c). The model ground was 270 mm in thickness with a 65% relative density by air pluviation method. An overburden pressure of 2.5 kPa was applied with stainless steel balls (with a diameter of 2 mm).

In order to investigate the movement of the ground from the x-ray CT results, a layer of iron particles (with a diameter of 0.3 mm) was used. The pile penetrated into the ground from the ground surface at a rate of 1 mm/min. The entire pile penetration experiment was conducted in the micro-focus x-ray CT scanner chamber, as shown in Figure 3. When the piles had penetrated to about 35 mm and 70 mm, the pile penetration was stopped, the load was released, and extension rods were added. To obtain test data, pile penetration was stopped at penetration intervals of 3 mm, and X-ray CT scanning was performed.

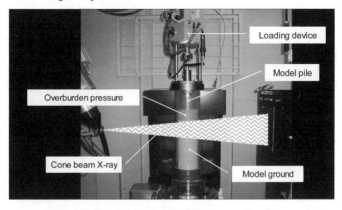

Figure 3. *Penetration test in CT scanner chamber*

The relationship between penetration resistance and depth is shown in Figure 4. A distinctive feature of the bearing capacity of the open-ended pile is that penetration resistance does not occur at the beginning of penetration, but decreases and increases in the middle of penetration. The increment of penetration resistance in both cases was almost equal after about 35 mm of penetration depth. This means that sufficient plugging of the open-ended pile may have developed. With the open-ended pile, resistance decreased and increased in the course of penetration at about 55 to 60 mm of penetration depth due to corresponding changes in the plugging effect. In other words, these results suggest that full plugging is not continuous but a plug is formed and broken repeatedly during pile penetration. This phenomenon was also reported through the research of a large-size laboratory experiment (Mizutani *et al.* 2003).

Figure 5 shows the movement of the particles during the pile penetration; the depth was from 42 mm to 81 mm, with points and lines extracted manually from the CT images. The points are the relative positions of the particles in the pile at each 3-mm step of penetration, and the lines are the particle routes. In the case of the closed-ended pile, the particles below the pile showed a tendency to be pushed to the outside of the pile toe. A clear wedge was constructed at this area, and the soil was unable to intrude there. Some of the particles below the pile were caught at the surface of the wedge, and some were discharged to the side of the pile at the edge and then moved along the pile. Because the wedge unified with the penetrating pile, the relative movement of soil at the surface of the wedge was greatly different. This implies that a shear zone may develop at the wedge surface. On the other hand, the particles below the pile toe were able to move upward and penetrate into the pile. The particles outside the pile were pushed to the outside of the pile toe.

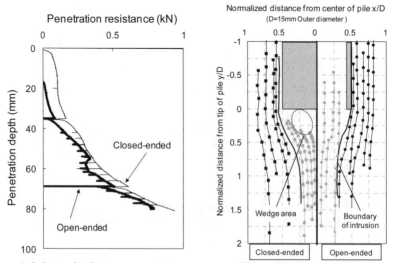

Figure 4. *Relationship between penetration resistance and depth*

Figure 5. *Movement of particles*

4. Deformation analysis using PIV

In order to examine in detail the ground behavior, the PIV method was applied to the CT images. This series of experiments was conducted to focus on the process of the plugging phenomenon evolution. In this series, the model pile was set up in the model ground at an initial penetration depth of 50 mm. The container and the loading device of this series of experiments were the same as the previous device. The pile used in this series was open-ended and 32 mm in diameter, 140 mm in length, and 1.5 mm in thickness. The pile was made of aluminum. The sand used was Souma sand #4 (0.7 mm of D_{50} and 1.6 of U_c). A larger-diameter model pile and larger-diameter sand were used to observe the ground behavior of the inner pipe pile by PIV in this series of experiments. The model ground was 270 mm in thickness with a 65% relative density by air pluviation method. An overburden pressure of 2.5 kPa was applied with stainless steel balls (with a diameter of 2 mm). The pile penetrated the ground at a rate of 1 mm/min from 50 mm to 98 mm in depth. The entire pile penetration experiment was conducted in the micro-focus x-ray CT scanner chamber.

The relationship between penetration resistance and depth is shown in Figure 6. Resistance occurred in the early stages of penetration in this experiment, because the soil had been packed in the pile at the start of penetration.

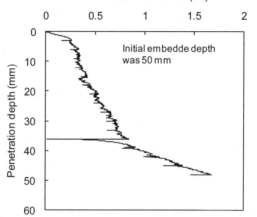

Figure 6. *Relationship between penetration resistance and depth*

The vectors of ground displacement that were measured by the 2D PIV method are shown in Figure 7. This analysis was conducted on the plane of the central pile axis to minimize the disappearance of the particles observed in the analysis. The displacement vectors presented were measured between each 3 mm of penetration. Numbers shown at the top of each figure are the penetration depths for each figure. The pile is shown as two white lines and gradations show displacement of the ground. Since it is difficult to recognize the deformation of the ground in this figure, the major displacements are presented by arrows in the figure.

The soil inside of the pile and below the pile moved downward at penetration depths from 0 to 3 mm. This is because that inner soil put in the pile in the initial state created frictional resistance and made a plug. While the penetration depth increased slightly, the rate of resistance increment went down immediately at penetration depths from 3 to 6 mm. Low rates of incremental resistance were observed at penetration depths from 3 to 33 mm. Movements of the soil inside and below the pile were small at this penetration depth. The transient process of the plugging effect occurred in this stage. A relatively large movement of the soil inside of the pile was observed at penetration depths from 18 to 21 mm. But little movement was observed in the next stage of penetration. It was confirmed that repeated production and destruction of plugging occurred at these steps. The rate of resistance increment rose again after 33 mm of penetration. The displacement of the soil inside of the pile and below the pile became larger at penetration depths ranging from 30 to 36 mm; in particular, a downward movement of the soil existed and was maintained at penetration depths from 36 to 42 mm. A sufficient plugging effect occurred at this step. In this way, the relationship between penetration resistance and ground deformation was observed by using the PIV method.

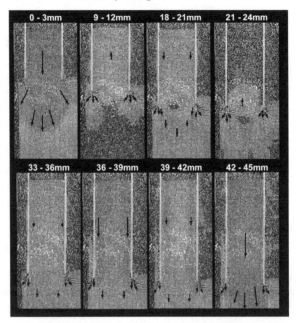

Figure 7. *Image of the ground displacement from CT images using PIV method. Arrows show the major direction and amount of displacement of the ground during the 3 mm penetration. The penetration depth is presented above each picture*

5. Expected plugging mechanism of open-ended piles

Based on these observations, the expected plugging mechanism is presented in Figure 8. The ground below the pile toe was deformed by pile penetration. The deformed and dilated soil intruded into the inside of the pile and it produced friction between the pile and the intruded soil. If the inner friction resistance and self-weight balanced with the bearing resistance of the ground below the pile toe, the plug was produced. Then, the area below the pile was compacted to form a soil wedge. However, if the bearing resistance of the ground below the pile overcame the resistance of the inner friction, the plug was destroyed and the wedge and the ground under the wedge intruded into the inside of the pile. This kind of mechanism in which the plug was produced and destroyed was repeated during the penetration of open-ended piles.

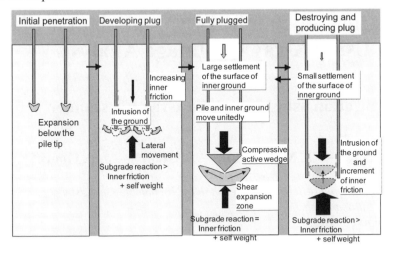

Figure 8. *Expected plugging mechanism of open-ended pile*

6. Conclusions

In this study, the behavior of the ground around a pile toe was analyzed based on a static penetration test. During the penetration test, the movement of the ground was observed using a micro-focus x-ray CT scanner. CT images of the experimental results showed that the condition of wedge formation below an open-ended pile was clearly different from that below a closed-ended pile. Although the penetration resistance of the open-ended pile and closed-ended pile was similar, the movement of soil inside the open-ended pile was not stopped but restricted, as shown by intermittent increases and decreases in penetration resistance during pile penetration. Conclusions about the expected plugging mechanism are presented in Figure 8.

7. References

Mizutani T., Kikuchi Y. and Taguchi H. "Cone Penetration Tests for the Examination of Plugging Effect of Open-ended Piles." *Proc. of IC on Foundations*, BGA, 655-664, 2003.

Kikuchi Y. "Investigation of Engineering Properties of Man-made Composite Geo-materials with Micro-focus X-ray CT". *International Workshop on X-ray CT for Geomaterials-GeoX2006-*, 53-78, 2006.

Development of a Bending Test Apparatus for Quasi-dynamical Evaluation of a Clayey Soil Using X-ray CT Image Analysis

Visualization using industrial x-ray CT scanner

T. Nakano* — T. Mukunoki* — J. Otani* — J. P. Gourc**

*Graduate School of Science and technology, Kumamoto University
2-39-1, Kurokami, Kumamoto City
Kumamoto 860-8555
Japan
081d8829@stud.kumamoto-u.ac.jp
mukunoki@kumamoto-u.ac.jp
junotani@kumamoto-u.ac.jp

**Lirigm, Université de Joseph Fourier, Grenoble, France
Grenoble Cedex9, France
gourc@ujf-grenoble.fr

ABSTRACT. *Local deformation of clay cover on the landfill for radioactive waste could cause serious damage to the environment. In order to evaluate the performance of clay barrier of the clay cover to local deformation, it is useful to conduct a bending test of the clay material. This bending test would facilitate understanding about the performance of clay barriers used to cover radioactive wastes. In this paper, a new testing method for obtaining the engineering properties of the cover soils was developed and the effectiveness of this method was evaluated. It was found that the local deformation of cover soil would cause catastrophic failure and that areas of low density would not as long as the clayey soil was compacted sufficiently.*

KEYWORDS: *compacted clay liner, landfill, industrial x-ray CT scanner, bending test*

1. Introduction

A compacted clay liner (CCL) is an impermeable barrier material that is placed at the bottom and top (cover) of landfill. As for use of the CCL to the cover system, there are potential issues such as aging of the CCL that result in local deformation of the CCL that occur as the settlement load becomes unbalanced. If this potential issue occurred in the actual site, serious environmental problems would result. Landfill for very low level radioactive waste was constructed in France Aube (Gourc and Camp, 2005 and Gourc and Oliver, 2006). The research group of Lirigm, Joseph Fourier University, Grenoble, in France has maintained a field monitoring problem and discovered that the localized settling of landfill surface. In order to observe the behavior of the CCL due to local settlement of the waste, an *in-situ* burst test was performed. The objective of the burst test was to simulate the critical force required to damage the CCL, namely to create cracks in this liner. This test helped display to visual observation the cracks on the surface of CCL. Meanwhile, the industrial x-ray computed tomography (CT) scanner installed at the X-Earth Center of Kumamoto University was used to visualize the crack propagation of the CCL at the each step during the loading level during the punching and bending tests. Authors developed a new bending test apparatus which can be operated in x-ray CT room. This test apparatus has made it possible to evaluate the crack generation in the specimen of CCL during incremental loading and without stress release between acquisitions of new images (Mukunoki *et al.*, 2008).

The objective of this paper is to confirm the test performance of the new bending test apparatus for x-ray CT scanner and to evaluate the bending property and the durability of the CCL used in the actual site.

2. Experimental method

2.1. *Materials*

Figure 1 shows a grain size distribution curve of CCL specimen tested. The compacted clayey specimen had a soil density of 2.72 t/m^3, optimum moisture content of 16%, its liquid limit and plastic index are 43.6% and 24.3%, respectively. In this study, the disk specimen with a diameter of 150 mm and the thickness of 25 mm was prepared for and subjected to a symmetric bending test with displacement control.

2.2. *New bending test apparatus*

The new test apparatus was used in this study. Figures 2(a) and (b) illustrate a schematic of the newly developed bending test apparatus that was used with the specimen with circular shape (referred as a disk specimen hereafter) and display a photograph of the installed test apparatus. The improved feature for this testing apparatus is that a new type of tester was developed. This testing apparatus has the loading system with an electrical actuator installed at the top of the test system as shown in Figures 2(a) and (b). The actuator can be operated by a remote controller. Additionally, the bottom plate was prepared especially for this CT system and an entire test apparatus can be fixed on the specimen table. Hence, all processes for testing the CCL can be conducted in the x-ray CT room. In this study, 300 kV was chosen. The total number of voxels in each x-ray image slice is 2048x2048 and the beam thickness is 1 mm (i.e. the dimension of a voxel is 0.195x 0.195x1 mm^3).

The features of this test apparatus can be drawn as follows:

1) to perform the symmetrical-loading test in the x-ray CT room; and,

2) to scan the specimen at each deformation level with x-ray CT scanner without stress release.

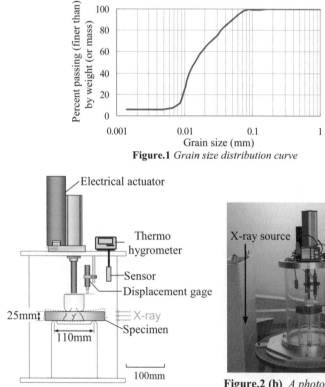

Figure.1 *Grain size distribution curve*

Figure.2 (a) *The schematic of the developed bending test apparatus*

Figure.2 (b) *A photograph of the developed bending test apparatus installed in the X-ray CT shield room*

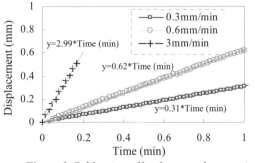

Figure.3 *Calibration of loading speed versus time*

3. Results and discussion

3.1. *Check the performance of the bending test apparatus*

Figure 3 shows displacement versus time due to loading with a speed of 0.3, 0.6 and 3.0 mm/min. The electrical actuator installed gave the linear relationship between displacement and time and it was confirmed that the loading speed was precisely maintained in this system. In this study, the loading speed was 0.6 mm/min.

3.2. *Load-displacement relationship*

Figure 4 shows the load-displacement curve at the center of the specimen with previous test apparatus (Case 1). In this case, the CCL specimen was loaded in the soil test room and the specimen had to be moved to the CT room. Hence, the sample rebound as shown in Figure 4. Eventually, at least 65% of the displacement was rebounded at the center of the specimen. This phenomenon probably resulted in close hair cracks, artefacts in the CCL specimen due to rebounding. Figure 5 shows the load-displacement curve at the center of the specimen with new test apparatus (Case 2) shown in Figures 2(a) and (b). In this newly developed case, the actuator that was used can control the displacement very precisely, there is no observation of displacement rebounded as shown in Figure 4.

The actuator motion must be stopped during the CT scanning. The scanning time is 2.5 minutes. Times the number of CT images, it took at least 62.5 minutes to scan one CCL. As the material is clayey soil, it is difficult to avoid the stress release. However, it is impossible to make the scanning time shorter than 62.5 minutes, so the issue of stress release is not discussed in this paper. These stress releases are evident in Figure 2 and displayed as a reduction in the load at increments of 62.5 minutes.

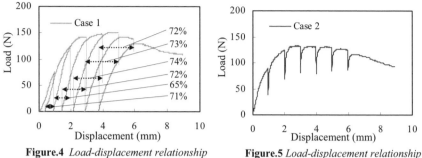

Figure.4 *Load-displacement relationship obtained from previous bending test apparatus (Case 1)*

Figure.5 *Load-displacement relationship obtained from the developed bending test apparatus (Case 2)*

3.3. X-ray CT image

Figures 6 and 7 (a), (b) show the x-ray CT images of vertical-cross section at the different steps from the initial condition for Case 1 and 2, respectively. At the initial condition of the CCL specimen: at the displacement level of 0 mm as shown in Figures 6(a) and 7(a), some cracks can be seen in the CT images of the CCL. Cracks were generated from the displacement of 3mm around the bottom surface and they grew with increase in the displacement level. Also, the point of the crack increased along the direction of propagation as keeping with the peak point of load-displacement curve for Case 1 and Case 2. Cracks generated in the CCL specimen seem not to have a relationship to the heterogeneity in the initial condition. The width equal of the crack in Case 2 was smaller than Case 1. It is conceivable that the crack was closed during rebound.

3.4. Visualization of 3D image of low density area extracted

Figure 8(a), (b) show the images that was extracted from the low density area for Case 1 and Case 2. Both images were seen from the view point of the bottom surface of CCL specimen. The crack which developed from the center of the specimen can be observed to have almost the radial spread at each step in the loading process. In the 3mm displacement the crack has occurred in both Cases, however in Case 1 the number and sized of the crack was smaller than in Case 2. It is difficult to judge a difference of both Cases in displacement of 4 mm, 5 mm and 6 mm. Therefore, Figure 9 shows the length of the crack to the direction of specimen thickness. Figure 10 shows the volume of low density area. These values could be obtained from Figure 8(a), (b) using the previous apparatus (Case 1) and in the newly developed one (Case 2). The length of the crack obtained from Figure 8(b) was slightly greater than that from Figure 8(a) as shown in Figure 9. Then, the volume of low density area shown in Figure 10 was reconstructed under the condition of less than 0 of CT-value. The volume of the low density area in the case of using the newly developed apparatus was approximately two times greater than the previously developed one. Likewise, the previous apparatus would underestimate the volume of low density area because of not avoiding the rebounding of specimen; namely, this result pointed out that the rebound effect was reduced.

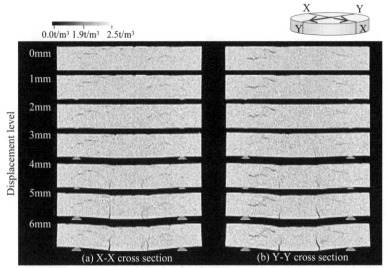

Figure.6 *X-ray CT images due to bending loading at each displacement level for Case 1*

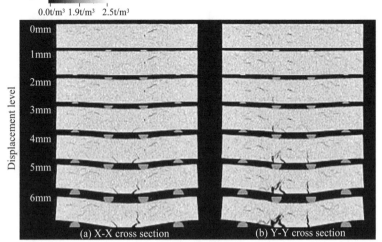

Figure.7 *X-ray CT images due to bending loading at each displacement level for Case 2*

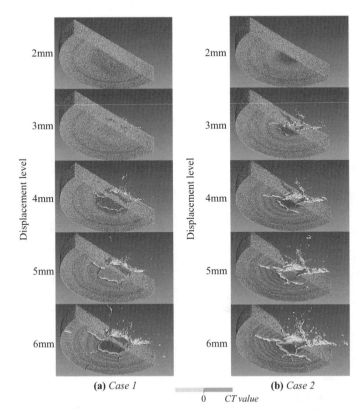

Figure.8 *3D image of the low density area for Case 1 and Case 2*

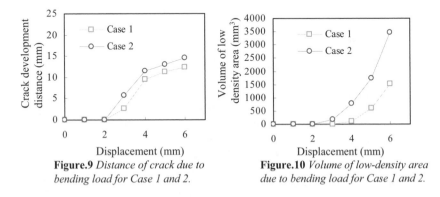

Figure.9 *Distance of crack due to bending load for Case 1 and 2.*

Figure.10 *Volume of low-density area due to bending load for Case 1 and 2.*

As in these results, to be able to perform the mechanical test in the X-ray CT room is a significant improvement and enables us to observe the phenomena while minimizing rebound artefacts in the CCL material that are caused by the unloading process.

4. Conclusions

The bending test apparatus for X-ray CT was newly developed to observe the deformation behavior of the compacted clay specimen due to symmetric loading in this paper. The conclusion can be summarized in the following:

(1) the developed bending test apparatus performed well in the x-ray CT shield room;

(2) use of the newly developed bending test apparatus made it possible to visualize the deformation process and the crack generation in the compacted clay specimen, because the sample could be symmetrically loaded with in the x-ray CT scanner;

(3) the obtained x-ray CT images of the compacted clay specimen that is subjected to various loads and load rates can be effectively evaluated with minimal rebound artefacts; and,

(4) as long as the clayey soil was compacted sufficiently, the local deformation would cause catastrophic failure of the clay cover and not areas of low density.

5. References

Gourc, J., and Camp, S. Proposal of joint program about the behavior of clay used as landfill cap cover and likely to crack under extension, Lirigm, University of Grenoble, France, 2005.

Gourc, J.P. and Olivier, F. "Overview of landfill instrumentation techniques for the monitoring of waste settlements", pp199-206, *Proceedings of the Sixth Japanese-Korean-French Seminar On Geo-Environmental Engineering*, 2006.

Mukunoki, T., Otani, J., Maekawa, A., Camp, S. "Investigation of Crack Behavior on Cover soils at Landfill using X-ray CT", pp213-219, *Advances in X-ray Tomography for Geomaterials-GeoX2006*, 2006.

Mukunoki, T., Nakano, T., Otani, J., A., Camp, S., Gourc, J.P. "Evaluation of Cracks in Cover Soils due to local deformation using X-ray CT", pp243-248, *Proceedings of the sixth Japanese-Korean-French Seminar On Geo-Environmental Engineering*, 2008.

Kenter, J.A.M., "Application of Computerized Topography in Sedimentology", vol. 8, pp. 201–211, *Marine Geotechnology*, 1989.

Author Index